Credit Risk

Contributions to Economics

http://www.springer.de/cgi-bin/search_book.pl?series=1262

Further volumes of this series can be found at our homepage.

Kirstin Hubrich
Cointegration Analysis in a German Monetary System
2001. ISBN 3-7908-1352-4

Nico Heerink et al. (Eds.)
Economic Policy and Sustainable Land Use
2001. ISBN 3-7908-1351-6

Friedel Bolle/Michael Carlberg (Eds.)
Advances in Behavioral Economics
2001. ISBN 3-7908-1358-3

Volker Grossmann
Inequality, Economic Growth, and Technological Change
2001. ISBN 3-7908-1364-8

Thomas Riechmann
Learning in Economics
2001. ISBN 3-7908-1384-2

Miriam Beblo
Bargaining over Time Allocation
2001. ISBN 3-7908-1391-5

Peter Meusburger/Heike Jöns (Eds.)
Transformations in Hungary
2001. ISBN 3-7908-1412-1

Claus Brand
Money Stock Control and Inflation Targeting in Germany
2001. ISBN 3-7908-1393-1

Erik Lüth
Private Intergenerational Transfers and Population Aging
2001. ISBN 3-7908-1402-4

Nicole Pohl
Mobility in Space and Time
2001. ISBN 3-7908-1380-X

Pablo Coto-Millán (Ed.)
Essays on Microeconomics and Industrial Organisation
2002. ISBN 3-7908-1390-7

Mario A. Maggioni
Clustering Dynamics and the Locations of High-Tech-Firms
2002. ISBN 3-7908-1431-8

Ludwig Schätzl/Javier Revilla Diez (Eds.)
Technological Change and Regional Development in Europe
2002. ISBN 3-7908-1460-1

Alberto Quadrio Curzio/Marco Fortis (Eds.)
Complexity and Industrial Clusters
2002. ISBN 3-7908-1471-7

Friedel Bolle/Marco Lehmann-Waffenschmidt (Eds.)
Surveys in Experimental Economics
2002. ISBN 3-7908-1472-5

Pablo Coto-Millán
General Equilibrium and Welfare
2002. ISBN 7908-1491-1

Wojciech W. Charemza/Krystyna Strzala (Eds.)
East European Transition and EU Enlargement
2002. ISBN 3-7908-1501-2

Natalja von Westernhagen
Systemic Transformation, Trade and Economic Growth
2002. ISBN 3-7908-1521-7

Josef Falkinger
A Theory of Employment in Firms
2002. ISBN 3-7908-1520-9

Engelbert Plassmann
Econometric Modelling of European Money Demand
2003. ISBN 3-7908-1522-5

Reginald Loyen/Erik Buyst/Greta Devos (Eds.)
Struggling for Leadership: Antwerp-Rotterdam Port Competition between 1870–2000
2003. ISBN 3-7908-1524-1

Pablo Coto-Millán
Utility and Production, 2nd Edition
2003. ISBN 3-7908-1423-7

Emilio Colombo/John Driffill (Eds.)
The Role of Financial Markets in the Transition Process
2003. ISBN 3-7908-0004-X

Guido S. Merzoni
Strategic Delegation in Firms and in the Trade Union
2003. ISBN 3-7908-1432-6

Jan B. Kuné
On Global Aging
2003. ISBN 3-7908-0030-9

Sugata Marjit, Rajat Acharyya
International Trade, Wage Inequality and the Developing Economy
2003. ISBN 3-7908-0031-7

Francesco C. Billari/Alexia Prskawetz (Eds.)
Agent-Based Computational Demography
2003. ISBN 3-7908-1550-0

Georg Bol
Gholamreza Nakhaeizadeh
Svetlozar T. Rachev
Thomas Ridder
Karl-Heinz Vollmer
Editors

Credit Risk

Measurement, Evaluation
and Management

With 85 Figures
and 47 Tables

Physica-Verlag
A Springer-Verlag Company

Series Editors
Werner A. Müller
Martina Bihn

Editors
Prof. Dr. Georg Bol
Prof. Dr. Svetlozar T. Rachev
Prof. Dr. Karl-Heinz Vollmer
Institut für Statistik
und Mathematische Wirtschaftstheorie
Universität Karlsruhe
76128 Karlsruhe
Germany

Prof. Dr. Gholamreza Nakhaeizadeh
DaimlerChrysler AG
RIC/AM
Postbox 2360
89013 Ulm
Germany

Dr. Thomas Ridder
DZ Bank AG
COR
Platz der Republik
60265 Frankfurt am Main
Germany

E-Mails:
georg.bol@statistik.uni-karlsruhe.de
zari.rachev@statistik.uni-karlsruhe.de
karl-heinz.vollmer@statistik.uni-karlsruhe.de
rheza.nakhacizadeh@daimlerchrysler.com
thomas.ridder@dzbank.de

ISSN 1431-1933
ISBN 3-7908-0054-6 Physica-Verlag Heidelberg New York

Cataloging-in-Publication Data applied for
A catalog record for this book is available from the Library of Congress.
Bibliographic information published by Die Deutsche Bibliothek
Die Deutsche Bibliothek lists this publication in the Deutsche Nationalbibliografie; detailed bibliographic data is available in the Internet at <http://dnb.ddb.de>.

This work is subject to copyright. All rights are reserved, whether the whole or part of the material is concerned, specifically the rights of translation, reprinting, reuse of illustrations, recitation, broadcasting, reproduction on microfilm or in any other way, and storage in data banks. Duplication of this publication or parts thereof is permitted only under the provisions of the German Copyright Law of September 9, 1965, in its current version, and permission for use must always be obtained from Physica-Verlag. Violations are liable for prosecution under the German Copyright Law.

Physica-Verlag Heidelberg New York
a member of BertelsmannSpringer Science+Business Media GmbH

http://www.springer.de

© Physica-Verlag Heidelberg 2003
Printed in Germany

The use of general descriptive names, registered names, trademarks, etc. in this publication does not imply, even in the absence of a specific statement, that such names are exempt from the relevant protective laws and regulations and therefore free for general use.

Softcover Design: Erich Kirchner, Heidelberg
SPIN 10924549   88/3130/DK-5 4 3 2 1 0 – Printed on acid-free paper

Preface

On March 13th - 15th 2002, the 8th Econometric Workshop in Karlsruhe was held at the University of Karlsruhe (TH), Germany. The workshop was organized jointly by the Institute for Statistics and Mathematical Economics and the Deutsche Zentral-Genossenschaftsbank, Frankfurt am Main (DZ-Bank AG, formerly SGZ-Bank AG). Almost 200 participants participated, discussing new developments in the measurement, evaluation and management of credit risk. This volume presents selected contributions to the conference, of which a short survey follows:

Arne Benzin, Stefan Trück and Svetlozar T. Rachev present the main features of the new Basel Capital Accord (Basel II). They include a survey of the two regulatory approaches to credit risk: the standardized and the internal rating based approach.

Christian Bluhm and Ludger Overbeck use the CreditMetrics/KMV one-factor model to discuss the quantification of systematic risk by estimation of asset correlations in homogeneous portfolios. Based on this concept the Basel-II proposal to fix asset correlation for corporate loans at the 20% level is compared with empirical data.

Dylan D'Souza, Keyvan Amir-Atefi and Borjana Racheva-Jotova investigate empirically how different distributional assumptions governing bond price uncertainty effect the price of a credit default swap. They use the two-factor Hull-White-model and the extension of the fractional recovery model of Duffie-Singleton given by Schönbucher. Prices are compared under the assumption of a Gaussian and a non-Gaussian distribution for the underlying risk factors.

Christoph Heidelbach and Werner Kürzinger point out the consequences of Basel II for the DaimlerChrysler Bank, a bank with a special profile as a provider of automotive financial services.

Alexander Karmann and Dominik Maltritz present a quantification of the probability that a nation will default on repayment obligations in foreign currency. They adopt the approach introduced by Merton, considering as underlying process the nation's ability-to-pay modeled by the sum of future payment surpluses. The method is successfully demonstrated on the examples of Latin America and Russia.

Rüdiger Kiesel, William Perraudin and Alex Taylor examine the practical aspects of using techniques from Extreme Value Theory (EVT) to estimate Value-at-Risk (VAR). They compare the performance of EVT estimators with estimators of VAR which are based on quantiles of empirical return distributions. The two types of estimators are shown to yield almost identical results for commonly used confidence levels.

Daniel Kluge and Frank Lehrbass consider default probabilities in structured commodity finance. While banks usually base investment decisions on cash flow modeling, they point out that the default risk of a project actually depends on the "combined downside risk" of spot price and production level movements. They present a workable approach to the measurement of this risk.

Filip Landskog, Alexander McNeil and Uwe Schmock start from the fact that some multivariate financial time series data are more plausibly modeled by an elliptical distribution other than the multivariate normal distribution. In that case, they propose estimation of the correlation by using Kendall's tau and the relation between tau and the linear correlation coefficient. They demonstrate that the well-known relation for the bivariate normal distribution generalizes to the class of elliptical distributions.

Marlene Müller and Wolfgang Härdle present semiparametric extensions of the logit-model which allow nonlinear relationships. The technique is based on the theory of generalized partial linear models and is used for credit risk scoring and the estimation of default probabilities.

Borjana Racheva-Jotova, Stoyan Stoyanov and Svetlozar T. Rachev illustrate a new approach for integrated market and credit risk management using Cognity software. The Cognity Credit Risk System comprises two models, the Asset Value Approach and the Stochastic Default Rate Model, both based on Stable distributions.

Thomas Rempel-Oberem, Rainer Klingeler and Peter Martin discuss a novel approach to the determination of default events which was developed for two different credit institutions. It is based on the CreditRisk$^+$ model, which was modified and extended to meet the particular requirements and standards of these institutions.

Ingo Schäl states principles of internal rating systems for corporate clients and presents his practical experience with building such systems. He proposes proceeding in two steps. First, each individual credit is evaluated with respect to expected exposure at default, expected default frequency and expected recovery rate. The second step is calculation of the correlations given the portfolio structure. This permits active portfolio management. He also describes how the quality of rating systems can be measured and gives basic requirements for backtesting.

Frank Schlottmann and Detlev Seese are concerned with computational aspects of the risk-return analysis of portfolios. They show that risk analysis requires the computation of a set of Pareto-efficient portfolio structures in a non-linear, nonconvex setting with additional constraints. They develop a new fast and flexible framework for solving this problem, using a hybrid heuristic method which combines a multi-objective evolutionary algorithm with a problem-specific local search. Empirical results demonstrate the advantage of the procedure.

Rafael Schmidt develops a model of dependence using elliptical copulae. He embeds the concepts of tail dependence and elliptical copulae into the framework of Extreme Value Theory and provides a parametric and a nonparametric estimator for the tail dependence coefficient.

Finally, Stefan Trück and Jochen Peppel give a review on credit risk models in practice. The first model is the structural model introduced by Merton and modified to the "distance-to-default" methodology of KMV. The second class are the Value-at-Risk models with CreditMetrics of JP Morgan, which use historical transition matrices. The third major approach is that proposed by Credit Suisse Financial Products with CreditRisk+. The paper closes with a brief description of CreditPortfolioView by McKinsey and the one-factor model proposed by the Basel Committee on banking supervision.

Many people have contributed to the success of the workshop: Irina Gartvihs did the major part of organizing the workshop. Bernhard Martin, Christian Menn, Theda Schmidt and Stefan Trück also proved indispensable for its organization. Tim Gölz and Matthias Rieber were responsible for the technical infrastructure while Thomas Plum prepared the layout of the book. All of their help is very much appreciated. The organization committee wishes also to thank the Fakultät für Wirtschaftswissenschaften, the dean Prof. Dr. Hartmut Schmeck and the managing director Dr. Volker Binder for their cooperation. Last but not least we thank the DaimlerChrysler AG for their sponsorship.

Karlsruhe, February 2003 *The Editors*

Contents

Approaches to Credit Risk in the New Basel Capital Accord
Arne Benzin, Stefan Trück, Svetlozar T. Rachev 1

Systematic Risk in Homogeneous Credit Portfolios
Christian Bluhm, Ludger Overbeck 35

Valuation of a Credit Default Swap: The Stable Non-Gaussian versus the Gaussian Approach
Dylan D'Souza, Keyvan Amir-Atefi, Borjana Racheva-Jotova 49

Basel II in the DaimlerChrysler Bank
Christoph Heidelbach, Werner Kürzinger 85

Sovereign Risk in a Structural Approach. Evaluating Sovereign Ability-to-Pay and Probability of Default
Alexander Karmann, Dominik Maltritz 91

An Extreme Analysis of VaRs for Emerging Market Benchmark Bonds
Rüdiger Kiesel, William Perraudin, Alex Taylor 111

Default Probabilities in Structured Commodity Finance
Daniel Kluge, Frank Lehrbass 139

Kendall's Tau for Elliptical Distributions
Filip Lindskog, Alexander McNeil, Uwe Schmock 149

Exploring Credit Data
Marlene Müller, Wolfgang Härdle 157

Stable Non-Gaussian Credit Risk Model; The Cognity Approach
Borjana Racheva-Jotova, Stoyan Stoyanov, Svetlozar T. Rachev 175

An Application of the CreditRisk⁺ Model
Thomas Rempel-Oberem, Rainer Klingeler, Peter Martin 195

Internal Ratings for Corporate Clients
Ingo Schäl 207

Finding Constrained Downside Risk-Return Efficient Credit Portfolio Structures Using Hybrid Multi-Objective Evolutionary Computation
Frank Schlottmann, Detlef Seese 231

Credit Risk Modelling and Estimation via Elliptical Copulae
Rafael Schmidt 267

Credit Risk Models in Practice - a Review
Stefan Trück, Jochen Peppel 291

List of Authors 331

Approaches to Credit Risk in the New Basel Capital Accord

Arne Benzin[1], Stefan Trück[2], and Svetlozar T. Rachev[3]

[1] Lehrstuhl für Versicherungswissenschaft, Universität Karlsruhe, Germany
[2] Institut für Statistik und Mathematische Wirtschaftstheorie, Universität Karlsruhe, Germany
[3] Institut für Statistik und Mathematische Wirtschaftstheorie, Universität Karlsruhe, Germany

Summary. We discuss the main features of the new Basel Capital Accord (Basel II) concerning the regulatory measurement of Credit Risk. After an overview of the basic ideas in the new accord the determining aspects of the approaches to Credit risk in the new capital accord are surveyed: the standardized approach (STD) as well as the two forms of the internal rating based (IRB) approach - foundation and advanced. We describe the issues of the second consultative document of the new accord and describe how to measure the required capital. Further the fair comment on several features of Basel II and its possible changes in the final version of the accord are illustrated.

1 Introduction

1.1 The History of the Basel Capital Standards

More than a decade has passed since the *Basel Committee on Banking Supervision*[4] (the Committee) introduced its 1988 Capital Accord (the Accord). The major impetus for this Basel I Accord was the concern of the governors of the central banks that the capital – as a "cushion" against losses – of the world's major banks had become dangerously low after persistent erosion through competition.

Since 1988 the business of banking, risk management practices, supervisory approaches and financial markets have undergone significant transformations. Consequently, the Committee released a proposal in June 1999 to replace the old Accord with a more risk-sensitive framework, the *New Basel Capital Accord* (Basel II). After receiving several comments by the industry and research

---

[4] The Basel Committee on Banking Supervision (BCBS) is a committee of central banks and bank supervisors from the major industrialised countries that meet every three months at the *Bank for International Settlements* (BIS) in Basel.

institutions in January 2001 the second consultative document was published. Again the suggestions were criticized a lot and according to the committee some features will be changed again. Reflecting the comments on the proposal and the results of the ongoing dialogue with the industry worldwide, the Committee will publish a revised version in 2003 with the new corrections. Therefore, the final version of the new Accord was postponed already several times. It is now expected to be implemented in the banking industry in 2006.

1.2 The Basel I Capital Accord

The 1988 Accord[5] requires internationally active banks to hold capital equal to at least 8% of a basket of assets measured in different ways according to their riskiness. The definition of capital is set in two tiers, whereby banks have to hold at least half of its measured capital in the tier one form:

Tier 1: The shareholders' equity and retained earnings.
Tier 2: The additional internal and external resources available to banks.

The Accord created capital requirements for credit risk – i.e. the risk of loss arising from the default by a creditor or a counterparty – in banking assets. A portfolio approach was taken to measure the risk, with assets classified into four risk buckets according to the debtor category. As a result, some assets have no capital requirement whereas other claims do have:

1. Risk Bucket: Generally consisting of claims on OECD[6] governments, has a 0% risk weight.
2. Risk Bucket: Generally consisting of claims on banks incorporated in OECD countries, has a 20% risk weight.
3. Risk Bucket: Generally consisting of residential mortgage claims, has a 50% risk weight.
4. Risk Bucket: Generally consisting of claims on consumers and corporates, has a 100% risk weight.

These *risk weights* multiplied with the respective exposure result in the so-called *risk-weighted assets* (RWA). As shown in the table below, the Accord requires that banks hold at least 8% of the RWA as a capital charge or as a *minimum capital requirement* (MCR) for protection against credit risk.

For instance, claims on banks have a 20% risk weight, which can be translated into a capital charge of 1.6% of the value of the claim. This means, that an exposure of 1 Mio. € is equivalent to RWA of 200.000 € and to MCR of 16.000€ due to the following calculations:

---

[5] See *Internal Convergence of Capital Measurement and Capital Standards*, BCBS, July 1988.
[6] Organisation for Economic Coordination and Development.

$$\text{RWA} = \text{Exposure} \cdot \text{Risk Weight} \tag{1}$$
$$= 1 Mio\ \text{€} \cdot 20\%$$
$$= 200.000\text{€}$$

$$\text{MCR} = \text{RWA} \cdot \text{Minimum Requirement of 8\%} \tag{2}$$
$$= 200.000\text{€} \cdot 8\%$$
$$= 16.000\text{€}$$

Comparatively, minimum capital requirements for claims on corporates of 1 Mio. € would be:

$$\text{MCR} = \text{RWA} \cdot \text{Minimum Requirement of 8\%} \tag{3}$$
$$= 1 Mio\ \text{€} \cdot 100\% \cdot 8\%$$
$$= 80.000\text{€}$$

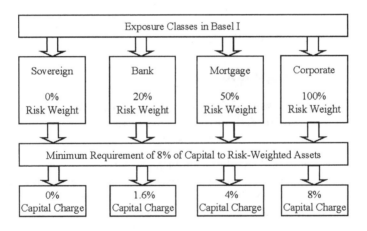

Fig. 1. Exposure classes of Basel I

However, virtually all claims on the non-bank private sector receive the standard 8% capital requirement, but there is also a scale of charges for off-balance sheet exposures through guarantees, bonds, etc. This is the only complex section of the 1988 Accord and requires a two-step approach:

Step 1: Banks convert their off-balance-sheet positions into a credit equivalent amount through a scale of *credit conversion factors* (CCF).

Step 2: These positions are weighted according to the counterparty's risk weight.

The 1988 Accord has been supplemented a number of times, most changes dealing with the treatment of the above mentioned off-balance-sheet activities. A significant amendment was enacted in 1996, when the Committee introduced a measure whereby trading positions – such as bonds or foreign exchange – were removed from the credit risk framework and given explicit capital charges related to the bank's open position in each instrument.

The two principal purposes of the Accord were to ensure an adequate level of capital in the international banking system and to create a "more level playing field" in competitive terms so that banks could no longer build business volume without adequate capital backing. These two objectives have been achieved, so that the Accord was widely recognised and became an accepted world standard during the 1990s.

However, there also have been some less positive features. The regulatory capital requirement has been in conflict with increasingly sophisticated internal measures of economic capital. The simple bucket approach with a flat 8% charge for claims on the private sector has given banks an incentive to move high quality assets off the balance sheet, thus reducing the average quality of bank loan portfolios. In addition, the 1988 Accord does not sufficiently recognise *credit risk mitigation* (CRM) techniques such as collateral and guarantees.[7] These are the principal reasons why the Committee decided to propose a more risk-sensitive framework.

1.3 The Basel II Capital Accord

In June 1999, the initial consultative proposal contained three fundamental innovations, each designed to introduce greater risk sensitivity into the Accord:

1) The current standard should be supplement with two additional "pillars" dealing with supervisory review and market discipline. They should reduce the stress on the quantitative pillar one by providing a more balanced approach to the capital assessment process.
2) Banks with advanced risk management capabilities should be permitted to use their own internal systems for evaluating credit risk – known as "internal ratings" – instead of standardised risk weights for each class of asset.
3) Banks should be allowed to use gradings provided by approved external credit assessment institutions to classify their sovereign claims into five risk buckets and their claims on corporates and banks into three risk buckets.

In addition, there were a number of other proposals like to refine the risk weightings and to introduce a capital charge for other risks. However, the

---

[7] Credit Risk Mitigation relates to the reduction of credit risk by - for example - taking collaterals, obtaining credit derivatives, guarantees or taking an offsetting position subject to a netting agreement.

basic definition of capital stayed the same. The comments on the June 1999 paper were numerous and reflected the important impact the old Accord has had. Nearly all welcomed the intention to refine the Accord supported by the three pillar approach due to safety and soundness in today's dynamic and complex financial system can be attained only by the combination of effective bank-level management, market discipline and supervision. Nevertheless, many details of the proposal were criticised. In particular, the threshold for the use of internal ratings should not be set so high as to prevent well-managed banks from using them.

The 1988 Accord focussed on the total amount of bank capital, which is vital in reducing the risk of bank insolvency and the potential cost of a bank's failure for depositors. Building on this, the new framework intends to improve safety and soundness in the financial system by placing more emphasis on banks' own internal control and management, the supervisory review process and the market discipline. Although the new framework's focus is primarily on internationally active banks, its underlying principles are suitable for application to banks of varying levels of complexity and sophistication, so that the new framework can be adhered by all significant banks within a certain period of time.

Table 1. Rationale for a new accord

Basel I Accord	Basel II Accord
Focus on a single risk measure	Emphasis on banks' own internal methodologies, supervisory review and market discipline
One size fits all	Flexibility, menu of approaches, incentives for better risk management
Broad brush structure	More risk sensitive

The 1988 Accord provided essentially only one option for measuring the appropriate capital of banks, although the way to measure, manage and mitigate risks differs from bank to bank. In 1996 an amendment was introduced which focussed on trading risks and allowed some banks for the first time to use their own systems to measure their market risks.

The new framework provides a spectrum of approaches from simple to advanced methodologies for the measurement of both credit risk and operational risk in determining capital levels. Therefore – due to the less prescriptive guidelines of the new Accord – capital requirements should be more in line with underlying risks and allow banks to manage their businesses more efficiently.

2 The Structure of the New Accord

The new Accord consists of three mutually reinforcing pillars, which together contribute to safety and soundness in the financial system. The Committee stresses the need for a rigorous application of all three pillars and plans to achieve the effective implementation of all aspects of the Accord.

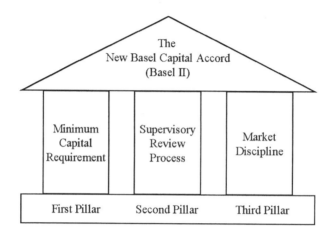

Fig. 2. The three pillars of the new accord

2.1 Minimum Capital Requirement

The first pillar sets out the MCR and defines the minimum ratio of capital to RWA. Therefore, it is necessary to know how the total capital is adequately measured by banks. The new framework maintains both the current definition of the total capital and the minimum requirement of at least 8% of the bank's capital to RWA:[8]

$$\text{Capital Ratio} = \frac{\text{Total Capital}}{\text{Credit Risk} + \text{Market Risk} + \text{Operational Risk}} \quad (4)$$

As you can see in formula 4, the calculation of the denominator of the capital ratio is dependent on three different forms of risk: Credit -, market - and operational risk. The credit risk measurement methods are more elaborate than those in the current Accord whereas the market risk measure remains unchanged. Nevertheless, the new framework proposes for the first time a measure for operational risk.

---

[8] To ensure that risks within the entire banking group are considered, the revised Accord is extended on a consolidated basis to holding companies of banking groups.

Credit Risk

For the measurement of credit risk two principal options are proposed and will be discussed later on. The first option is the *standardised* (STD) approach[9] and the second the *internal ratings-based* (IRB) approach[10] from which two variants exist, a foundation and an advanced IRB approach. The use of the IRB approach is subject to an approval by the supervisors, based on standards established by the Committee.

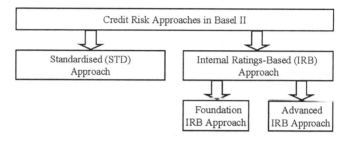

Fig. 3. Credit risk approaches in Basel II

The STD Approach: This approach is conceptually the same as the present Accord, but it is more risk sensitive. The bank allocates a risk weight to each of its assets and off-balance-sheet positions and produces a sum of RWA values. A risk weight of 100% means that an exposure is included in the calculation of RWA at its full value, which translates into a capital charge equal to 8% of that value. Similarly, a risk weight of 20% results in a capital charge of 1.6% (i.e. 20% of 8%). Individual risk weights currently depend on the broad category of borrower which are sovereigns, banks and corporates. Under the new Accord, the risk weights are refined by the reference to a rating provided by an *external credit assessment institution* (ECAI), such as rating agencies. For example, for corporate lending, the existing Accord provides only one risk weight category of 100%, the new Accord provides four categories: 20%, 50%, 100% and 150%.

The IRB Approach: Under this approach, banks are allowed to use their internal estimates of borrower creditworthiness to assess credit risk in their portfolios, subject to strict methodological and disclosure standards. Distinct analytical frameworks are provided for different types of loan exposures whose loss characteristics are different. Under the IRB approach, banks estimate each borrower's creditworthiness and translate the results into estimates of a potential future loss amount, which forms the basis of MCR. The framework allows on the one hand a foundation method and

---
[9] See section 3: *The STD Approach to Credit Risk*.
[10] See section 4: *The IRB Approach to Credit Risk*.

on the other hand more advanced methodologies for corporate, sovereign and bank exposures. In the foundation methodology, banks estimate the probability of default associated with each borrower, and the supervisors supply the other inputs. In the advanced methodology, a bank with a sufficiently developed internal capital allocation process is permitted to supply other necessary inputs as well. Under both IRB approaches, the range of risk weights are far more diverse than those in the STD approach, resulting in greater risk sensitivity.

Market Risk

The 1988 Accord set a capital requirement simply in terms of credit risk – the principal risk for banks – though the overall capital requirement (the 8% minimum ratio) was intended to cover other risks as well. In 1996, market risk exposures – i.e the risk of losses in trading positions when prices move adversely – were removed and given separate capital charges. The menu of approaches to measure market risk in the revised framework stays unchanged and contains beside a standardised approach an internal model approach.

Operational Risk

In its attempt to introduce a greater credit risk sensitivity, the Committee has cooperated with the industry to develop a suitable capital charge for operational risk. Operational risk is the risk of direct or indirect loss resulting from inadequate or failed internal processes, people and systems or from external events like the risk of loss from computer failures. Many major banks presently allocate 20% or more of their internal capital to operational risk. The work on operational risk is in a developmental stage, but three different approaches of increasing sophistication have been identified:

1) The basic indicator approach utilises one indicator of operational risk for a bank's total activity.
2) The standardised approach specifies different indicators for different business lines.
3) The internal measurement approach requires banks to utilise their internal loss data in the estimation of required capital.

As a result, the Committee expects operational risk on average to constitute approximately 20% of the overall capital requirements under the new framework.

Overall Capital

Concerning the overall capital, the Committee's goal remains to neither raise nor to lower the aggregate regulatory capital – inclusive of operational risk

– for internationally active banks using the STD approach. With regard to the IRB approach, the ultimate goal is to ensure that the regulatory capital requirement is sufficient to address underlying risks and contains incentives for banks to migrate from the STD to the IRB approach.

Furthermore, the new framework introduces more risk sensitive approaches to the treatment of CRM – a range of techniques whereby a bank can partially protect itself against a counterparty default – and asset securitisation – the packaging of assets or obligations into securities for sale to third parties – under both the STD and the IRB approach.

2.2 The Second Pillar – Supervisory Review Process

The supervisory review pillar requires supervisors to undertake a qualitative review of their bank's capital allocation techniques and compliance with relevant standards.[11] Supervisors have to ensure that each bank has sound internal processes to assess the adequacy of its capital based on a thorough evaluation of its risks. The new framework stresses the importance of bank management developing an internal capital assessment process and setting targets for capital that are commensurate with the bank's particular risk profile and control environment. Thus, supervisors are responsible for evaluating how well banks are assessing their capital adequacy needs relative to their risks. This internal process is – where it is appropriate – subject to supervisory review and intervention.

2.3 The Third Pillar – Market Discipline

The third pillar aims to bolster market discipline through enhanced disclosure requirements by banks which facilitate market discipline.[12] Effective disclosure is essential to ensure that market participants do better understand banks' risk profiles and the adequacy of their capital positions. The new framework sets out disclosure requirements and recommendations in several areas, including the way a bank calculates its capital adequacy and risk assessment methods. The core set of disclosure recommendations applies to all banks with more detailed requirements for supervisory recognition of internal methodologies for credit risk, CRM techniques and asset securitisation.

3 The STD Approach to Credit Risk

This section describes the STD approach to credit risk in the banking book which is the simplest of the three broad approaches to credit risk and is not based on banks' internal rating systems like the other two approaches.[13]

---

[11] See *Consultative Document*, BCBS, January 2001.
[12] See *Consultative Document*, BCBS, January 2001.
[13] See section 4: *The IRB Approach to Credit Risk*.

Compared to the present Accord, the STD approach aligns regulatory capital requirements more closely with the key elements of banking risk by introducing a wider differentiation of risk weights and a wider recognition of CRM techniques, while avoiding excessive complexity. Accordingly, the STD approach produces capital ratios more in line with the actual economic risks that banks are facing. This should improve the incentives for banks to enhance their risk management and measurement capabilities and reduce the incentives for regulatory capital arbitrage. In this review we will concentrate on the most discussed feature - the assignment of risk weights for sovereigns, banks and corporates.

3.1 Risk Weights in the STD Approach

Along the lines of the proposals in the consultative paper to the new capital adequacy framework,[14] the RWA in the STD approach continue to be calculated as the product of the amount of exposures and supervisory determined risk weights:

$$RWA = E \cdot r \tag{5}$$

where: E is the value of the exposure
r is the risk weight of the exposure

As in the current Accord, the risk weights are determined by the category – sovereigns, banks and corporates – of the borrower. However, there is no distinction on the risk weighting depending on whether the country is a member of the OECD. Instead the risk weights for exposures depend on external credit assessments like e.g. rating agencies.

3.2 Risk Weights for Sovereigns and for Banks

Despite the concerns regarding the use of external credit assessments – especially credit ratings – the old Accord (with the 0% risk weight for all sovereigns) was replaced by an approach that relies on sovereign assessments of eligible ECAI. Following the notation,[15] the risk weights of sovereigns and their central banks are as follows:

The assessments used should generally be in respect of the sovereign's long-term rating for domestic and foreign currency obligations. At national

---
[14] See *A New Capital Adequacy Framework*, BCBS, June 1999.
[15] The notation follows the methodology used by Standards & Poor's as an example only.

Table 2. Risk weights of sovereigns – option 1

Rating	Risk Weights
AAA to AA-	0%
A+ to A-	20%
BBB+ to BBB-	50%
BB+ to B-	100%
Below B-	150%
Unrated	100%

discretion, a lower risk weight[16] may be applied to banks' exposures to the sovereign or central bank of incorporation denominated in domestic currency and funded in that currency.[17]

To address at least in part the concern expressed over the use of credit ratings and to supplement private sector ratings for sovereign exposures, there is also the possibility of using country risk ratings assigned to sovereigns by *export credit agencies* (ECA). The key advantage of using publicly available ECA risk scores for sovereigns is that they are available for a far larger number of sovereigns than private ECAI ratings. Banks may then choose to use the risk scores produced by an ECA recognised by their supervisor. As detailed in the below table, each of these ECA risk scores corresponds to a specific risk weight category.[18]

Table 3. Risk weights of sovereigns – option 2

Risk Scores	Risk Weights
1	0%
2	20%
3	50%
4 to 6	100%
7	150%

---

[16] The Committee is no longer requiring adherence to the International Monetary Fund's Special Data Dissemination Standards, the Basel Committee's Core Principles for Effective Banking Supervision or the International Organisation of Securities Commissions' Objectives and Principles of Securities Regulation as preconditions for preferential risk weights.

[17] This is to say that banks also have liabilities denominated in the domestic currency.

[18] The Bank for International Settlements, the International Monetary Fund, the European Central Bank and the European Community receive the lowest risk weight applicable to sovereigns and central banks.

Further there are two options for deciding the risk weights on exposures to banks, but national supervisors have to apply one option to all banks in their jurisdiction. As a general rule for both options, no claim on an unrated bank may receive a risk weight less than that applied to its sovereign of incorporation.

Under the first option – as shown in table 4 – all banks incorporated in a given country are assigned a risk weight one category less favourable than that assigned to claims on the sovereign of incorporation. Therefore, the Committee left the fixed 20 % risk weight of the old Accord for all bank claims. However, for claims to banks in sovereigns rated BB+ to B- and to banks in unrated countries the risk weight is capped at 100%, except for banks incorporated in countries rated below B-, where the risk weight is capped at 150%.

Table 4. Risk weights of banks – option 1

Rating	Sovereign Risk Weights	Bank Risk Weights
AAA to AA-	0%	20%
A+ to A-	20%	50%
BBB+ to BBB-	50%	100%
BB+ to B-	100%	100%
Below B-	150%	150%
Unrated	100%	100%

The second option bases the risk weighting on the external credit assessment of the bank itself. Under this option, a preferential risk weight that is one category more favourable than the risk weight shown in table 5 is applied to all claims with an original maturity[19] of three months or less, subject to a "floor" of 20%. This treatment is available to both – rated and unrated bank claims – but not to banks risk weighted at 150%.

Table 5. Risk weights of banks – option 2

Rating	Risk Weights	Short-Term Claim Risk Weights
AAA to AA-	20%	20%
A+ to A-	50%	20%
BBB+ to BBB-	50%	20%
BB+ to B-	100%	50%
Below B-	150%	150%
Unrated	50%	20%

---

[19] Supervisors should ensure that claims with an original maturity under three months which are expected to be rolled over – i.e. where the effective maturity is longer than three months – do not qualify for this preferential treatment.

In order to maintain liquidity in local inter-bank markets, there are proposals to extend the preferential treatment of domestic government exposures to domestic short-term inter-bank exposures. Accordingly, when the national supervisor has chosen to apply the preferential treatment for claims on the sovereign as described above, it can also assign – under both options – a risk weight that is one category less favourable than that assigned to claims on the sovereign of incorporation – subject to a floor of 20% – to bank claims of an original maturity of three months or less denominated and funded in the domestic currency.

3.3 Risk Weights for Corporates

The maybe most-discussed feature of the new Basel Capital Accord is the assignment of different risk weights on corporate claims. Table 6 illustrates the risk weighting of rated corporate claims according to the second consultive document in 2001, including claims on insurance companies. As a general rule, no claim on an unrated corporate may be given a risk weight preferential to that assigned to its sovereign of incorporation and the standard risk weight for unrated claims on corporates is 100%. As with the case of exposure to banks, there is no sovereign floor, recognising that there are legitimate cases where a corporate can have a higher assessment than the sovereign assessment of its home country.

Table 6. Risk weights of corporates

Rating	Risk Weights
AAA to AA-	20%
A+ to A-	50%
BBB+ to BB-	100%
Below BB-	150%
Unrated	100%

Assigning 100% risk weights to unrated companies is questionable and therefore, still discussed. Obviously, if the risk weighting of unrated exposures is lower than that for low-rated exposures, borrowers with a low rating have an incentive to give up their solicited rating. Therefore, there is a risk of *adverse selection*: For example, if many low-rated corporates give up their ratings, the quality of the average unrated borrower deteriorate to the extent that a 100% risk weight no longer offers sufficient protection against credit risk. As a consequence, it is indispensable to balance awareness of this incentive with consideration of the fact that in many countries corporates and banks do not need to acquire a rating in order to fund their activities. For this reason the assignment of 100% risk weight to unrated companies is still discussed and subject to possible changes. However, the fact that a borrower is not rated does not signal low credit quality. In balancing these conflicting considerations, a 100% risk weight to unrated corporates was assigned. This is the same risk weighting that all corporate exposures receive under the present Accord in order to do not cause an unwarranted increase in the costs of funding for small and medium-sized businesses, which in most countries are a primary source of job creation and of economic growth. However, in countries with higher default rates, national authorities may increase the standard risk weight for unrated claims where they judge that a higher risk weight is warranted by their overall default experience.

3.4 Maturity

Although maturity is a relevant factor in the assessment of credit risk, it is difficult to pursue greater precision in differentiating among the maturities of claims within the STD approach given the broad-brush nature of the counterparty risk weighting. The STD approach is designed to be suitable for application by banks of varying degrees of size and sophistication. However, the costs of increasing the complexity of the STD approach are relatively high. In general, the benefits of improved risk-sensitivity would be outweighed by the costs of greater complexity. Despite its improved risk sensitivity, the new STD approach remains intentionally simple and broad-brush. Therefore, a maturity dimension is not incorporated throughout the STD approach in contrast to the IRB approach. As set out above, the only maturity elements which are included are the distinction between short and long-term commitments, lendings and the use of short-term assessments as it is discussed below.

3.5 Credit Risk Mitigation

Credit Risk Mitigation (CRM) relates to the reduction of credit risk by – for example – taking collateral, obtaining credit derivatives or guarantees or taking an offsetting position subject to a netting agreement.

The current Basel I Accord recognises only collateral instruments and guarantees deemed to be identifiably of the very highest quality. This led to an "all-or-nothing" approach to credit risk mitigants: Some forms were recognised while others were not. Since 1988, the markets for the transfer of credit risk have become more liquid and complex and thus, the number of suppliers of credit protection has increased. New products such as credit derivatives have allowed banks to unbundle their credit risks in order to sell those risks that they do not wish to retain. These innovations result in greater liquidity in itself, reduce the transaction costs of intermediating between borrowers and lenders and also encourage a more efficient allocation of risks in the financial system. In designing a new framework for CRM, three main aims were pursued:

- Improving incentives for banks to manage credit risk in an effective manner.
- Offering an approach that may be adopted by a wide range of banks.
- Relating capital treatments to the economic effects of different CRM techniques and greater consistency in the treatment of different forms of CRM.

The revised approach allows a wider range of credit risk mitigants to be recognised for regulatory capital purposes and depart from the "all-or-nothing" approach. It also offers a choice of approaches that allow different banks to strike different balances between simplicity and risk-sensitivity. As a result, there are three broad treatments to CRM depending on which credit risk approach is used by the banks. However, the treatment of CRM in the STD and in the foundation IRB approach is very similar. While CRM techniques generally reduce credit risk, they do not fully eliminate it. In such transactions, banks – often for good business reasons – leave some residual risks unhedged. Therefore, three forms of residual risk are explicitly addressed: Asset-, maturity- and currency mismatch. As a consequence the determination of CRM numbers offers a lot of options and is too manifold to be described in this review paper. For a detailed description we therefore refer to *Credit Risk mitigation in the Standardised Approach* in *The Standardised Approach to Credit Risk* of the Basel committee on Banking Supervision.

4 The IRB Approach to Credit Risk

4.1 Key Elements and Risk Components of the IRB Approach

In this section we will give an overview of the Internal Ratings-Based Approach (IRB Approach) to capital requirements for credit risk by the Basle Committee on Banking Supervision. The approach relies - in opposite to the Standardised Approach described above heavily upon a bank's internal assessment of its counterparties and exposures. Since the most discussed topic were the risk weights for corporate exposures we will concentrate on this feature of the new Capital Accord in the IRB approach. The assignment of risk weights to banks and sovereign exposures are very similar and based mainly on the same function as calculating the risk weights for corporate exposures.

According to the consultative document, the IRB approach has five key elements:

- A classification of exposures by broad exposure type.
- For each exposure class, certain risk components which a bank must provide, using standardized parameters or its internal estimates.
- A risk-weight function which provides risk weights (and hence capital requirements) for given sets of these components.
- A set of minimum requirements that a bank must meet in order to be eligible for IRB treatment for that exposure.

- Across all exposure classes, supervisory review of compliance with the minimum requirements.

The capital charge for exposures within each of the three exposure classes depends on a set of four risk components (inputs) which are provided either through the application of standardised supervisory rules (foundation methodology) or internal assessments (advanced methodology), subject to supervisory minimum requirements.

Probability of Default (PD)

All banks – whether using the foundation or the advanced methodology – have to provide an internal estimate of the PD associated with the borrowers in each borrower grade. Each estimate of PD has to represent a conservative view of a long-run average PD for the grade in question and has to be grounded in historical experience and empirical evidence. The preparation of the estimates, the risk management processes and the rating assignments that lay behind them, have to reflect full compliance with supervisory minimum requirements to qualify for the IRB recognition.

Loss Given Default (LGD)

While the PD – associated with a given borrower – does not depend on the features of the specific transaction, LGD is facility-specific due to losses are generally understood to be influenced by key transaction characteristics such as the presence of collateral and the degree of subordination.

The LGD value can be determined in two ways: In the first way – respectively under the foundation methodology – LGD is estimated through the application of standard supervisory rules. The differentiated levels of LGD are based upon the characteristics of the underlying transaction, including the presence and the type of collateral. The starting point is a value of 50% for most unsecured transactions whereas a higher value of 70% is applied to subordinated exposures, but the percentage can be scaled to the degree to which the transaction is secured. If there is a transaction with financial collateral, a haircut methodology adapted from the STD approach is used.[20] In the advanced methodology, LGD – which is applied to each exposure – is determined by the banks themselves. Thus, banks using internal LGD estimates for capital purposes are able to differentiate LGD values on the basis of a wider set of transaction and borrower characteristics.

Exposure at Default (EAD)

As with LGD, EAD is also facility-specific. Under the foundation methodology, EAD is estimated through the use of standard supervisory rules and is

---

[20] A separate set of LGD values is applied to transactions with real estate collateral.

determined by the banks themselves in the advanced methodology. In most cases, EAD is equal to the nominal amount of the facility but for certain facilities – e.g. those with undrawn commitments – it includes an estimate of future lending prior to default.

Maturity (M)

Where maturity is treated as an explicit risk component like in the advanced approach, banks are expected to provide supervisors with the effective contractual maturity of their exposures. Where there is no explicit adjustment for maturity, a standard supervisory approach is presented for linking effective contractual maturity to capital requirements.

4.2 Risk Weights for Corporate Exposures

After introducing the entering risk components we will now investigate how the risk weights for corporate exposures are calculated in the IRB approach.

Formula for RWA

The derivation of risk weights is dependent on the estimates of PD, LGD and M that are attached to any exposures. Where there is no explicit maturity dimension in the foundation approach, corporate exposures receive a risk weight (RW_C) that depends on PD and LGD after recognizing any credit enhancements from collateral, guarantees or credit derivatives. The average maturity of all exposures is assumed to be three years, RW_C is expressed as a function of PD and LGD[21] according to the formula below:[22]

$$RW_C = min\left[\frac{LGD}{50} \cdot BRW_C(PD) \; ; \; 12.50 \cdot LGD\right] \quad (6)$$

where: RW_C is the risk weight associated with given PD and LGD
            $BRW_C(PD)$ is the corporate benchmark risk weight associated with a given PD

In the advanced approach – or where there is an explicit maturity dimension in the foundation approach – for an exposure with an effective maturity

---

[21] PD, LGD and EAD are expressed as whole numbers rather than decimals. The only exception are the benchmark risk weight and maturity slope calculations.
[22] The cap is to ensure that no risk weight can be more penal than the effect of deducting from capital the exposure's expected loss in the event of default.

M different from three years, an asset's *maturity-adjusted* risk weight is calculated by scaling up or down the BRW$_C$ for a hypothetical three-year loan having the same PD and LGD. Thus, RW$_C$ in the advanced approach is expressed as a function of PD, LGD and M according to the following formula:

$$RW_C = min\left[\frac{LGD}{50} \cdot BRW_C(PD) \cdot \left(1 + b(PD) \cdot (M-3)\right) \,;\, 12.50 \cdot LGD\right] (7)$$

where: $b(PD)$ is the maturity adjustment factor dependent of PD

$1 + b(PD) \cdot (M-3)$ is the multiplicative scaling factor linear in M

For maturities ranging from one to seven years, a linear relationship between maturity and credit risk is viewed as a reasonable approximation to both industry-standard mark-to-market (MTM) credit risk models – such as J.P. Morgan's CreditMetrics^{TM} (CM) and KMV's Portfolio Manager^{TM} (PM) – and multi-period default-mode (DM) models.[23] Research undertaken to calibrate the maturity adjustment factor indicates that this factor is very sensitive to whether the underlying credit risk modelling approach is based on a MTM or a DM framework. Thus, for each framework an alternative has been developed for the calibration of b(PD).

Calibration of BRW

The *Value at Risk* (VaR) approach recognizes that trading activities generate a return for banks but there is still a range of possible outcomes that includes possible operating losses. The market risk approach considered capital as being required to insulate the institution against a very severe negative event or events in its trading portfolio. Thus, the VaR approach relates capital to some target level of confidence that capital for market risk will not be exhausted.

Since total regulatory capital includes at least a portion of a bank's general loan loss reserves, IRB risk weights have been developed within the context of achieving adequate coverage of total credit losses – i.e expected and unexpected losses (EL and UL) – over an assumed one-year time horizon. Therefore, the concept of economic capital is based on the idea that future gains or losses of a portfolio of credit exposures can be described by its *probability density function* over a specified time horizon. In theory, a bank which knows this function can assign capital that will reduce its probability of failure over the appropriate time horizon to any desired confidence level. This can be seen as the *target solvency probability* or *loss coverage target* for the bank.

---

[23] See *CreditMetrics-Technical Document*, Bathia, M. and Finger, C.C. and Gupton, G.M., April 1997, as well as *The Merton/KMV Approach to Pricing Credit Risk*, Sundaram, R.K., January 2001.

The calibration methods described below are based on the same credit risk modelling framework but modified to ensure coverage of both, EL and UL. Under this framework, the risk weights are implicitly calibrated so that with a specified minimum probability – i.e. the target solvency probability – capital will cover total credit losses. An appropriate balance between two empirical approaches – one direct and another survey-based or indirect – was applied for calibrating risk weights for corporate exposures under the IRB standard which involves the following:

- Estimating the volatility of credit losses in a portfolio over a one-year time horizon.
- Determining the level of capital needed to achieve possible target solvency probability.

Survey-Based Risk Weights: The implementation of the indirect approach for estimating risk weights is based on detailed information about internal economic capital allocations against large corporate loans. The estimated weights imply relative economic capital requirements which are attributed by each bank to corporate loans having particular configurations of PD, LGD and M. The survey data indicated broad comparability across banks in the relationship between PD and relative economic capital levels, holding LGD and M fixed. In contrast to the current 100% risk weight for all corporate loans, the survey evidence highlighted the greater relative credit risk attributed to higher PD borrowers and the proportionality of economic capital levels and banks' LGD assumptions. This suggests, that for large corporate portfolios, a capital standard could be constructed to generate risk weights whose sensitivity with respect to PD would be broadly compatible with existing internal economic capital processes. In opposite to the effects of PD and LGD, M was much less consistent across surveyed banks and depended critically on the underlying credit risk model.

Direct Estimates of Risk Weights: The results of exercises in credit risk modelling are comparable to those obtained from the survey evidence. In particular, while the sensitivity of economic capital levels to PD and LGD were well defined across a range of modelling frameworks and parameter specifications, the effects of M on economic capital were quite sensitive to particular modelling choices.

Derivation of Benchmark Risk Weights

The equation of BRW_C for corporate exposures – dependent on a given PD – was developed on the basis of survey and model-based evidence. Finally, a continuous function was selected to represent the relationship between a corporate borrower's PD and the associated risk weight for a benchmark loan to that borrower having a three-year maturity and LGD equal to 50%. The formula presented here is the BRW function that was usually suggested in the second consultive document of the Basel committee:

$$BRW_C(PD) = 976.5 \cdot \Phi\left(1.118 \cdot \Phi^{-1}(PD) + 1.288\right) \cdot \left(1 + 0.047 \cdot \frac{1-PD}{PD^{0.44}}\right) \quad (8)$$

where: $\Phi(x)$ is the cumulative distribution function for a standard normal random variable.

Obviously the above expression consists of three separate factors. The first term 976.5 is a constant scaling factor to calibrate the BRW_C to 100% for PD = 0.7% and LGD = 50% as shown below:

$$BRW_C(0.7\%) = 976.5 \cdot \Phi(1.118 \cdot \Phi^{-1}(0.007) + 1.288)(1 + 0.047 \cdot \frac{1-0.007}{0.007^{0.44}})$$
$$= 976.5 \cdot \Phi(1.118 \cdot (-2.455) + 1.288) \cdot 1.414$$
$$= 1380.96 \cdot \Phi(-1.457)$$
$$= 100$$

The second term $\Phi(1.118 \cdot \Phi^{-1}(PD) + 1.288)$ represents the sum of EL and UL and is associated with a hypothetical, infinitely-granular portfolio of a one-year loan having a LGD of 100%, based on the idea of only one systematic risk factor X. The systematic risk factor stands e.g. for the current state of the economy that next to firm-specific risks has an influence on the fact whether a firm defaults or not. Such models are in the literature generally called *one-factor models*.[24] The committee emphasizes that this approach includes two industry-standard credit risk models (PM and CM) and provides a reasonable approximation to a third, Credit Suisse Financial Product's *CreditRisk^{+TM}* (CR$^+$).[25]

In the one-factor model, the borrowers' asset-change Y_i can be divided into two components: A systematic (X) and a idiosyncratic (U_j) component whereby the idiosyncratic risk factor is in contrast to the systematic risk factor dependent on every single borrower j (with j = 1,...,K). Concerning these two variables X and U_j, the first two of five assumptions are met:

1) $X, U_1, ..., U_K \sim N(0,1)$.
2) $\forall$ j = 1,...,K: X and U_j are stochastically independent as well as all U_j.

Due to these assumptions, X and U_j (for j = 1,...,K) are standard normal distributed and in pairs stochastically independent but not the sum of X and

---

[24] See *A one-parameter representation of credit risk and transition matrices*, Belkin, B., Forest, L.R., Suchower, S., CreditMetrics Monitor, Third Quarter, 1998 as well as *The one-factor CreditMetrics model in the New Basel Capital Accord*, Finger, C.C., RiskMetrics Journal 2, 2001.

[25] See Credit Risk$^+$: A credit risk management framework, Credit Suisse Financial Products, 1997.

U_j. Therefore, a weight-factor w (with $w \in [0;1)$) has to be introduced with w_j for the systematic risk factor and with $\sqrt{1-w_j^2}$ for the idiosyncratic risk factor as shown in the third assumption:

3) ∀ borrowers j = 1,...,K: The modelling relationship between the borrowers' asset-change Y_i - which is lognormally distributed - the systematic risk factor X and the idiosyncratic risk factors U_j is as follows:

$$Y_j = w_j \cdot X + \sqrt{1-w_j^2} \cdot U_j \qquad (9)$$

In addition, if there are identically correlations ρ, the equation 9 can be simplified to the following:[26]

4) ∀ borrowers j = 1,...,K: There are identical correlations with $\rho = \rho(Y_j, Y_k) = w_j \cdot w_k > 0$.

$$Y_j = \rho \cdot X + \sqrt{1-\rho} \cdot U_j \qquad (10)$$

From this assumption follows, that all Y_i are standard normal distributed with the same multi-normal distribution and the same correlation ρ. Concerning the correlation ρ, the fifth and last assumption is met. This asset return correlation assumption (better known as asset value correlation) is underpinning the proposed corporate credit risk charges and is consistent with the overall banks' experience.[27]

5) The asset return correlation ρ is constant over time, independent of any risk factor and assumed to be 0.2.

Given all these assumptions, a *default-point model* is introduced. In this model, there is a default for borrower j exactly when Y_i is equal or less than a certain default point d_j which – in turn – is defined as:

$$PD_j = P(D_j = 1) = P(Y_j \le d_j) \qquad (11)$$

where: d_j is the default point with: $d_j = \Phi^{-1}(PD_j)$
D_j is a bernoulli distributed default indicator with:
$$D_j = \begin{cases} 1 & , \text{ for default by obligor j} \\ 0 & , \text{ otherwise} \end{cases}$$

---

[26] See *Die Risikogewichte der IRB-Ansätze: Basel II und "schlanke" Alternativen*, Wehrsohn, U. et al, risknews, 11/2001.
[27] See *ISDA's Response to the Basel Committee on Banking Supervision's Consultation on the New Basel Capital Accord*, International Swaps and Derivates Association (ISDA), 2001, S.13.

To go back to the modelling relationship, it is necessary to abstract from a single borrower to the whole homogeneous portfolio. Therefore, in the event of a default – i.e $Y_j \leq d_j$ – the conditional PD for a fixed realisation x of the same risk factor X is used:

$$PD_j(x) := P(D_j = 1 | X = x)$$

$$= P\left[\sqrt{\rho} \cdot x + \sqrt{1-\rho} \cdot U_j \leq \Phi^{-1}(PD_j)\right]$$

$$= P\left[U_j \leq \frac{\Phi^{-1}(PD_j) - \sqrt{\rho} \cdot x}{\sqrt{1-\rho}}\right] \quad (12)$$

As it can be seen from the assumptions above, all U_j are standard normal distributed and the average asset correlation coefficient ρ is 0.20. Furthermore, the Committee decided to calibrate the coefficients within this expression to an assumed loss coverage target of the 99.5% quantile of the standard normal distribution. Thus, the 1 - 99.5% = 0.5% quantile of $N(0,1)$ is $\Phi^{-1}(0.005) = -2.576$. Therefore, $PD_j(x)$ can be calculated as follows:

$$PD_j(z) := \Phi\left(\frac{\Phi^{-1}(PD_j) - \sqrt{0.20} \cdot (-2.576)}{\sqrt{0.80}}\right)$$

$$= \Phi\left(1.118 \cdot \Phi^{-1}(PD_j) + 1.288\right) \quad (13)$$

The last factor $\left(1 + 0.047 \cdot (1 - PD)/(PD^{0.44})\right)$ is an adjustment to reflect that the BRW_C are calibrated to a three-year average maturity.

This construction is based on survey evidence and simulation results which were pooled judgementally to develop a smooth functional relationship between the values of PD and b(PD).

A graphical depiction of IRB risk weights for a hypothetical corporate exposure having an LGD equal to 50% without an explicit maturity dimension is presented below. Note that for given PD, the corresponding BRW_C can be calculated by using formula 8 as presented in Table 7 below. Furthermore, the minimum BRW_C is 14 due to a 0.03%-floor which was imposed for PD values. The floor is due to the committees evaluation of banks difficulties in measuring PDs adequately. As described in the minimum risk weight function 6, the RW_C are capped at 625% which corresponds to a PD of exactly 17.15%. From this point on, the BRW_C function runs above the RW_C function.

Maturity-Adjustments to Corporate Risk Weights

Especially in the advanced IRB approach maturity is treated as an explicit risk component. Therefore, the approach to maturity in the Basel Capital Accord

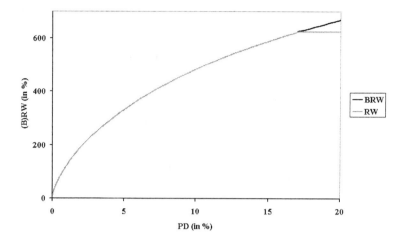

Fig. 4. (Benchmark) risk weights as a function of PD

Table 7. Benchmark risk weights for corporate exposures

PD (in %)	BRW_C (in %)	PD (in %)	BRW_C (in %)
0.03	14.1	1.0	125.0
0.05	19.1	2.0	192.4
0.1	29.3	3.0	246.0
0.2	45.1	5.0	331.4
0.4	69.9	10.0	482.4
0.5	80.6	15.0	588.0
0.7	99.8	20.0	668.2

shall also be discussed here. Generally within the banking industry two classes of credit risk models tend to be most prevalent, the mark-to-market (MTM) and default-mode (DM) models.

Both models are generally employed under the assumption of a one-year time horizon and both capture credit losses associated with defaults that occur within this time horizon. However, the two approaches differ with regard to how they deal with credit deterioration short of default. While in default-mode models there are only two states of an exposure, default and non-default, in MTM models changes in a loan's credit quality are also considered. Usually in a MTM model the probability of an upgrade or downgrade - and therefore of an improvement or an deterioration of the credit quality - of a loan is modelled by a transition matrix.

The sensitivity of economic capital to maturity depends critically on the choice of the credit risk model. In this light, two alternative schedules of maturity adjustment factors b(PD) were developed, one based on a MTM another on a DM framework.

While both approaches were described in the second quantitative impact study we will illustrate here only one of the two options: the MTM approach to maturity. The reason is that in the new suggestions of the Basel committee probably only this option will be considered, while the DM option was excluded from the accord.

Changes in a loan's credit quality over the time horizon are translated into changes in the loan's economic value based on an assumed valuation relationship. This links the loan's risk rating to an assumed market-based credit spread that is used to value the loan at the end of the horizon. A credit deterioration short of default is presumed to reduce a loan's value, generating an implicit credit loss. The sensitivity of a loan's end-of-horizon value to a credit quality deterioration short of default is dependent on its maturity. As a consequence, maturity has a substantial influence on economic capital within MTM models, with longer-maturity loans requiring greater economic capital. The schedule of maturity adjustment factors is based on an underlying MTM calibration approach. The calibration of b(PD) according to the committee reflects survey evidence and simulation results and is a smooth functional relationship between PD and b(PD):

$$b(PD) = \frac{0.0235 \cdot (1 - PD)}{PD^{0.44} + 0.047 \cdot (1 - PD)} \tag{14}$$

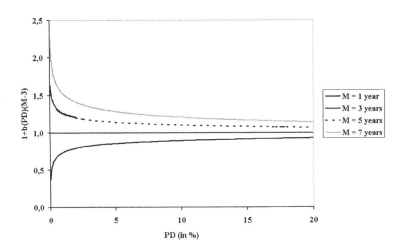

Fig. 5. MTM-based maturity adjustment for different maturities

As it can be seen from the chart above, the MTM-based maturity adjustment factors are a decreasing function of PD. This inverse relationship reflects the fact that maturity has a greater proportional effect on economic capital the greater the probability of downward credit quality migrations short of default relative to PD. For very low PD values – i.e. very high credit quality – the likelihood of a downgrade short of default within one year is high relative to the likelihood of a default. Consequently, the effect of maturity on economic capital is relatively large. In contrast, as PD values increase, the likelihood of default within one year increases more rapidly than the likelihood of a downgrade short of default, implying a reduced sensitivity of economic capital to maturity. In the limit as PD approaches 100%, it becomes certain that the borrower will default within the one-year horizon and the likelihood of a downgrade short of default tends to zero, implying that maturity has little or no effect on economic capital, i.e. b(PD) equals 0.

At the end of this section, a graphical representation of risk weights is given for corporate exposures with a maturity of 1, 3, 5 and 7 years setting LGD to 50% as it is assigned in the foundation approach. Especially for exposures with a long maturity and a high PD estimate, the ceiling of a 625% is the limit for the assigned risk weight. Obviously for longer maturity horizons the assigned risk weights are clearly higher for the same probability of default.

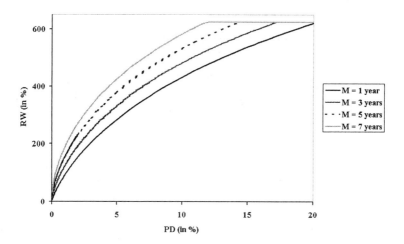

Fig. 6. Risk weights as a function of PD for different maturities

5 Modification of the Approaches

5.1 Severe Criticism of the Suggestions

The suggestions both for the standardised and IRB approach of the second consultative document were subject to extensive discussions. Especially small and medium-sized companies (SMEs) were afraid of higher capital costs for banks that would lead to worse credit conditions for these companies. Also the desired incentive character of the IRB approach for banks was very questionable, since risk weights in many cases were rather higher for the IRB approach than for the STD approach.

Table 8 shows a comparison of the assigned risk weights for the old accord and both approaches of the new capital accord for different risk classes. Obviously there is a clear tendency in the IRB of assigning lower risk weights to companies with a very good rating and much higher risk weights to such companies with a rating worse than BB-. On the one hand this is consistent with recommendations e.g. by Altman/Saunders (2000) where is stated that risk weights for AAA or AA bonds in the ST approach are chosen too high while for BB- or worse rated companies the risk weights may be chosen too low. However, considering especially SMEs that will probably not obtain a rating in one of the first three categories, banks will have to hold a higher amount of capital for such companies.

Table 8. Comparison of risk weights

Rating	AAA to AA-	A+ to A-	BBB+ to BB-	below BB-	without
Old Accord	100%	100%	100%	100%	100%
ST	20%	50%	100%	150%	100%
IRB	14%	14%	36%-200%	200%-625%	–

Thus, especially small and medium sized companies and companies with a rating rather in the "speculative grade" area may suffer from higher capital costs for the banks that will be passed down to the companies.

One can further see that there is not a real incentive for banks with many exposures in the "higher risk" area to use the internal rating based approach, since the capital to hold may be much higher than so far. Therefore, it is questionable whether the Incentive Compatibility which shall encourage banks to continue to improve their internal risk management practices is really existing.

Also the relationship between assigned risk weights and maturity of the second consultative document is questionable. The assumed linear relationship between maturity and assigned risk weights can be observed in these curves.

Comparing the assigned Risk Weights to actually observed spreads in the market, one can find that for high rated bonds credit spreads rise - maybe even in a linear relationship - with longer maturities. This behavior matches

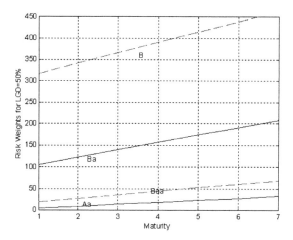

Fig. 7. Assigned risk weights as function of maturity for different ratings

the model of the Basel committee where higher risk weights are assigned to exposures with longer maturities. However, especially for lower rated bonds credit spreads do not show a positive correlation with maturity. For Ba rated bonds the spreads are constant while for single B rated bonds the spreads fall from year one. Obviously the assigned Risk Weights according to the maturity adjustments are not really matching the market Credit Spreads for corporate exposures rated from B to Baa. The reason for the falling credit spreads in lower rating categories can be explained empirically by the fact that as the threat of default recedes, risk neutral investors require a smaller yield spread to compensate them for expected default loss.

It seems as if this market behaviour of credit spreads is not incorporated in the MTM model also considering maturities. Even worse, many SMEs were afraid that due to higher risk weights for long-term loans, banks could even refuse to make such contracts anymore.

The problems and criticism mentioned above was also confirmed by so-called quantitative impact studies (QIS) conducted by banks for the Basel committee. According to the second quantitative impact study (QIS2) by the Basel Committee requirements for banks would increase by the suggestions of the STD approach. For the IRB approach capital requirements would even be higher. According to a study for several of the G10 major banks the capital requirements would be between 6% and 14% higher. Therefore, the goals of the accord to keep the overall capital unchanged were not satisfied.

Thus, obviously to match the goals set by the committee there was need for a revision of some features in the accord - especially for the BRW function in the IRB approach that should provide an incentive for banks to use the IRB and refine their risk management procedure.

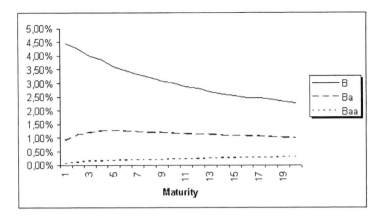

Fig. 8. Maturity and credit spreads for bonds rated Baa-B

5.2 Possible Changes in the Final Version

According to the Basel Committee there will be several changes in the final version of the accord. After a third quantitative impact study that is conducted in late 2002 the final version will probably be available in 2003. Still, there has already been an announcement of possible changes by the committee and further suggestions for a refined risk weight function and different treatment of maturity, SMEs and overall capital. In the following section some of the possible changes will be briefly described according to a press release by the committee.

Treatment of SMEs

One of the major criticism was the treatment of more risky loans or companies especially small and medium sized companies (SMEs). In recognition of the different risks associated with SME borrowers, under the IRB approach for corporate credits, banks will now be permitted to separately distinguish loans to SME borrowers from those to larger firms. SME borrowers are defined as companies with less than Euro 50 Million in annual sales - thus, e.g. in Germany more than 90% of the companies will fall into this class. Under the proposed treatment, exposures to SMEs will be able to receive a lower capital requirement than exposures to larger firms. The reduction in the required amount of capital will be up to twenty percent, depending on the size of the borrower. Further, banks that manage small-business-related exposures in a manner similar to retail exposures will be permitted to apply the less capital requiring retail IRB treatment[28] to such exposures, provided that the

---

[28] For a detailed description of the treatment of retail exposures see *Consultative Document - The IRB approach to Credit Risk*, BCBS, 2001

total exposure of a bank to an individual SME is less than Euro 1 Million. Such exposures are then treated the same way as credits to private customers. A similar threshold is also assumed to be established in the standardised approach. The committee assumes that this should result in an average reduction of approximately ten percent across the entire set of SME borrowers in the IRB framework for corporate loans.

Treatment of Maturity

A second criticised drawback was the assigned maturity adjustment in the second consultative document - especially the linear relationship between maturity horizon and the assigned risk weights for more risky loans. Banks using the advanced IRB framework for corporate lending, as well as banks using the foundation IRB approach, will be required to incorporate maturity adjustments calculated using a mark-to-market methodology. However, in recognition of the unique characteristics of national markets, supervisors will have the option of exempting smaller domestic firms from the maturity framework. In this framework smaller domestic firms are defined as those with consolidated sales and consolidated assets of less than Euro 500 Million. If the exemption is applied, all exposures to qualifying smaller domestic firms will then be assumed to have an average maturity of 2.5 years, as under the foundation IRB approach. With these new regulations the committee hopes to counter the fears that due to higher risk weights, banks would refuse to give long-term loans to SMEs.

Treatment of Overall Capital

One concern that has been identified in the Committee's prior impact surveys has been the potential gap between the capital required under the foundation and advanced IRB approaches. To modestly narrow this gap, the average maturity assumption in the foundation approach will be modified from 3 years to 2.5 years, and the majority of the supervisory "loss-given-default" (LGD) values in the foundation IRB approach will be reduced by five percentage points (e.g. for senior unsecured exposures from 50% to 45%). These changes will be combined with offsetting changes to the IRB risk-weight function for corporate lending.

More fundamentally, the Committee is also proposing to alter the structure of the minimum floor capital requirements in the revised Accord. Under the new approach, there will be a single capital floor for t2006 and 2007 following implementation of the new Accord. This floor will be based on calculations using the rules of the existing Accord. Beginning year-end 2006 and during the first year following implementation, IRB capital requirements for credit risk together with operational risk capital charges cannot fall below 90% of the current level required, and in the second year, the minimum will be 80% of this minimum. Should problems emerge during this period, the Committee will seek to take appropriate measures to address them, and, in particular, will be prepared to keep the floor in place beyond 2008 if necessary.

The capital requirements for the various exposures included in QIS 3 have been designed to be consistent with the Committee's goal of neither significantly decreasing nor increasing the aggregate level of regulatory capital in the banking system. Nevertheless, it is possible that the QIS 3 study may indicate the need for some adjustments. It is important to note that - if required - the Committee is prepared to make both upward and downward adjustments to the amount of required capital.

The Refined BRW Function

As a consequence of the fair comment on the risk weight function assigned in the IRB approach the Basel committee will suggest a refined BRW function. The modified formula relating probability of default (PD) to capital requirements differs from the formula proposed in January in several ways.

One major change is that there is no explicit scaling factor in the formula anymore. Also the confidence level that was implicit in the formula has been increased from 0.995 to 0.999, to cover some of the elements previously dealt with by the scaling factor. The January formula incorporated an implicit assumption that asset correlation is equal to 0.20. The new formula assumes that asset correlation declines with PD according to the following formula:

$$\rho(PD) = 0.1 \cdot \frac{1 - e^{-50 \cdot PD}}{1 - e^{-50}} + 0.2 \cdot (1 - \frac{(1 - e^{-50 \cdot PD})}{1 - e^{-50}}) \tag{15}$$

For the lowest PD value it is equal to 0.20 and for the highest PD value it is equal to 0.10. There is at this stage no modification to the proposed inclusion of an implicit maturity adjustment for all exposures in the foundation IRB approach based on the assumption of an average three-year maturity. The modified formula can then be calculated by first calculating the correlation value that corresponds to the appropriate PD value. This value enters then the main formula for the capital requirement. Capital requirements and risk-weighted assets are related in a straightforward manner. With the given confidence level and $\Phi^{-1}(0.999) = 3.090$, the resulting formula is the following:

$$BRW_C(PD) = \Phi\left(\frac{\Phi^{-1}(PD) + \sqrt{\rho} \cdot 3.090}{\sqrt{1 - \rho}}\right)\left(1 + 0.047 \cdot \frac{1 - PD}{PD^{0.44}}\right) \tag{16}$$

As it can be seen in figure 9 the combined impact of these changes is a risk-weight curve that is generally lower and flatter than that proposed in January, which are the directions suggested by industry feedback and the Committees own quantitative efforts.

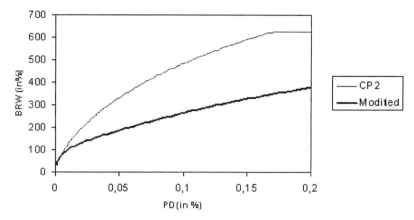

Fig. 9. Modified corporate risk weight curve compared to CP2

With this altered risk weight functions assigned BRWs are clearly lower than before and especially for more risky loans the required capital will decrease by up to 40%.

6 Summary

In 1988 when the first Capital Accord was publicized, there was only one option for measuring the appropriate capital of internationally active banks. Since then the business of banking and the financial markets have undergone significant changes. Therefore, the Committee was obliged to develop a new Accord which should be more comprehensive and more risk sensitive to the default risk of the obligor than the old one. As a consequence, from 2006 on banks will have the possibility to choose one of three approaches to credit risk for their portfolios. The choice wether a bank uses the STD, the foundation IRB or the advanced IRB approach depends critically on the ability of estimating own risk components and on meeting supervisory requirements. However, in the long run all internationally active banks should use the advanced IRB approach or at least the foundation IRB approach in line with improvements in their risk management practices and in line with the benefit in holding their intrinsic amount of credit risk.

Nevertheless, there were also some undesirable drawbacks in the new Accord and not all of them have been changed yet. For instance, there was no incentive for banks dealing with low-rated companies to use one of the IRB approaches to calculate their risk weights. In the version that was suggested in the second consultative document, in the STD approach the risk weights of good-rated companies were chosen too high while for worse-rated companies they were too low. Low-rated companies may still have an incentive to give up their solicited rating if the 100% risk weight for non-rated companies remains. Another point of critique are the requirements for using one of the IRB approaches. They were originally set too high, so that it might still be questionable whether all banks will be willing to use the IRB approach by the beginning of 2006.

By reading this survey paper, one should always keep in mind that the described version of the new Basel II Accord is probably not the final one which will be implemented. There ares still many features and several changes in discussion, before the final version will be provided in 2003. However, what remains is the evolution of a more risk sensitive and comprehensive regulatory approach to credit risk. Also the way the BIS is dealing with suggested changes and criticism is commendable: the final version is likely to be a result of a lively discussion between regulators, practitioners from banks and researchers in the field of credit risk.

References

1. Aleksiejuk, A. and Holyst, J.A. (2001). A simple model of bank bankruptcies.
2. Basel Committee on Banking Supervision (November 2001) Potential Modifications to the Committee's Proposal.
3. Basel Committee on Banking Supervision (June 1999). A New Capital Adequacy Framework.
4. Basel Committee on Banking Supervision (April 1999). Credit Risk Modelling: Current Practices and Applications.
5. Basel Committee on Banking Supervision (July 1988). Internal Convergence of Capital Measurement and Capital Standards.
6. Basel Committee on Banking Supervision (January 2001). Consultative Document – Overview of the New Basel Capital Accord.
7. Basel Committee on Banking Supervision (January 2000). Range of Practice in Banks' Internal Ratings Systems.
8. Basel Committee on Banking Supervision (January 2001). Consultative Document – The Internal-Ratings-Based Approach to Credit Risk.
9. Basel Committee on Banking Supervision (January 2001). Consultative Document – The Standardised Approach to Credit Risk.
10. Basel Committee on Banking Supervision (January 2001). Consultative Document – The New Basel Capital Accord.
11. Bathia, M. and Finger, C.C. and Gupton, G.M. (April 1997). CreditMetrics-Technical Document. J.P. Morgan.
12. Baule, R. and Entrop, O. and Wilkens, M. (December 2001). Basel II – Berücksichtigung von Diversifikationseffekten im Kreditportfolio durch das Granularity Adjustment. Kreditwesen.
13. Belkin, B. and Forest, L.R. and Suchower, S. (Third Quarter 1998). A one-parameter representation of credit risk and transition matrices. CreditMetrics Monitor.
14. Belkin, B. and Forest, L.R. and Suchower, S. (First Quarter 1998). The effect of systematic credit risk on loan portfolio value-at-risk and loan pricing. CreditMetrics Monitor.

15. Boos, K.-H. and Schulte-Mattler, H. (May 2001). Basel II: Externes und internes Rating. Die Bank.
16. Credit Suisse Financial Products (1997). Credit Risk$^+$: A credit risk management framework.
17. Crosby, P.J. (1999). Modelling Default Risk. KMV Corporation.
18. Deutsche Bundesbank (April 2001). Die neue Baseler Eigenkapitalvereinbarung (Basel II). Monatsbericht.
19. Entrop, O. and Völker, J. and Wilkens, M. (April 2001). Strukturen und Methoden von Basel II – Grundlegende Veränderung der Bankenaufsicht. Kreditwesen.
20. Europäische Zentralbank (EZB) (May 2001). Die neue Baseler Eigenkapitalvereinbarung aus Sicht der EZB. Monatsbericht.
21. Finger, C.C. (2000). A comparison of stochastic default rate models. RiskMetrics Journal 1.
22. Finger, C.C. (April 1999). Conditionla approaches for CreditMetrics portfolio distributions. CreditMetrics Monitor.
23. Finger, C.C. (2001). The one-factor CreditMetrics model in the new Basel Capital Accord. RiskMetrics Journal 2.
24. Frye, John (2000), Depressing Recoveries, Working Paper
25. Gordy, M. B. (2000). A comparative anatomy of credit risk models. Journal of Banking & Finance 24.
26. Gordy, M. B. (February 2001). A risk-factor model foundation for ratings-based bank capital rules.
27. International Swaps and Derivates Association (ISDA) (2001). ISDA's Response to the Basel Committee on Banking Supervision's Consultation on the New Basel Capital Accord.
28. J.P. Morgan & Co.Incorporated (April 1997). Credit Metrics.
29. Link, T. and Rachev, S. and Trück, S. (October 2001). New Tendencies in Rating SMES with Respect to Basel II.
30. Sundaram, R.K. (January 2001). The Merton/KMV Approach to Pricing Credit Risk.
31. Trück, S. (September 2001). Basel II and the Consequences of the IRB Approach for Capital Requirements. Proceedings of the METU International Conference in Econometrics, Ankara, Turkey.
32. Wehrspohn, U. and Gersbach, H. (November 2001). Die Risikogewichte der IRB-Ansätze: Basel II und "schlanke" Alternativen. risknews.
33. Wilde, T.(May 2001). IRB approach explained. Risk.
34. Wilde, T.(August 2001). Probing Granularity. Risk.
35. Zentraler Kreditausschuss (2001a). Kurzfassung der Stellungnahme des Zentralen Kreditausschusses zum Konsultationspapier des Baseler Ausschusses zur Neuregelung der angemessenenen Eigenkapitalausstattung von Kreditinstituten vom 16.Januar 2001 ("Basel II")..
36. Zentraler Kreditausschuss (2001b). Stellungsnahme des Zentralen Kreditausschusses zum Konsultationspapier des Baseler Ausschusses zur Neuregelung der angemessenen Eigenkapitalausstattung von Kreditinstituten vom 16.Januar 2001 ("Basel II").

Systematic Risk in Homogeneous Credit Portfolios*

Christian Bluhm[1] and Ludger Overbeck[2]

[1] HypoVereinsbank, Munich, Germany
[2] Deutsche Bank, Frankfurt, Germany

1 Systematic Risk in Credit Portfolios

In credit portfolios (see [5] for an introduction) there are typically two types of counterparties: Listed firms and non-listed borrowers. For the first type, a time series of the firm's equity values can be used to derive an *Ability-to-Pay Process* (APP), showing for every point in time the firm's ability to pay, see e.g. [6]. For the second type, equity processes are not available, but still every borrower somehow admits an APP, depending on the customer's assets and liabilities, sometimes known by the lending institute, but in any case imposed as an unobservable *latent variable*. In general, we can expect that correlations between the obligor's APPs strongly influence the portfolio's credit risk. The calculation of *APP correlations* usually is based on a suitable factor model, e.g., a (single-beta) linear model

$$r_i = \beta_i \Phi_i + \varepsilon_i , \qquad (1)$$

where r_i denotes the standardized log-return of the i-th borrower's APP, Φ_i denotes the *composite factor* of borrower i, and ε_i denotes the *residual* part of r_i, which can not be explained by the customer's composite factor. Usually the composite factor of a borrower is itself a weighted sum of country- and industry-related indices, see e.g. [5], Chapter 1. Along with representation (1) comes a decomposition of variance,

$$\mathbb{V}[r_i] = 1 = \underbrace{\beta_i^2 \mathbb{V}[\Phi_i]}_{=R_i^2, \text{ systematic}} + \underbrace{\mathbb{V}[\varepsilon_i]}_{=1-R_i^2, \text{ specific}} \qquad (2)$$

in a *systematic* and an *idiosyncratic* effect. The systematic part of variance is the so-called *coefficient of determination*, denoted by R_i^2, implicitly determined by the regression (1). It can be seen as a quantification of the *systematic risk* of borrower i and is an important input parameter in credit portfolio

---
* The contents of this paper reflects the personal view of the authors and not the opinion of HypoVereinsbank or Deutsche Bank.

management tools, heavily driving the portfolio's *Economic Capital*[3] (EC). For example, the following chart shows CEC, the *contributory EC* (w.r.t. a reference portfolio of corporate loans to middle-size companies) as a function of R^2 for a loan with a *default probability* of 30bps, a *severity*[4] of 50%, a 100% country weight in Germany, and a 100% industry weight in automotive industry:

CEC in % Exposure

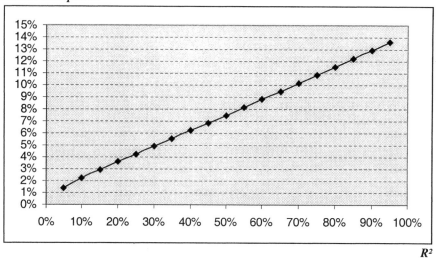

Fig. 1. CEC as a function of R^2

The chart shows that the increase in contributory EC implied by an increase of systematic risk (quantified by R^2) is significant.

This note has a two-fold intention: First, we want to present a simple approach for estimating the systematic risk, that is, the parameter R^2, for a homogeneous credit portfolio. Second, we discuss the proposal of the *Basel Committee on Banking Supervision*, see [1], to fix the *asset correlation* respectively the systematic risk for the calibration of the benchmark risk weights for corporate loans at the 20% level. In our discussion we apply the method introduced in the first part of this note to Moody's corporate bond default statistics and compare our estimated APP correlation with the correlation level suggested in the Basel II consultative document. On one hand, our findings show that the average asset correlation within a rating class is close to the

---

[3] The Economic Capital w.r.t. a level of confidence α of a credit portfolio is defined as the α-quantile of the portfolio's loss distribution minus the portfolio's expected loss (i.e. the mean of the portfolio's loss distribution).

[4] A severity of 50% means that in case of default the recovery rate can be expected to be (1-severity)=50%.

one suggested by the Basel II approach. On the other hand, the assumption of a *one-factor model* with a uniform asset correlation of 20% as suggested in the current draft of the new capital accord turns out to be violated as soon as we consider the correlation between different segments, e.g., rating segments as in our case.

2 Homogeneous Credit Portfolios

The simplest way to model default or survival of borrowers in a credit portfolio is by means of binary random variables, where a '1' indicates default and a '0' means survival w.r.t. to a certain valuation horizon.

2.1 A General Mixture Model for Uniform Portfolios

We start with a standard mixture model of *exchangeable binary random variables*, see [10], 7.1.3. More precisely, we model our credit portfolio by a sequence of Bernoulli random variables $L_1, ..., L_m \sim B(1;p)$, where the default probability p is random with a distribution function F (with support in $[0,1]$), such that given p, the variables $L_1, ..., L_m$ are conditionally i.i.d. The (unconditional) joint distribution of the L_i's is then determined by the probabilities

$$\mathbb{P}[L_1 = l_1, ..., L_m = l_m] = \int_0^1 p^k(1-p)^{m-k} dF(p), \; k = \sum_{i=1}^m l_i, \; l_i \in \{0,1\}. \quad (3)$$

The uniform default probability of borrowers in the portfolio is given by

$$\bar{p} = \mathbb{P}[L_i = 1] = \int_0^1 p \, dF(p) \quad (4)$$

and the uniform default correlation between different counterparties is

$$r = \text{Corr}(L_i, L_j) = \frac{\mathbb{P}[L_i = 1, L_j = 1] - \bar{p}^2}{\bar{p}(1-\bar{p})} = \frac{\int_0^1 p^2 dF(p) - \bar{p}^2}{\bar{p}(1-\bar{p})}. \quad (5)$$

Therefore, $\text{Corr}(L_i, L_j) = \mathbb{V}[Z]/(\bar{p}(1-\bar{p}))$, where Z is a random variable with distribution F, showing that the dependence between the L_i's is either positive or zero. Moreover, $\text{Corr}(L_i, L_j) = 0$ is only possible if F is a Dirac distribution (degenerate case). The other extreme, $\text{Corr}(L_i, L_j) = 1$, can only occur if F is a Bernoulli distribution, $F \sim B(1;\bar{p})$.

2.2 Construction of a Homogeneous Portfolio

Being started from a general perspective, we now briefly elaborate one possible approach to construct a mixture distribution F reflecting the APP-model indicated in the introduction. Following the classical *Asset Value Model* of MERTON [11] and BLACK / SCHOLES [4], we model the borrower's APPs as correlated geometric Brownian motions,

$$dA_t(i) = \mu_i A_t(i)dt + \sigma_i A_t(i)dB_t(i) \qquad (i = 1, ..., m), \qquad (6)$$

where $(B_t(1), ..., B_t(m))_{t \geq 0}$ is a multivariate Brownian motion with correlation ϱ (the *uniform APP-correlation*). Assuming a one-year time window, the vector of asset returns at the valuation horizon, $\left(\ln \frac{A_1(1)}{A_0(1)}, ..., \ln \frac{A_1(m)}{A_0(m)} \right)$, is multivariate normal with mean vector $\left(\mu_1 - 0.5\sigma^2, ..., \mu_m - 0.5\sigma^2 \right)$ and covariance matrix $\Sigma = (\sigma_i \sigma_j \varrho_{ij})_{1 \leq i,j \leq m}$ where $\varrho_{ij} = \varrho$ if $i \neq j$ and $\varrho_{ij} = 1$ if $i = j$. A standard assumption in this context is the existence of a so-called *default point* $\tilde{c}_i$ for every borrower i such that i defaults if and only if its APP at the valuation horizon falls below $\tilde{c}_i$, see CROSBIE [6] for more information about the calibration of default points. So we can define binary variables by a *latent variables* approach,

$$L_i = 1 \quad \Longleftrightarrow \quad A_1(i) < \tilde{c}_i \qquad (i = 1, ..., m).$$

As a consequence of the chosen framework we obtain

$$\bar{p} = \mathbb{P}[L_i = 1] = \mathbb{P}[A_1(i) < \tilde{c}_i] = \mathbb{P}[X_i < c_i] = N[c_i], \qquad (7)$$

where N denotes the standard normal distribution function, the variables X_i are standard normal with uniform correlation ϱ, and $c_i = (\ln \tilde{c}_i - \ln A_0(i) - \mu_i + 0.5\sigma_i^2)/\sigma_i$. Moreover, (7) shows that the c_i's must be equal to a constant c, namely the $\bar{p}$-quantile of the standard normal distribution, $c = N^{-1}(\bar{p})$.

Because the distribution of a Gaussian vector is uniquely determined by their expectation vector and covariance matrix, we can parametrize the variables X_i by means of a *one-factor model*

$$X_i = \underbrace{\sqrt{\varrho}\, Y}_{\text{systematic}} + \underbrace{\sqrt{1-\varrho}\, Z_i}_{\text{specific}} \qquad (i = 1, ..., m), \qquad (8)$$

where $Y, Z_1, ..., Z_m$ are independent standard normal random variables. Equation (8) is obviously a linear regression equation, and based on (1) and (2) we see that the systematic risk or R^2 of the regression is given by the APP-correlation ϱ.

Therefore, estimating systematic risk within our parametric framework means estimating the asset respectively APP correlation ϱ. As soon as ϱ is determined, the *default correlation* r is also known, because based on equation (5) we only need to know the joint default probability $\mathbb{P}[L_i = 1, L_j = 1]$. Because the X_i's are standard normal, the *Joint Default Probability* (JDP) is given by the bivariate normal integral

$$\mathbb{P}[X_i < c, X_j < c] = \frac{1}{2\pi\sqrt{1-\varrho^2}} \int_{-\infty}^{N^{-1}(\bar{p})} \int_{-\infty}^{N^{-1}(\bar{p})} e^{-\frac{1}{2}(x_1^2 - 2\varrho x_1 x_2 + x_2^2)/(1-\varrho^2)} dx_1 dx_2 \qquad (9)$$

So for fixed $\bar{p}$ we can derive r from ϱ and vice versa by evaluating formulas (5) and (9).

At this point we come back to the distribution F in our mixture model (3). From (8) we derive

$$\bar{p} = \mathbb{P}[L_i = 1] = \int_{-\infty}^{\infty} \mathbb{P}[L_i = 1 \mid Y = y] \, dN(y) = \int_{-\infty}^{\infty} g(y) \, dN(y)$$

where

$$g(y) = \mathbb{P}[L_i = 1 \mid Y = y] = \mathbb{P}\left[\sqrt{\varrho}\, Y + \sqrt{1-\varrho}\, Z_i < c_i \mid Y = y\right] \quad (10)$$

$$= \mathbb{P}\left[Z_i < \frac{c - \sqrt{\varrho}\, Y}{\sqrt{1-\varrho}} \mid Y = y\right] = N\left[\frac{N^{-1}(\bar{p}) - \sqrt{\varrho}\, y}{\sqrt{1-\varrho}}\right],$$

because Z_i is standard normal. We therefore obtain equation (4) with F being the distribution function of the random variable $g(Y)$, $Y \sim N(0,1)$,

$$F = N(0,1) \circ g^{-1}. \quad (11)$$

Note that this is just one possible approach to realize a mixture model of exchangeable binary variables. The fundamental assumption here is the lognormality of APPs. For related work regarding homogeneous or uniform portfolios we refer to BELKIN ET. AL. [2]-[3], FINGER [9], and VASICEK [13]. For a more detailed investigation of mixture models applied to credit risk modelling we refer to FREY AND MCNEIL [8], and to Chapter 2 in [5].

3 Estimation of Correlation

In this section we fix F as in (11) and assume the underlying model. The (percentage) portfolio loss is given by $L = \frac{1}{m} \sum_{i=1}^{m} L_i$, and its distribution is determined by (3). We assume that we observed a time-series of vectors of default events $(\hat{L}_{j1}, ..., \hat{L}_{jm_j})_{j=1,...,n}$ where j refers to the year of observation and m_j denotes the number of counterparties in the portfolio in year j. The write-offs immediately imply default frequencies

$$\hat{p}_j = \frac{1}{m_j} \sum_{i=1}^{m_j} \hat{L}_{ji} \quad (j = 1, ..., n).$$

According to our model assumption and Equation (10) we can also write

$$g(y_j) = \hat{p}_j = \frac{1}{m_j} \sum_{i=1}^{m_j} \hat{L}_{ji},$$

where y_j denotes the (unknown!) realization of the factor Y in year j. Conditional on y_j the variables L_{ji} are i.i.d. Bernoulli for fixed j. The observed default frequency $\hat{p}_j$ therefore constitutes the standard maximum-likelihood estimate for the default probability $g(y_j)$ of year j. Assuming $y_1,...,y_n$ to be realizations of i.i.d. copies $Y_1,...,Y_n$ of Y, we obtain

$$\frac{1}{n}\sum_{j=1}^n g(Y_j) \stackrel{n\to\infty}{\longrightarrow} \mathbb{E}[g(Y)] = \bar{p} \quad \text{a.s.}$$

$$\frac{1}{n-1}\sum_{j=1}^n \left(g(Y_j) - \overline{g(Y)}\right)^2 \stackrel{n\to\infty}{\longrightarrow} \mathbb{V}[g(Y)] \quad \text{a.s.} \quad (12)$$

where $\overline{g(Y)} = \sum g(Y_j)/n$. Therefore, the sample mean and variance

$$m_p = \frac{1}{n}\sum_{j=1}^n \hat{p}_j \quad \text{and} \quad s_p^2 = \frac{1}{n-1}\sum_{j=1}^n (\hat{p}_j - m_p)^2$$

are reasonable estimates the mean and variance of $g(Y)$. The underlying unknown asset correlation ϱ is the only 'free' parameter in the variance of $g(Y)$. Then, (5) and (9) yield

$$\mathbb{V}[g(Y)] = \mathbb{E}[g(Y)^2] - \mathbb{E}[g(Y)]^2 = \int_0^1 p^2 dF(p) - \bar{p}^2 =$$

$$= \frac{1}{2\pi\sqrt{1-\varrho^2}} \int_{-\infty}^{N^{-1}(\bar{p})} \int_{-\infty}^{N^{-1}(\bar{p})} e^{-\frac{1}{2}(x_1^2 - 2\varrho x_1 x_2 + x_2^2)/(1-\varrho^2)} dx_1 dx_2 \; - \bar{p}^2 \; . \quad (13)$$

Estimating $\mathbb{V}[g(Y)]$ by s_p^2 and $\bar{p}$ by m_p, we can now determine ϱ by solving the equation

$$s_p^2 = N_2\left[N^{-1}(m_p), N^{-1}(m_p); \varrho\right] - m_p^2 \quad (14)$$

for ϱ. Here, $N_2(\cdot,\cdot,\rho)$ denotes the standard bivariate normal distribution function with correlation ρ. Note again that in (14) only ϱ is unknown.

Equation (14) represents a very simple method for estimating asset respectively APP correlations for homogeneous credit portfolios. Because in many cases portfolios admit an *analytical approximation* by a suitably calibrated uniform portfolio, *systematic risk* can be estimated for portfolios admitting a representation by a synthetic homogeneous reference portfolio.

4 Beyond Models with Uniform R-Squared

By a similar approach we can derive asset/APP correlations between different *segments* (e.g., rating classes; see the next sections) from the time series of default rates in the considered segments.

4.1 Correlation Between Segments - Basic Version

This section presents a straightforward application of Equation (14), interpreted in a slightly different manner. In our example we define two segments: Segment 1 consists of Moody's universe of Baa-rated corporate bonds, whereas Segment 2 consists of Ba-rated bonds. The idea now is to pick a 'typical' bond from every segment and to calculate the asset correlation ϱ between these bonds. Because segments are assumed to behave like a uniform portfolio, the so calculated correlation must be equal to the correlation between the segments.

More explicitely, we proceed as follows. Denote the covariance of the default event of an obligor in class Baa and an obligor in class Ba by $\text{Cov}_{Baa,Ba}$. Our model assumptions yield

$$\text{Cov}_{Baa,Ba} = \mathbb{P}[L_{Baa,i} = 1, L_{Ba,j} = 1] - \bar{p}_{Baa}\bar{p}_{Ba} = \qquad (15)$$

$$= N_2\left[N^{-1}(\bar{p}_{Baa}), N^{-1}(\bar{p}_{Ba}); \varrho\right] - \bar{p}_{Baa}\bar{p}_{Ba} \ .$$

Here, $L_{Baa,i}$ and $L_{Ba,j}$ are loss variables referring to bonds in rating class Baa and Ba. The parameters $\bar{p}_{Baa}$ and $\bar{p}_{Ba}$ are the corresponding default probabilities. By a result similar to Equation (12) we can estimate this covariance by the sample covariance of the time series of default rates, see also Equation (20). Replacing the default probabilities by the corresponding sample means and solving (15) for ϱ yields the correlation between rating classes Baa and Ba.

4.2 Correlation Between Segments - Multi Index Approach

In this section we follow a slightly more complex approach. Let us assume that we have m different segments, for example, rating classes or industry buckets. Every segment k will be considered as a uniform portfolio with default probability p_k and asset correlation ϱ_k. Equation (8) can then be rewritten by

$$X_{ki} = \sqrt{\varrho_k}\, Y_k + \sqrt{1-\varrho_k}\, Z_{ki} \qquad (k=1,...,m;\ i=1,...,m_k), \qquad (16)$$

where Y_k denotes a segment-specific index, and Z_{ki} is the specific effect of obligor i in segment k. The number of obligors in segment k is given by m_k. Additionally we introduce a global factor Y by means of which all segments are correlated,

$$Y_k = \sqrt{\varrho}\, Y + \sqrt{1-\varrho}\, Z_k \qquad (k=1,...,m), \qquad (17)$$

where ϱ is the uniform R^2 of the segment indices w.r.t. the global factor Y. The variables Z_k are the segment-specific effects. It is assumed that the variables Y, Z_k, Z_{ki} are independent standard normal random variables. The correlation ϱ is the unknown quantity we want to determine in the sequel; see Equation (20).

To give an example, let us consider the two extreme cases regarding ϱ. In case of $\varrho = 0$, the segments are uncorrelated. In case of $\varrho = 1$, the segments are perfectly correlated, such that the union of the segments yields an aggregated uniform portfolio. In both cases, the R^2 of obligors depends on the obligor's segment k and is given by ϱ_k. The correlation matrix $C = (c_{\sigma\tau})_{1 \leq \sigma, \tau \leq m_1 + \ldots + m_m}$ of the portfolio consisting of the union of all segments is given by

$$c_{\sigma\tau} = \mathrm{Corr}[X_{k_\sigma i_\sigma}, X_{k_\tau i_\tau}] = \sqrt{\varrho_{k_\sigma} \varrho_{k_\tau}}\, \varrho + \sqrt{\varrho_{k_\sigma} \varrho_{k_\tau}}\, (1 - \varrho)\, \mathrm{Corr}[Z_{k_\sigma} Z_{k_\tau}] +$$

$$+ \sqrt{(1-\varrho_{k_\sigma})(1-\varrho_{k_\tau})}\, \mathrm{Corr}[Z_{k_\sigma i_\sigma} Z_{k_\tau i_\tau}] = \begin{cases} \varrho_k & \text{if } k_\sigma = k_\tau = k, i_\sigma \neq i_\tau \\ 1 & \text{if } k_\sigma = k_\tau = k, i_\sigma = i_\tau \\ \sqrt{\varrho_{k_\sigma} \varrho_{k_\tau}}\, \varrho & \text{if } k_\sigma \neq k_\tau. \end{cases}$$
(18)

Equation (18) confirms ϱ_k as a *segment intra-correlation*, whereas the correlation between counterparties from different segments k_σ and k_τ is given by $\sqrt{\varrho_{k_\sigma} \varrho_{k_\tau}}\, \varrho$.

By arguments analogous to the one in Section 3, one can see that the empirical covariance of the default rates of different segments over time converges against the theoretical covariance

$$\mathrm{Cov}[p_{k_\sigma}(Y_{k_\sigma}), p_{k_\tau}(Y_{k_\tau})] = \int_{\mathbb{R}^2} p_{k_\sigma}(y_{k_\sigma}) p_{k_\tau}(y_{k_\tau})\, dN_2(y_{k_\sigma}, y_{k_\tau} | \varrho) - \overline{p}_{k_\sigma} \overline{p}_{k_\tau},$$
(19)

where the functions $p_k(\cdot)$, $k = 1, \ldots, m$, are defined by

$$p_k(y_k) = N\left[\frac{N^{-1}(\overline{p}_k) - \sqrt{\varrho_k}\, y_k}{\sqrt{1 - \varrho_k}}\right],$$

reflecting the same arguments as presented in (10).

Comparing the empirical with the theoretical covariance, we obtain the following Equation, where n refers to the number of considered years:

$$\frac{1}{2\pi\sqrt{1-\varrho^2}} \int_\mathbb{R} \int_\mathbb{R} N\left[\frac{N^{-1}(\overline{p}_{k_\sigma}) - \sqrt{\varrho_{k_\sigma}}\, y_{k_\sigma}}{\sqrt{1 - \varrho_{k_\sigma}}}\right] N\left[\frac{N^{-1}(\overline{p}_{k_\tau}) - \sqrt{\varrho_{k_\tau}}\, y_{k_\tau}}{\sqrt{1 - \varrho_{k_\tau}}}\right] \times$$

$$\times\, e^{-\frac{1}{2(1-\varrho^2)}(y_{k_\sigma}^2 - 2\varrho y_{k_\sigma} y_{k_\tau} + y_{k_\tau}^2)}\, dy_{k_\sigma}\, dy_{k_\tau} - \overline{p}_{k_\sigma} \overline{p}_{k_\tau}$$

$$\overset{!}{=} \frac{1}{n} \sum_{j=1}^{n} (p_{k_\sigma j} - \overline{p}_{k_\sigma})(p_{k_\tau j} - \overline{p}_{k_\tau}),$$
(20)

where p_{kj} denotes the default frequency of segment k in year j. Replacing the $\overline{p}_k$'s by sample means, the only unknown parameter in Equation (20) is the correlation ϱ between segments k_σ and k_τ. Therefore, we can solve (20) in order to get an estimate for ϱ.

5 The 20% Correlation Assumption of Basel II

As already mentioned in the introduction, the new Basel capital accord in its recent version suggests a 20% -level of systematic risk for the calibration of the benchmark risk weights for corporate loans, see [1].

In Section 5.1 we apply Equation (14) to Moody's historic default data for corporate bonds in order to estimate the asset/APP correlation for every rating class, assuming that the underlying corporate bond portfolios can be analytically approximated by a homogeneous reference portfolio; see the beginning of Section 3.

We will also estimate a *systematic APP* process; see Section 5.2.

5.1 Example (Part I): APP Correlations from Moody's Data

The following Table 1 shows the relative default frequency of corporate bonds according to the Moody's report [12] from 2002, including default data from 1970 to 2001.

Table 1. Moody's corporate bond defaults [12]

Rating	1970	1971	1972	1973	1974	1975	1976	1977	1978	1979
Aaa	0,00%	0,00%	0,00%	0,00%	0,00%	0,00%	0,00%	0,00%	0,00%	0,00%
Aa	0,00%	0,00%	0,00%	0,00%	0,00%	0,00%	0,00%	0,00%	0,00%	0,00%
A	0,00%	0,00%	0,00%	0,00%	0,00%	0,00%	0,00%	0,00%	0,00%	0,00%
Baa	0,28%	0,00%	0,00%	0,47%	0,00%	0,00%	0,00%	0,28%	0,00%	0,00%
Ba	4,19%	0,43%	0,00%	0,00%	0,00%	1,04%	1,03%	0,53%	1,10%	0,49%
B	22,78%	3,85%	7,14%	3,77%	6,90%	5,97%	0,00%	3,28%	5,41%	0,00%
Caa	53,33%	13,33%	40,00%	44,44%	0,00%	0,00%	0,00%	50,00%	0,00%	0,00%
	1980	1981	1982	1983	1984	1985	1986	1987	1988	1989	1990
	0,00%	0,00%	0,00%	0,00%	0,00%	0,00%	0,00%	0,00%	0,00%	0,00%
	0,00%	0,00%	0,00%	0,00%	0,00%	0,00%	0,00%	0,00%	0,69%	0,00%
	0,00%	0,00%	0,27%	0,00%	0,00%	0,00%	0,00%	0,00%	0,00%	0,00%
	0,00%	0,00%	0,31%	0,00%	0,37%	0,00%	1,36%	0,00%	0,00%	0,61%	0,00%
	0,00%	0,00%	2,78%	0,94%	0,87%	1,80%	1,78%	2,76%	1,26%	3,00%	3,37%
	5,06%	4,49%	2,41%	6,31%	6,72%	8,22%	11,80%	6,27%	6,10%	9,29%	16,18%
	33,33%	0,00%	27,27%	44,44%	100,00%	0,00%	23,53%	20,00%	28,57%	33,33%	53,33%
	1991	1992	1993	1994	1995	1996	1997	1998	1999	2000	2001
	0,00%	0,00%	0,00%	0,00%	0,00%	0,00%	0,00%	0,00%	0,00%	0,00%	0,00%
	0,00%	0,00%	0,00%	0,00%	0,00%	0,00%	0,00%	0,00%	0,00%	0,00%	0,00%
	0,00%	0,00%	0,00%	0,00%	0,00%	0,00%	0,00%	0,00%	0,00%	0,00%	0,17%
	0,29%	0,00%	0,00%	0,00%	0,00%	0,00%	0,00%	0,12%	0,11%	0,39%	0,30%
	5,43%	0,31%	0,57%	0,24%	0,70%	0,00%	0,19%	0,64%	1,03%	0,91%	1,19%
	14,56%	9,05%	5,86%	3,96%	4,99%	1,49%	2,16%	4,15%	5,88%	5,42%	9,35%
	36,84%	27,91%	30,00%	5,26%	12,07%	13,99%	14,67%	15,09%	20,05%	18,15%	32,50%

The table shows observed default frequencies per rating class and year. Table 2 shows the result when applying the estimation procedure from Section 3.

Recall that for every rating class - according to (13) - the asset correlation ϱ is determined by equation (14) with m_p and s_p^2 being the mean value and variance according to the default history as given in Table 1. Hereby, the upper table reports on the result when smoothing the historic means and standard deviations by a linear regression on a logarithmic scale. The lower table shows the result of the same calculation but with the original sample moments. For Aaa-rated bonds no defaults have been observed, such that the lower table shows 'not observed' for the Aaa-asset correlation estimate.

Table 2. First and second moments according to Table 1 and estimated asset correlations

Rating	Mean	Stand.Dev.	Impl.Ass.Corr.
Aaa	0,0005%	0,0031%	21,17%
Aa	0,0030%	0,0134%	20,17%
A	0,0194%	0,0585%	18,80%
Baa	0,1263%	0,2557%	17,23%
Ba	0,8228%	1,1180%	15,90%
B	5,3623%	4,8879%	16,41%
Caa	34,9453%	21,3696%	32,08%
Mean	5,8971%		20,25%

Rating	Mean	Stand.Dev.	Ass.Corr.Orig.
Aaa	0,0000%	0,0000%	not observed
Aa	0,0216%	0,1220%	31,50%
A	0,0138%	0,0556%	22,89%
Baa	0,1528%	0,2804%	15,95%
Ba	1,2056%	1,3277%	13,00%
B	6,5256%	4,6553%	11,77%
Caa	24,7322%	21,7857%	42,51%
Mean	4,6645%		22,94%

Our conclusion from the result of our calculations (Table 2) is as follows:

Given that our model assumptions are not taking us too far away from the 'real world', our calculations show that the Basel II level of 20% correlation is often close to the estimated correlation. However in more than half of the rating classes 20% correlation is conservative.

5.2 Example (Part II): Implied Systematic APP Process

One assumption which could at first sight seem to be critical, is the way we treated the underlying systematic APP process $Y_1, Y_2, Y_3, ...$; see Section 3. There, we assumed these variables to be independent. In a more realistic approach one would probably prefer to model these systematic variables by means of an *autoregressive process*, e.g., with time lag 1 (i.e. an AR(1)-process). However, in our model we are *not* thinking in terms of Y being a *macroeconomic* factor, for which an autoregressive modelling would be recommended. Our Y reflects the 'instantanous dependency' between borrower's ability to pay and does not refer to some time-lagged macroeconomic effect.

Moreover, we can get the process of realizations $Y_1, Y_2, Y_3, ...$ of the APP-factor Y back by a simple least-squares fit. For this purpose, we used an L^2-solver for calculating $y_1, ..., y_n$ with

$$\sqrt{\sum_{j=1}^{32}\sum_{i=1}^{7}|p_{ij}-g_i(y_j)|^2} = \min_{(v_1,...,v_n)} \sqrt{\sum_{j=1}^{32}\sum_{i=1}^{7}|p_{ij}-g_i(v_j)|^2},$$

where p_{ij} refers to the observed historic default frequency in rating class i in year j, according to Table 1, and $g_i(v_j)$ is defined by

$$g_i(v_j) = N\left[\frac{N^{-1}[\overline{p}_i] - \sqrt{\varrho_i}v_j}{\sqrt{1-\varrho_i}}\right] \quad (i=1,...,7;\ j=1,...,32),$$

reflecting Equation (10) with i denoting the respective rating class i. Note that, ϱ_i refers to the just estimated asset correlations for the rating classes according to Table 2, lower table.

Figure 1 shows the resulting 'APP-factor cycle' and the time-dependent overall mean of the default frequencies in Moody's corporate bond universe.

The result is very intuitive:

Comparing the APP-factor cycle $y_1, ..., y_n$ with the historic mean default path, one can see that any systematic 'APP-downturn' corresponds to an increase of default frequencies.

5.3 Example (Part III): Correlation between Segments

Recalling our results from Section 4, we can consider every rating class as a segment and apply Equations (15) ('basic version') and (20) ('multi index approach') in order to estimate the segment correlation ϱ between rating classes.

Basic Version

Based on Equation (15), we can calculate the asset/APP correlation between rating classes Baa and Ba. From Table 2 we have

- $\overline{p}_{Baa} = 15{,}28\text{bps}$ and $\overline{p}_{Ba} = 120{,}56\text{bps}$.

The empirical covariance can be obtained from the time series in Table 1:

- $\text{Cov}[(p_{Baa,j})_{j=1,...,32}, (p_{Ba,j})_{j=1,...,32}] = 0{,}00104\%$.

We then apply Equation (15) and obtain

- $\varrho = 5{,}60\%$.

This example indicates that the Basel II assumption of a one-factor model with a uniform asset correlation of 20% is violated as soon as we consider correlations between different segments.

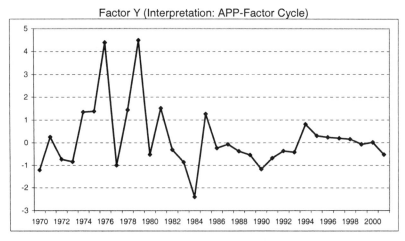
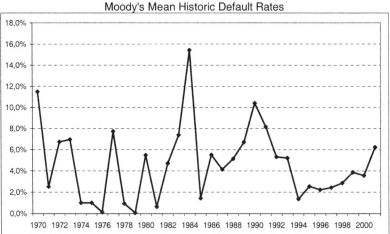

Fig. 2. Systematic APP process and underlying mean default frequency path

Multi Index Approach

Using the same notation as in Section 4, the intra-segment correlation ϱ_k for segment k, where k ranges over all seven rating classes, is given in Table 2. In our example, we work with the lower table in Table 2, which is based on the original moments (without regression).

As an example, consider rating classes 4 (Baa) and 5 (Ba). From Table 2 we have

- $\bar{p}_{Baa} = 15{,}28$bps and $\bar{p}_{Ba} = 120{,}56$bps;
- $\varrho_{Baa} = 15{,}95\%$ and $\varrho_{Ba} = 13{,}00\%$.

For calculating ϱ, we first of all need to calculate the empirical covariance of the default frequency time series of rating classes Baa and Ba. In the previous section, the covariance of the time series of default rates in Table 1 has been estimated as 0,00104%.

Dividing the covariance by the respective standard deviations yields a correlation between the two time series of about 28%. Next, we solve Equation (20) for ϱ and get

- $\varrho = 38{,}7\%$

as the correlation between the two factors. So much regarding an example calculation. Now let us interpret our result in terms of the 20%-correlation assumption of Basel II.

Following Basel II, a 'pure' one-factor approach is claimed to be sufficient for capturing *diversification effects*. Under this hypotheses, the correlation between systematic factors Y_k must be equal to $\varrho = 1$; cp. Equations (16) and (17). In contrast, our calculations above indicate that ϱ in fact is much lower than 100%. It is easily verified by means of analogous calculations, that this observation remains true even when dropping the multi-segment approach (allowing for different R^2's in different segments) by assuming ϱ_k to be constant for all segments k.

The assumption of a uniform asset correlation of 20% as made in the current draft of the new capital accord underestimates diversification benefits and does not provide any incentive to optimise the portfolio's risk profile by investing in different 'risk segments' like countries or industries.

References

1. Basel Committee on Banking Supervision; *The Internal Ratings-Based Approach*; Consultative Document, January (2001)
2. Belkin, B., Suchower, S., Forest, L. R. Jr.; *The effect of systematic credit risk on loan portfolio value-at-risk and loan pricing*; CreditMetrics Monitor, Third Quarter (1998)
3. Belkin, B., Suchower, S., Forest, L. R. Jr.; *A one-parameter representation of credit risk and transition matrices*; CreditMetrics^{TM} Monitor, Third Quarter (1998)
4. Black, F., Scholes, M.; *The Pricing of Options and Corporate Liabilities*; Journal of Political Economy **81**, 637-654 (1973)
5. Bluhm, C., Overbeck, L., Wagner, C.; *An Introduction to Credit Risk Modeling*; Chapman & Hall/CRC Financial Mathematics; CRC Press (2002)
6. Crosbie, P.; *Modelling Default Risk*; KMV Corporation (1999) (www.kmv.com)
7. Embrechts, P., McNeil, A., Straumann, D.; *Correlation and Dependence in Risk Management: Properties and Pitfalls*, Preprint, July 1999.
8. Frey, R., McNeil, A. J.; *Modelling Dependent Defaults*; Preprint, March (2001)
9. Finger, C. C.; *Conditional Approaches for CreditMetrics Portfolio Distributions*; CreditMetrics Monitor, April (1999)
10. Joe, H.; *Multivariate Models and Dependence Concepts*; Chapman & Hall (1997)

11. Merton, R.; *On the Pricing of Corporate Debt: The Risk Structure of Interest Rates*; The Journal of Finance **29**, 449-470 (1974)
12. Moody's Investors Service; *Default & Recovery Rates of Corporate Bond Issuers*; February (2002)
13. Vasicek, O. A.; *Probability of Loss on Loan Portfolio*; KMV Corporation (1987)

Valuation of a Credit Default Swap: The Stable Non-Gaussian versus the Gaussian Approach *

Dylan D'Souza[1], Keyvan Amir-Atefi[2], and Borjana Racheva-Jotova[3]

[1] Department of Economics, University of California, Santa Barbara, USA
[2] Department of Economics, University of California, Santa Barbara, USA
[3] Bravo Risk Management Group and Faculty of Economics and Business Administration, Sofia University, Bulgaria.

Summary. This empirical paper investigates the effect of different distributional assumptions governing defaultable bond price uncertainty on the price of a credit default swap. We value a credit default swap using the two-factor Hull-White (1994) model for the term structure of default-free spot interest rates and the credit spread process of a Baa-rated bond index and use the fractional recovery model of Duffie-Singleton (1999) and its multiple default extension as given in Schönbucher (1996,1998). The model is implemented using a tree algorithm outlined in Schönbucher (1999) where one factor is used for the spot interest rate and the other factor is for the credit spread or the intensity rate of a Cox process. This tree representation permits correlation between the spot rate and the credit spread dynamics of the Baa bond index, enabling us to provide a model that better fits the term structure of default-free and defaultable bond prices. We compare the values of a credit default swap obtained when the underlying risk factors are modeled with Gaussian and stable non-Gaussian distributions.

1 Introduction

Credit derivatives are over the counter contracts developed in the early nineties to meet investors' needs for more efficient and effective tools for the management of credit risk exposures. By employing credit derivatives a counter party to the contract may alter his risk/return profile by trading in these synthetic assets. Credit derivatives, unlike traditional credit securi-

---

* We thank Stephen LeRoy, John Marshall and Svetlozar Rachev for many helpful comments. We also thank Stoyan Stoyonov from Bravo Risk Management Group for his computational assistance and participants at the 8th Econometrics Workshop on Credit Risk Management, March 2002 at Karlsruhe. We are responsible for all errors. Please send correspondence to Dylan D'Souza, Department of Economics, University of California, Santa Barbara, CA 93106. Email: dylan@econ.ucsb.edu

ties,[4] permit the transfer of an asset's credit risk from one counter party to another without trading the underlying asset. Although the global market for credit derivatives accounts for only one per cent of the "notional" value of the global market for over-the counter derivatives,[5] the demand for such assets is growing rapidly.

The increasing popularity of these securities stems from the desire of commercial banks to have a flexible tool that diversifies portfolios of commercial loans and other credit risky assets in a cost-effective manner. These synthetically created credit assets diversify portfolios by making previously untradeable risks tradeable, thus facilitating market completion and efficient risk sharing. In other words these assets have created a secondary market, which allows the risks of corporate loans and bonds to be traded. The main credit derivatives traded in financial markets are default swaps, also known as credit default swaps or credit swaps, total-rate-of return swaps and credit spread options.

The focus of this paper is on the most popular of these credit derivatives, the credit default swap.[6] A credit default swap insures an investor, party **A**, against a potential loss on a reference asset resulting from adverse movements in the credit quality of the issuer, party **C** of the reference asset.[7] In a default swap contract "protection buyer", **A**, buys insurance from a "protection seller", party **B**. Alternatively, **A** sells its credit risk exposure to **B**. Typical cashflows of a plain vanilla default swap are shown in Figure 1. Assuming the asset is a corporate bond, **A** has the right to sell the defaultable bond issued by **C** to **B** for its face value if the credit event occurs. In return for this right, **B** is compensated with either an upfront fee or an annuity premium and only makes a payment if a default event occurs, otherwise **B** has no obligation. This paper assumes that **B** is not subject to default risk, there is no uncertainty about **B** meeting its financial obligations.

The default swap contract terminates at the time of the pre-specified credit event or at the stated maturity of the contract if the credit event has not occurred. Upon the occurrence of a credit event, **A** is obliged to exercise the contract immediately. Typical credit events include bankruptcy, insolvency, a credit rating downgrade or failure to make an obligated scheduled payment.

If the credit event occurs before the maturity of the swap, **B** is obliged to make a termination payment. A typical termination payment is the par

---

[4] Examples of traditional credit securities are letters of credit, mortgage guarantees, and private sector bond reinsurance.

[5] See Tavakli (1998).

[6] Surveys by the British Bankers' Association (2000) and Hargreaves (2000) both conclude that default swaps constituted the largest single component of the credit derivatives market at the end of 1999, based on outstanding principal amounts of underlying reference security.

[7] Reference assets include corporate bonds and bank loans. However, bank loans are not commonly used as reference assets because of their price intransparency, heterogeneity and illiquidity.

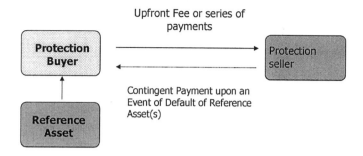

Fig. 1. The payoff structure of a plain vanilla credit default swap

value minus the recovery value of the underlying credit. The recovery value is computed based on a recovery rate which is specified in the contract. The recovery rate contracted is related to the average recovery rate of firms in the same industry or of firms with the same credit rating. The recovery value is also based on either the reference asset's pre-default price, post-default price or on its par value. For instance the termination payment could be the par value minus a pre-determined fraction of the pre-default price. Thus the recovery value is negotiable, it depends on the provisions of the contract.

Most models of credit derivatives valuation assume Brownian motion as the driver of uncertainty for the underlying asset returns. The discrete time counterpart of Brownian motion is a random walk with Gaussian innovations. However, empirical evidence suggest the distribution of asset returns are non-Gaussian. Many asset returns exhibit heavy tails, skewness, excessive kurtosis, and stochastic volatility. As a result of the non-Gaussian nature of asset returns, Mandelbrot (1963,1967) and Fama (1965) were among the first to propose the use of stable Paretian distributions as an alternative to model these empirical properties.[8] The advantage of employing stable Paretian distributions over other non-Gaussian distributions is that sums of stable random variables are also stable distributed, which is convenient for portfolio choice theory. Stable distributions have heavy tails and have a higher probability mass around the mean. Figure 2 depicts the probability density function for various indices of stability: $\alpha = 2$ (i.e. Gaussian distribution), $\alpha = 0.5, 1$ and $\alpha = 1.5$. The smaller the index of stability the heavier the tail. Figure 3 shows the flexibility of stable non-Gaussian distributions in modeling various levels of skewness. This flexibility permits stable non-Gaussian distributions to provide a better fit to the empirically observed distributions of asset returns.[9]

---

[8] Rachev and Mittnik (2000), Rachev, Schwartz and Khindanova (2000) are among others that have applied stable non-Gaussian distributions to model financial asset returns.

[9] Mittnik and Rachev (2000) discuss this in detail.

The Gaussian distribution is a special case of stable distributions. It is the only stable distribution that has a finite second moment.

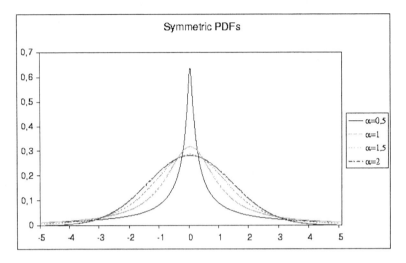

Fig. 2. The probability density function for a standard symmetric α-stable random variable, varying the index of stability α

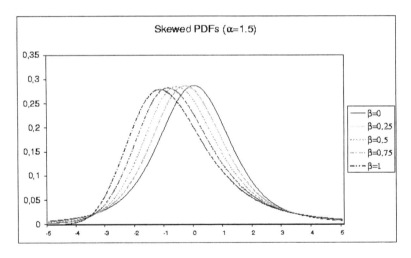

Fig. 3. Probability density functions for skewed stable random variables, varying the skewness parameter beta, for a given index of stability $\alpha = 1.5$

This paper, building on the work of Schönbucher (June 1999), values a credit default swap using the two-factor extended Vasicek Hull-White (1994a,1994b) term structure model for the default-free spot rate and the intensity process of a defaultable bond. Additionally, the paper models recovery of defaultable bonds using the fractional recovery model of Duffie-Singleton (1999) and is extended to allow for multiple defaults following Schönbucher (1996,1998). The recovery value is computed on the basis of the asset's pre-default price using a constant recovery rate.[10] The additional feature of our paper is that we compare the values of a default swap obtained when spot rate is modeled with Gaussian versus stable non-Gaussian innovations. We compute the value of the default swap on a lattice, a trinomial tree. The tree algorithm we implement to value a default swap is outlined in Schönbucher (June, 1999), which uses the Hull and White (1994a, 1994b) two factor numerical procedure for modelling the term structure of a defaultable bond's return, where one factor is used for the spot rate and the other factor is for the credit spread or the intensity rate governing the time of default for a defaultable asset. This procedure allows for any degree of correlation between the spot rate and the intensity rate process. This generalization enables us to obtain a better fit to the term-structures of default-free and defaultable bond prices.

We value a 3-year default swap written on a 10-year coupon bond with Baa rating.[11] The rationale of obtaining default swap prices using the stable non-Gaussian distribution is to obtain more realistic market valuations since stable non-Gaussian distributions are capable of providing a better fit to the underlying empirical distribution of asset returns.[12]

Section 2 provides a brief literature review on credit risk modeling. Section 3 describes the credit risk model for the defaultable reference asset. The models for the spot and intensity rate processes are discussed in Section 4. Stable non-Gaussian distributions and its application to the return process is described in Section 5. Section 6 discusses the computational procedures used to implement the model framework. Section 7 describes the default swap payoff structure and valuation on the trinomial tree. Section 8 provides the numerical results and Section 9 concludes.

2 Literature Review

Currently there are two main approaches used to the model credit risk. The first approach is based on structural models while the second is based on intensity or hazard rates, commonly called reduced form models. Both approaches

---

[10] The recovery rate is assumed to be 50%.
[11] The rating is obtained from Moody's Investor Service rating agency.
[12] See Rachev and Mittnik (2000) for applications of stable non-Gaussian distributions to model financial asset returns.

address the two sources of uncertainty associated with a defaultable bond, which are the probability of default and the loss incurred by an investor if the bond defaults.

The structural approach is derived from the pioneering work of Black and Scholes (1973) and Merton (1974). In these papers a defaultable security is a contingent claim on the value of the issuing firm's assets and is valued according to option pricing theory. The reason these models are described as structural is because these models make explicit assumptions on the firm's capital structure. Structural models take as a primitive the evolution of firm's value (i.e. the market value of the firm's total assets), in order to determine the probability of default and the recovery rates, or equivalently the loss rate given default.

In the structural form framework, diffusion processes describe the evolution of a firm's value. Additionally a default event is modeled explicitly in terms of the issuing firm's value process and its capital structure. Default occurs as soon the firm's value hits a pre-specified (exogenous) boundary, defined by the level of debt in the firms capital structure. Because of the continuity of the diffusion process, the time of default is a predictable stopping time. Thus within structural models, default is an endogenous and predictable event. The payoff at default is usually a cash payment representing the proceeds from liquidating the firm's assets (possibly after bankruptcy costs). In this approach recovery of the defaulted security is endogenous to the model. Models by Merton (1974), Black and Cox (1976), Geske (1977), Longstaff and Schwartz (1993) and Das (1995) are representative of the structural approach. The disadvantage with the structural approach is that the firm value process is not easily observable. The primary reason for this is because corporate debt relative to equity is illiquid. Furthermore, although the diffusion approach is theoretically insightful as it explains the economic mechanism leading to default, it is shown to be inadequate as it cannot explain the level of credit spreads observed empirically on defaultable assets.

While structural models develop a probability of default based on a firm's value process, the reduced form or intensity-based approach makes no explicit reference to the value process. Instead, the reduced form or intensity-based approach, models the default process as an exogenous Poisson event involving a sudden loss in market value of the firms assets. As a result default events can never be expected, default events are unpredictable. The time of default is modeled as a totally inaccessible stopping time with a stochastic intensity rate and the recovery rate is exogenous to the model. Because of the tractability they offer and the relative ease which they can be calibrated to empirically observed credit spreads, reduced form models attractive to researchers and practitioners. Examples of papers employing these models are Artzner and Delbaen (1992; 1994), Jarrow and Turnbull (1995), Lando (1994;1998), Jarrow, Lando and Turnball (1997), Madan and Unal (1998), Flesaker et al. (1994), Duffie and Singleton (1997;1999), Duffie, Schroder and Skiadas (1994), Duffie and Huang (1996) and Duffie (1994).

3 Credit Risk Model Structure

There are two elements of risk associated with a defaultable security. They are the risk of default and the uncertainty of the loss rate given default. The credit risk model will therefore have to specify the default time and a recovery rate in the event of default. Additionally, a dynamic model framework for credit derivative valuation requires a description of the dynamics of these risk factors associated with the reference asset used for the derivative. In this paper the credit spread of the reference asset is a measure of the credit risk and we model the dynamics of the credit spread.

In modeling credit risk we make the following assumptions. We model the risk-neutral world with the probability space $(\Omega, \mathcal{F}, \mathcal{Q})$ where $\mathcal{Q}$ is the martingale or pricing measure. The filtration $F = (\mathcal{F}_t)$ satisfies the usual conditions and there is a finite time horizon $\bar{T}$. Default-free and defaultable discount bonds of all maturities are traded continuously between $[0, \bar{T}]$. Markets are assumed to be frictionless with no arbitrage opportunities and are complete.

3.1 Default Time Model

A defaultable bond *promises* to pay the principal value borrowed at the bond's maturity, thus allowing for the likelihood of default. Let the price of a defaultable discount bond at time t that promises to pay \$1 at time T be denoted by $\bar{P}(t, T)$. We assume that the default counting process $N(t)$

$$N(t) = \max\{i \mid \tau_i \leq t\} = \sum_{i=1}^{\infty} 1_{\{\tau_i \leq t\}}$$

is a Cox process with the intensity process $\lambda(t)$. Here τ_i is the time of the ith default. The increasing sequence of defaults are modeled by the Cox process. Recall that $N = (N(t), t \geq 0)$ is called a Cox process, if there is a non-negative $\mathcal{F}_t$ adapted stochastic process $\lambda(t)$, the intensity rate of the Cox process, with $\int_0^t \lambda(s)ds < \infty$, $\forall t$, and conditional on the realization $\{\lambda(t)\}_{\{t>0\}}$ of the intensity, $N(t)$ is a time-inhomogeneous Poisson process with intensity $\lambda(t)$. The events of default occur at the times when the point process N, jumps in value. For an time inhomogeneous Poisson process, given the realization of λ, the probability of having exactly n jumps in the interval $[t, T]$ is

$$\Pr[N(T) - N(t) = n \mid \{\lambda(s)\}_{\{T \geq s \geq t\}}] = \frac{1}{n!} \left(\int_t^T \lambda(s)ds\right)^n \exp\left\{-\int_t^T \lambda(s)ds\right\}$$

Give the assumption that the Cox process is the default triggering mechanism, we get the intensity rate of the Cox process does not depend upon previous defaults and therefore the default times become a totally inaccessible stopping time. This stochastic intensity rate specification for the default process permits rich dynamics for the intensity process, thus allowing for the flexibility of capturing empirically observed stochastic credit spreads.

3.2 Fractional Recovery and Multiple Default Model

In practice, a default event does not terminate the debt contract, rather firms reorganize and re-float their debt.[13] Debt holders recover a positive amount of their assets either through cash or in terms of the newly floated debt. This structure allows for subsequent defaults and hence multiple defaults can occur with debt restructuring at each default event. This model is more realistic than other recovery models since it allows a firm to reorganize its debt instead of liquidating each time an event that triggers default occurs. Liquidating may be a very costly option for a firm or impossible in the case of sovereign debt. As stated earlier the fractional recovery model that we employ is identical to the recovery of market value (RMV) of Duffie and Singleton (1999) and is extended to allow for multiple defaults following Schönbucher (1996, 1998, 1999). In the RMV scheme the payoff to bond holders in the event of default is based on a fraction of the value of the same asset directly before it defaulted. Thus, the defaulted asset is worth only a fraction of its pre-default value but continues to trade.

The increasing sequence of default times $\{\tau_i\}_{i \in N}$ is driven by a Cox process. At each default τ_i the defaultable bond's face value is reduced by a factor q_i, the fractional loss rate, which could be a random variable. In the multiple default framework, a defaultable discount bond that promises to pay \$1 at its maturity T has a final payoff at time T given by

$$Q(T) := \prod_{\tau_i \leq T} (1 - q_i)$$

which is a product of the face value reductions given all defaults events that occurred until the bond's maturity T. For simplicity we assume that the fractional loss rate associated with the defaultable bond[14], is a constant, say $q = 50\%$.

With the assumption of arbitrage-free and complete markets, which implies the existence of a unique equivalent martingale measure $\mathcal{Q}$, permits the use of arbitrage pricing theory. From the APT, the price of a default-free discount bond is computed by discounting the final claim using the default-free spot interest rate and then taking the expectation with respect to the martingale measure $\mathcal{Q}$. Let $P(t,T)$ denote the t-value of a default-free discount bond with maturity T then its value can be expressed as,

---

[13] See Franks and Torous (1994).
[14] The defaultable bond we are interested in is a coupon bond with has Baa rating given by Moody's Investor Services.

$$P(t,T) = E_t[e^{(-\int_t^T r(u)du)}] \quad (1)$$

where $r(t)$ is the instantaneous default-free spot interest rate at time t. The value of a defaultable discount bond in the fractional recovery, multiple default framework is given by

$$\bar{P}(t,T) = E_t[Q\bar{P}(\tau-,T)e^{-\int_t^{\tau-} r(u)du}1_{\{t\leq \tau \leq T\}} + e^{-\int_t^T r(u)du}1_{\{T\leq\tau\}}] \quad (2)$$

where $E_t[\bullet]$ is the conditional expectation with respect to the martingale measure Q given all available information at time t and τ is the time of default.

From Equation (2) we see that there are two components to a defaultable bond's value. The first component is the present value of the promised payment in default and the second part is the present value of the promised payment if default does not occur.

In the fractional recovery, multiple default approach we can express the price of a defaultable discount bond as

$$\bar{P}(t,T) = Q(t)E_t[e^{(-\int_t^T \bar{r}(s)ds)}]$$

where $Q(t)$ is $\mathcal{F}_t$ measurable.[15] The process $\bar{r}(t)$ is called the defaultable adjusted spot rate and it is defined by

$$\bar{r}(t) = r(t) + \lambda(t)q(t) \quad (3)$$

Here $r(t)$ is the default free spot rate, $\lambda(t)$ is the intensity rate of the default and $q(t)$ is the fractional loss rate each time the bond defaults.[16] We can decompose the defaultable bond's price into the product of the face value reductions up to time t, $Q(t)$, the default free bond's price at time t and the survival probability in the time interval $[t,T]$,

$$\bar{P}(t,T) = Q(t)P(t,T)\Pr(t,T) \quad (4)$$

Observing Equation (3), we note it is possible to model the term structure of defaultable bond prices by modelling the default-free spot rate and the intensity rate process separately using two factors, one factor for each process and also permitting correlation between the two processes.

4 Model for the Spot and Intensity Rate Processes

We value defaultable bonds and credit derivatives, using the no-arbitrage extended Vasicek Hull-White (1994a,1994b) model for the term structure of

---

[15] Duffie and Singleton (1999) discuss the technical conditions when this can be done.
[16] We have assumed the fractional loss rate is constant at 50%.

default-free spot interest rates and credit spreads of a Baa bond index and their evolution over time. In the Hull-White (1994a,1994b) term structure model, the default-free spot rate has a mean reverting process, given by the following SDE under the risk neutral probability measure $\mathcal{Q}$,

$$dr(t) = [\theta(t) - ar(t)]dt + \sigma dW(t), 0 \leq t \leq T, \qquad (5)$$

where $r(t)$ is the default-free spot rate, $\theta(t)$ is a deterministic function of time t, a is a positive constant called the mean reversion rate and σ is the instantaneous standard deviation of the spot rate. $dW(t)$ is a standard Wiener process under the risk-neutral probability measure that drives the evolution of the term structure. The model is Gaussian and is consistent with the current term structure of default-free discount bonds where the parameter $\theta(t)$ is given as,

$$\theta(t) = F_t(0,t) + aF(0,t) + \frac{\sigma^2}{2a}(1-e^{[-2at]}), t \geq 0 \qquad (6)$$

where $F(t,T)$ is the forward rate of the default-free discount bonds. The drift process for r at time t is approximately equal to $F_t(0,t) + aF(0,t)$. On average the default-free rate r, follows the slope of the initial instantaneous forward rate curve. However, if it deviates from the forward rate it reverts back to it with a mean reversion rate a. This is an affine term structure model, where the price of a default-free discount bond at time t maturing at time T is given by,

$$P(t,T) = e^{[A(t,T) - B(t,T)r(t)]} \qquad (7)$$

where

$$B(t,T) = \frac{1 - e^{[-a(T-t)]}}{a} \qquad (8)$$

and

$$A(t,T) = \frac{1}{2}\int_t^T \sigma^2(s)B(t,s)^2 ds - \int_t^T B(t,s)\theta(s)ds \qquad (9)$$

The credit spread of the Baa-rated bond index is defined as the difference between the yield on the index and the yield on an equivalent 30-year treasury security. We model this term structure of credit spreads and their dynamics by modelling the intensity rate of the Cox process, $\lambda(t)$, that determines the instantaneous conditional default probability of the defaultable bond. The reason we model the intensity rate process instead of the credit spread process is because we have assumed the fractional loss rate to be a constant so the stochastic behavior of the credit spread is derived from the stochastic behavior of the intensity rate. To model the intensity of the Cox process, we use the same extended Vasicek Hull-White (1994a, 1994b) term structure model. The stochastic intensity is interpreted as the likelihood of default of the bond and under the risk-neutral probability measure this stochastic intensity $\lambda(t)$ evolves according to the SDE

$$d\lambda(t) = [\bar{\theta}(t) - \bar{a}\lambda(t)]\,dt + \bar{\sigma}d\bar{W}(t),\ 0 \le t \le T \quad (10)$$

According to Fons (1994), there is evidence that default probabilities are related to the firm's life cycle. This suggests that the behavior in default probabilities should be modeled as a mean reverting stochastic process given as in Equation (10). $\bar{\theta}(t)$ is determined so that the model is consistent with the current term structure of credit spreads making it flexible enough to describe any observed term structure of credit spreads.

From the Hull-White (1994a, 1994b) term structure model, the price at time t of a defaultable discount bond maturing at time T, is expressed as

$$\bar{P}(t,T) = P(t,T)e^{[\bar{A}(t,T) - \bar{B}(t,T)\lambda(t)]} \quad (11)$$

where

$$\bar{B}(t,T) = \frac{1}{a}[1 - e^{(-\bar{a}(T-t))}] \quad (12)$$

$$\bar{A}(t,T) = \frac{1}{2}\int_t^T \sigma^2(s)\bar{B}(t,s)^2 ds - \int_t^T \bar{B}(t,s)\tilde{\theta}(s)ds \quad (13)$$

and

$$\tilde{\theta}(s) = \bar{\theta}(t) + \rho\bar{\sigma}(t)\sigma(t)B(s,T) \quad (14)$$

where $\bar{a}$ is the mean reversion rate and $\bar{\sigma}$ is the instantaneous standard deviation of the intensity rate process. $d\bar{W}(t)$ is a standard Wiener process driving the term structure of the credit spread evolution.

The two factor model for the yield of a defaultable discount bond permits us to incorporate correlation, between the spot and the intensity rate processes, ρ, between dW and $d\bar{W}$. There is evidence that aggregate default failures depend on macro economic variables and several studies show that credit spreads are affected by common economic variables.[17] We therefore assume the spot and intensity rate processes are correlated, which is captured by assuming a correlation between the two Wiener processes, dW and $d\bar{W}$, where the correlation coefficient, $\rho = dW * d\bar{W}$.

5 Stable Paretian Modeling of the Spot Rate and Credit Spread Processes

The Hull-White (1994a, 1994b) term structure model uses Gaussian innovations to model asset returns. However in examining empirical data, asset returns show a clear departure from Gaussian properties. These empirical properties include heavy tails, skewness, excess kurtosis and stochastic volatility. We therefore use a more general distribution than the Gaussian distribution to model these empirical properties.

---

[17] See Altman (1983,1990), Wilson (1997a,b) and Pedrosa and Roll (1998).

Stable non-Gaussian processes are an important class of stochastic processes used to model extreme events. The Gaussian distribution model for asset returns is inadequate to model crashes and upturns in financial markets because these extreme financial market conditions occur more frequently than implied by the Gaussian assumption.[18] There are several reasons why stable non-Gaussian distributions are used over other non-Gaussian distributions to model extreme events. A very desirable property of stable distributions is that they are the only limiting distribution of properly normalized and centered partial sum processes for i.i.d. random variables X_i.[19] They have the ability to capture well documented[20] empirical properties of financial time series data better than other non-Gaussian distributions. A distribution in the domain of attraction of a stable distribution will have properties close to that of a stable distribution.

Typically credit returns have distributions that are characterized by skewed returns with heavy downside tails. The heavy-tailed nature of credit returns implies that there is a greater probability for large outlier events to occur than thin-tailed distributions. Defaultable bonds typically have returns that are left skewed. The reason that these returns are left skewed is because the probability of a defaultable bond earning a substantial price appreciation is relatively small. However, there is large probability of receiving a small profit through interest rate earnings. The distribution tends to be skewed around a positive value with a very small positive tail reflecting the limited upside potential. Adverse movements in credit quality occur with small probability but can have a significant negative impact on the value of asset. These credit quality migrations have the ability to generate significant losses thus producing skewed distributions with large downside tails. Additionally these skewed returns with heavy downside tails are characteristic of portfolios of defaultable bond as well.

The motivation for employing stable distributions to model the default-free spot interest rate can be seen from the following figures. The distribution of the daily changes in the 1-year Treasury bill returns, shown in Figure 4, are characterized by heavy tails. Figure 5 shows the inadequacy of the Gaussian distribution's ability to fit the left tail of the empirical distribution. As can be seen from these two figures, the stable non-Gaussian distribution provides a much better fit to the empirical distribution of this return process.

The degree to which a stable distribution is heavy tailed depends on its index of stability. As the index of stability, α becomes smaller than 2 ($\alpha = 2$, represents a Gaussian distribution) the distribution gets more peaked around its mean and more heavy tailed. In empirical analysis, high frequency financial

---

[18] See Embrechts et al. (1997) and Rachev and Mittnik (2000).
[19] See Embrechts et al. (1997) and Rachev and Mittnik (2000).
[20] In the empirical finance literature Rachev and Mittnik (2000) model financial assets returns which exhibit properties such as heavy-tails, asymmetries, volatility clustering, temporal dependence of tail behavior and short and long range dependence.

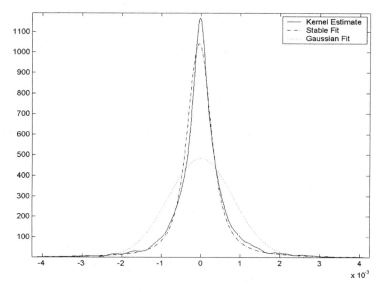

Fig. 4. Comparison of the Gaussian and stable non-Gaussian fit for the residuals of the changes in the 1-year Treasury bill returns

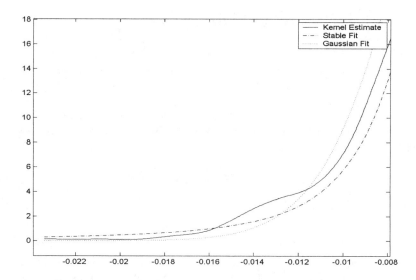

Fig. 5. Comparison of the Gaussian and stable non-Gaussian left tail fit for residuals of the change in 1-year Treasury bill returns

data have indices smaller than 2 while low frequency data have indices close to 2. To determine whether a distribution is in the domain of attraction of a stable distribution requires an examination of the tails of the distribution, which influences the properties of the distribution.

We model the default-free spot rate using a symmetric α-stable distribution, denoted by $S\alpha S$. We briefly discuss the properties of a stable non-Gaussian distribution and the estimation method employed to obtain the stable parameters that characterize the stable non-Gaussian distribution for the default-free spot return process in the next section. We also describe the properties of symmetric α-stable random variables, $S\alpha S$.

5.1 Description of Stable Distributions

Stable distributions are characterized by the equality in distribution between a random variable and the normalized sum of any number of i.i.d. replicas of the same variable. The Gaussian distribution is a special case of the stable distribution, it is the only stable distribution with finite second moments, with an index of stability, α, equal to 2.

Definition 1. *A random variable X (or a distribution function, F_X) is called stable if it satisfies*

$$c_1 X_1 + c_2 X_2 \stackrel{d}{=} b(c_1, c_2) X + a(c_1, c_2) \tag{15}$$

for all non-negative numbers c_1, c_2 and appropriate real numbers $b(c_1, c_2) > 0$ and $a(c_1, c_2)$, where X_1 and X_2 are independent random variables, $X_i \stackrel{d}{=} X_1$, $i = 1, 2$.

$\stackrel{d}{=}$ denotes equality in distribution. Consider the sum S_n of i.i.d. stable random variables. From Equation (15) we have for some real constants a_n and $b_n > 0$ and $X = X_1$

$$S_n = X_1 + \ldots + X_n \stackrel{d}{=} b_n X + a_n, \; n \geq 1 \tag{16}$$

which can be rewritten as

$$b_n^{-1}(S_n - a_n) \stackrel{d}{=} X \tag{17}$$

then it can be concluded that if a distribution is stable, then it is the unique limit distribution for sums of i.i.d. random variables. In general the stable distribution has no explicit form for the distribution function but can be expressed with its characteristic function which is as follows,

$$\Phi_X(t) = \exp\left\{-\sigma^\alpha \, |t|^\alpha \, (1 - i\beta sign(t) \tan \frac{\pi\alpha}{2}) + i\delta t\right\} \text{ if } \alpha \neq 1 \tag{18}$$

$$\Phi_X(t) = \exp\left\{-\sigma \, |t| \, (1 - i\beta \frac{2}{\pi} sign(t) \ln t) + i\delta t\right\} \text{ if } \alpha = 1$$

The characteristic function is described by the following parameters: $\alpha \in (0, 2]$, called the index of stability, $\beta \in [-1, 1]$, the skewness parameter, $\delta \in \Re$, is the location parameter and $\sigma \in [0, \infty)$, is the scale parameter. A stable random variable is represented as $X \sim S_{\alpha,\beta}(\delta, \sigma)$. The Gaussian distribution is a special case of the stable distribution with $\alpha = 2$. As the stability index gets smaller in value, the distribution becomes more leptokurtotic, i.e. with a higher peak and a heavier tail. If $\beta = 0$ the distribution is symmetric like the Gaussian. When $\beta > 0 (\beta < 0)$ the distribution is skewed to the right(left). The distribution is called symmetric α-stable when $\delta = 0$ and $\beta = 0$. The scale parameter generalizes the definition of standard deviation. The stable analog of variance is variation, denoted by σ^α. The pth absolute moment of X, $E|X|^p = \int_0^\infty P(|X| > y) dy$, is finite if $0 < p < \alpha$ and infinite otherwise. Hence when $\alpha \leq 1$ the first moment is infinite and when $\alpha < 2$ the second moment is infinite. The Gaussian distribution is the only stable distribution that has finite second moments.

Mittnik et al. (1996) use an approximate conditional maximum likelihood (ML) procedure to estimate the parameters of the stable distribution. The unconditional (ML) estimate $\Theta = (\alpha, \beta, \mu, \sigma)$ is obtained by maximizing the logarithm of the likelihood function

$$L(\Theta) = \prod_{t=1}^{T} S_{\alpha,\beta}\left(\frac{r_t - \mu}{\sigma}\right) \sigma^{-1} \tag{19}$$

The estimation is similar to DuMouchel (1973) but differs in that the stable density function, is numerically approximated using Fast Fourier Transforms (FFT) of the stable characteristic function in equation (18). Since the stable density is approximated the estimation of stable parameters is approximate. For further details on stable maximum likelihood estimation see Mittnik et al. (1996), Paulauskas and Rachev (1999) and Rachev and Mittnik (2000).

The stable non-Gaussian and Gaussian parameter estimated for the changes in the daily returns of the 1-year treasury bill and the credit spread of the long term Baa rated index are presented in Table 1.

Table 1. Gaussian and stable non-Gaussian parameters for 1-year Treasury bill returns and the credit spread of the Baa–rated bond index

	Gaussian parameters		Stable Non-Gaussian parameters			
	mu	sigma	alpha	beta	sigma	delta
1-year Treasury bill returns	0.00	0.00083	1.3145	0.0591	0.000306	0.00
Credit spread on Baa bond index	0.00	0.00078	1.6070	0.00	0.003	0.00

A symmetric α-stable random variable Z can be constructed where $Z = A^{1/2}G$, where G is a zero mean Gaussian random variable, i.e. $G \sim N(0,\sigma)$ and A is an $\alpha/2$-stable random variable totally skewed to the right and independent of G. A has the following distribution, $A \sim S_{\alpha/2}(1,0,\sigma)$; A is called the $\alpha/2$ stable subordinator. A symmetric α-stable random variable can be interpreted as a random rescaling transformation of a Gaussian random variable. The result can be extended to a two dimensional case[21] where the two stable random variables are the default-free spot interest rate and the intensity rate default driving the default process of the defaultable bond. A two dimensional vector $X = (X_1, X_2)$ is defined by

$$X = (A_1^{1/2}G_1, A_2^{1/2}G_2) \qquad (20)$$

The vector X is called a sub-Gaussian[22] $S\alpha S$ vector in R^2 under the Gaussian vector $G = (G_1, G_2)$. As the covariance of stable distributed random variables with $\alpha < 2$ is infinite, the sub-Gaussian $S\alpha S$ vector can be employed to incorporate the Gaussian dependence structure among different stable random variables. The Gaussian dependence is easy to compute and can easily be transferred to the sub-Gaussian case.

We generate independent and dependent $S\alpha S$ random variables that use truncated Gaussian covariances from the empirical data. Generating two-dimensional random vector X is performed by simulating a two-dimensional Gaussian random vector G with correlated elements G_1 and G_2 and a stable subordinator A independent of G_1 and G_2.[23]

6 Computational Procedures

6.1 Model Inputs

The Hull-White (1994a, 1994b) extended Vasicek model describes the dynamics of the entire term structure in a manner that is consistent with the initial or observed market data. To price the default swap we require default-free and defaultable discount bond prices on each node of the tree. The Hull-White (1994a, 1994b) trinomial tree algorithm is a discrete time approximation to the two mean reverting continuous processes given in equations (5) and (10). Using the trinomial tree algorithm we are able to describe the discrete-time

---

[21] In general it can be extended to d- dimenisonal case.
[22] A_i are α_i-stable subordinators, $i = 1, 2$, $1 < \alpha_i < 2$, and independent of the Gaussian vector $G = (G_1, G_2)$. If $A_i = A$, $i = 1, 2$, then X is sub-Gaussian. In general $X = (X_1, X_2)$, $X_i = A_i^{1/2}G_i$, is only a bivariate infinitely divisible vector. In fact $X = (X_1, X_2)$ is (α_1, α_2)-stable. See Rachev and Mittnik (2000), page 729. The class of infinitely divisible random variables is larger than the class of stable random variables.
[23] See Rachev, Schwartz and Khindanova (2000) for details.

dynamics of the spot rate and the intensity rate process that exactly fits the term structures of default-free and defaultable discount bond prices. Uncertainty in the trinomial tree is captured through branching. The tree is three-dimensional with one time dimension and two state space dimensions representing the states, $r(t)$, the spot rate and $\lambda(t)$, the intensity rate of default.

We calibrate the Hull-White (1994a, 1994b) model to market data using a least squares approach together with the assumption that the fractional loss rate in the event of default, is constant, set at 50 %, i.e. the defaultable bond loses 50 % of its pre-default value. We obtain estimates of the parameters that characterize the two mean-reverting spot and intensity rate processes, using the closed form expressions given in equations (7 - 14) and using a time series dataset of daily returns for 1-year Treasury bills for the period July/1959:April/2001 and credit spreads of a Baa rated long term bond index for the period January/1986:April/2001 obtained from the Federal Reserve dataset. Additionally, the correlation coefficient is estimated using this calibration procedure. From these estimated parameters we obtain discount curves for default-free and defaultable bond prices.

The advantage of using the closed form expressions for discount bond prices given in Equations (7-11) can be appreciated when valuing short term derivatives written on long term bonds. For example if we want to value a credit derivative that expires in three years written on a thirty year coupon bond it is very inefficient to build a thirty year tree. Instead, the closed form expressions for discount bond prices provides the coupon bond price at each node of the tree for the three years needed to find the payoff of the derivative under consideration, without the need to construct a thirty year tree.

6.2 Default Branching

The next two sections discuss the default branching and recovery mechanism of the tree. From the closed form expressions for default-free and defaultable discount bond prices, we have information of the current defaultable and default-free discount curves at each node in the tree. Thus we know the current default intensity $\lambda(t)$. The survival probability in the period from t to $t + \Delta t$ is given by

$$1 - p = E_t[e^{(-\int_t^{t+\Delta t} \lambda(s)ds)}] \qquad (21)$$

Over this small time interval $[t, t + \Delta t)$, the default intensity rate is constant, the survival probability can then be written as

$$1 - p = e^{(-\lambda(t)\Delta t)} \qquad (22)$$

where p is the probability of default over the time interval $[t, t + \Delta t)$. If the time step Δt is not large, we can assume with little loss in accuracy that if default occurs, it occurs at the left end of the time interval, i.e. default occurs at $\tau = t$.

As Figure 6 indicates, an additional branch is added to the each node of the tree to incorporate default. Thus at each node of the tree, if default has occurred the tree branches down to default, otherwise, the tree branches across as depicted in the figure. Default is an absorbing state and the tree terminates there if default occurs. If the bond survives, the tree continues with the evolution of the spot rates and default intensities. At the default node the tree terminates but payoffs in the event of default are computed.

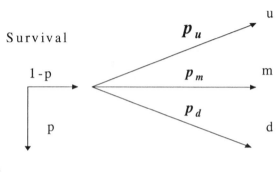

Fig. 6. Default branching at each node of the trinomial tree

From Figure 6, the conditional probability of surviving and reaching nodes u, m, d, the up, middle and down nodes, over the next time interval Δt are $(1-p) \cdot p_u$, $(1-p) \cdot p_m$ and $(1-p) \cdot p_d$ respectively. A security with survival contingent payoffs x_u, x_m and x_d in the up, middle and down nodes and zero at default, will have an expected value $x' = x_u p_u + x_m p_m + x_d p_d$, if default is not considered. However in the case of default the expected value is $x = (1-p) \cdot x'$.

6.3 Recovery Modelling in the Default Branch

Similar to Schönbucher (June 1999) we permit multiple defaults.[24] In the discrete-time fractional recovery model, we can approximate multiple defaults in two alternative ways.[25] The number of defaults is restricted to a single default in each time interval $[t, t + \Delta t)$ or multiple defaults are allowed in this time interval. If V_n denotes the value of the defaultable asset at time $t = n \Delta t$, and V_n^* its value if it survived until time $t = n \Delta t$ and if only a

---
[24] See Schönbucher (1996,1998) for more details on multiple defaults.
[25] See Schönbucher (June 1999).

single default is allowed each period, the following recursion will hold for V_n (ignoring discounting by default-free interest rates)

$$V_n = e^{(-\lambda_n \Delta t)} V_{n+1} + (1 - e^{(-\lambda_n \Delta t)})(1-q) V^*_{n+1}$$
$$= (1 - q(1 - e^{(-\lambda_n \Delta t)})) V^*_{n+1} \qquad (23)$$

The above equations states that the value of the defaultable security at time period $n\Delta t$ is the expected value of the value of the defaultable security in the next time period $(n+1)\Delta t$ depending on survival or default at period $(n+1)\Delta t$. Alternatively, allowing for multiple defaults in the interval $[t, t+\Delta t)$ and assuming $q * \Delta t$ is small, the value is given by,

$$V_n = e^{(-q\lambda_n \Delta t)} V^*_{n+1} \qquad (24)$$

The dynamics of equation (23) converge to equation (24) as $\Delta t \to 0$, and for reasonably small time step sizes the difference is negligible. If the time step is large (e.g. larger than $\frac{1}{12}$) the approach in (24) is more appropriate. We do not consider stochastic recovery rates in this paper however it can be incorporated in this framework[26].

6.4 Independence between the Spot and Intensity Rate Processes

The first case that we examine, we assume the spot rate is independent of the intensity rate process of the defaultable bond. Thus from equation (4), the price of the defaultable discount bond is given by

$$\bar{P}(t,T) = E_t[Q(T) e^{(-\int_t^T r(s)ds)}] =$$
$$= E_t[e^{(-\int_t^T r(s)ds)}] E_t[Q(T)] = P(t,T)\Pr(t,T) \qquad (25)$$

conditional on no default up to time t, where

$$\Pr(t,T) = E_t[e^{(-\int_t^T q\lambda(s)ds)}] \qquad (26)$$

$\Pr(t,T)$ is the survival probability from time t to T. Because the spot rate process and the intensity process are independent we can model them separately.

Hull-White Trinomial Tree Algorithm for the Spot and Intensity Rate Processes

The Hull and White (1994a, 1994b) numerical method constructs a discrete-time trinomial tree that approximates the mean reverting spot rate process, given as in equation (5).

---

[26] Schönbucher (June 1999) discusses stochastic recovery rates.

$$dr(t) = [\theta(t) - ar(t)]dt + \sigma dW(t),\ 0 \leq t \leq T$$

The Hull-White algorithm is a robust and efficient numerical procedure that reproduces in discrete-time the mean reverting property of the continuous spot rate process and is capable of exactly fitting the current term structure of the default-free discount bond prices and its volatility. The trinomial tree describes the spot rate suitably because it is capable of duplicating the expected drift and the instantaneous standard deviation of the continuous-time process at each node of the tree.[27] The spot rate is in terms of an unknown function of time, $\theta(t)$, which is implicitly obtained to exactly match the current term structure of default-free discount bond prices. The spot rate tree is constructed using the estimated parameters of the Hull-White (1994a, 1994b) model that describe the continuous spot rate stochastic process expressed in Equation (5).

The trinomial tree represents movements in the state variable r, by approximating the continuous-time model using discrete time intervals, $\triangle t$. The time interval $[0, T]$ is made up of n subintervals, so each time step is $n\triangle t$. The interest rate at each time interval is $r_0 + j\triangle r$ where j is the index indicating the position or the node of the tree, where j can take positive and negative integers. Each node on the spot rate trinomial tree is summarized by the coordinates (n, j). From each node there are three alternative branching processes given in Figures 7, 8 and 9. The geometry of the tree is constructed so that the middle branch corresponds the expected value of r, making tree construction faster and permitting more accurate pricing and better values for hedge parameters.

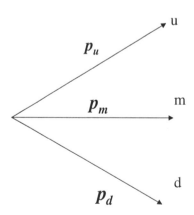

Fig. 7. Scheme A: Standard trinomial tree branching

---

[27] This is in general not possible to do with a constant step binomial tree without increasing the number of nodes exponentially with the number of time steps.

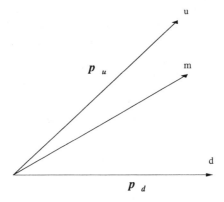

Fig. 8. Scheme B: "Up" branching in the trinomial tree

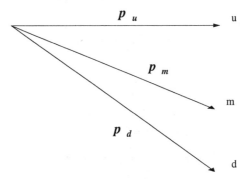

Fig. 9. Scheme C: "Down" branching in the trinomial tree

The tree in constructed in two stages. In the first stage the tree is constructed assuming $\theta(t) = 0$ and the initial value of r, $r(0)$ is zero. With this assumption the dynamics for r^* is given by the SDE

$$dr^* = -ar^* dt + \sigma dW \qquad (27)$$

In discrete time, r^* corresponds to the continuously compounded Δt period interest rate. $r^*(t + \Delta t) - r^*(t)$ is normally distributed with mean $r^*(t)M = r^* a \Delta t$ and variance $\sigma^2 \Delta t$. The central node of the tree at each time step corresponds to $r^* = 0$. Error minimization in the numerical procedures suggests the size of the interest rate step or the vertical distance between nodes at each time step be $\Delta r^* = \sqrt[2]{3V} = \sigma \sqrt[2]{3\Delta t}$. At each time step Δt, the variance in r^* is σ^2. Therefore node (n,j) represents the node where $t = n\Delta t$ and $r^* = j\Delta r^*$. Let p_u, p_m and p_d denote the probabilities of the up, middle and down branches respectively emanating from a node. These probabilities are chosen to match the expected change and variance of the change in r^* over the time interval Δt and they should add up to one. These conditions give us three equations in three unknowns, given as follows,

$$E[r^{*n+1} - r^{*n}] = p_u \Delta r_u^* + p_m \Delta r_m^* + p_d \Delta r_d^* = -ar_j^{*n}\Delta t \qquad (28)$$

$$E[(r^{*n+1} - r^{*n})^2] = p_u(\Delta r_u^*)^2 + p_m(\Delta r_m^*)^2 + p_d(\Delta r_d^*)^2 = \sigma^2 \Delta t + a^2(r_j^{*n})^2 \Delta t^2 \qquad (29)$$

and

$$p_u + p_m + p_d = 1 \qquad (30)$$

The tree is constructed so that at each node the first two moments of the discrete time process coincide with the continuous-time process up to terms of order Δt^2. Changes in the spot rate from node (n,j) are given by $\Delta r_u, \Delta r_m$ and Δr_d where the spot rate goes to an upper, middle and lower node respectively. The solutions to these equations depends on the branching from node (n,j). If the branching from node (n,j) is as in Scheme A (Figure 7) then the solution to the above equations is

$$p_u = \frac{1}{6} + \frac{1}{2}(j^2 M^2 + jM)$$

$$p_m = \frac{2}{3} - j^2 M^2$$

$$p_d = \frac{1}{6} + \frac{1}{2}(j^2 M^2 - jM)$$

If the branching from node (n,j) is as in Scheme B (Figure 8) then the solution is given by

$$p_u = \frac{1}{6} + \frac{1}{2}(j^2 M^2 - jM) \qquad (31)$$

$$p_m = -\frac{1}{3} - j^2 M^2 + 2jM \qquad (32)$$

$$p_d = \frac{7}{6} + \frac{1}{2}(j^2 M^2 - 3jM) \qquad (33)$$

If the branching from node (n,j) is as in Scheme C (Fig 9) then the probabilities are given as

$$p_u = \frac{7}{6} + \frac{1}{2}(j^2 M^2 + 3jM) \qquad (34)$$

$$p_m = -\frac{1}{3} - j^2 M^2 - 2jM \qquad (35)$$

$$p_d = \frac{1}{6} + \frac{1}{2}(j^2 M^2 + jM) \qquad (36)$$

The trinomial branch in scheme A is appropriate most of the time. However to ensure the probabilities are positive, branching switches from scheme A to scheme C when j is large and when $a > 0$. For the same reason it is necessary to switch from branching as shown in scheme A to scheme B when j is small. Switching from one branching alternative to another truncates the tree to keep the tree from growing very large as the number of time periods increases. Let j_{max} be the value of j where the branching is switched from scheme A to scheme C and j_{min} be the value of j when the branching is switched from scheme A to scheme B. The probabilities p_u, p_m and p_d are always positive when j_{max} is between $-0.184/M$ and $-.816/M$ provided $a > 0$ and $M > 0$. The probabilities at each node are a function of j, additionally the tree structure branching is symmetrical.

In the second stage of tree construction we employ forward induction. Starting from time period zero to the end of the tree, each node of the tree is displaced by an amount α_i, in order to match the initial term structure and simultaneously preserving the probabilities. The displacement magnitude is the same for all nodes at a particular time $n\Delta t$. The new tree for r will be for

$$r(t) = \alpha(t) + r^*(t)$$

and the initial discount curve is obtained

$$P(0,T) = E[e^{(-\int_0^T r(s)ds)}]$$

The new displaced tree models the mean-reverting stochastic process

$$dr(t) = [\theta(t) - ar(t)]dt + \sigma dW(t)$$

If $\hat{\theta}(t)$ is defined to be the discrete-time estimate of $\theta(t)$ between time t and $t + \Delta t$, and the drift in r at time $i\Delta t$ at the midpoint of the tree is $\hat{\theta}(t) - a\alpha_i$ such that

$$[\hat{\theta}(t) - a\alpha_n]\Delta t = \alpha_n - \alpha_{n-1}$$

therefore

$$\hat{\theta}(t) = \frac{\alpha_n - \alpha_{n-1}}{\Delta t} + a\alpha_n$$

In the limit as $\Delta t \to 0$, $\hat{\theta}(t) \to \theta(t)$.

Let $Q_{i,j}$ denote the present value of a security that pays off \$1 if node (n,j) is reached and zero otherwise. $Q_{i,j}$ is then the state price of node (n,j). The α_i's and $Q_{i,j}$'s are calculated using forward induction. The initial value of the state price, $Q_{0,0}$ is one. The value of α_0 is chosen to match the discount bond price that matures in time Δt. So α_0 is set to the initial Δt period interest rate given as

$$\alpha_0 = -\frac{1}{\Delta t}\ln P(0,1) \tag{37}$$

The next step is to compute $Q_{1,1}$, $Q_{1,0}$, $Q_{1,-1}$. This procedure is expressed formally as the following. Suppose the $Q_{i,j}$'s have been determined up to time $n \leq m$ ($m \geq 0$). The next step is to determine α_m so the tree correctly prices a discount bond maturing at time $(m+1)\Delta t$. The interest rate at (m, j) is $\alpha_m + j\Delta r$. The price of a discount bond maturing at time $(m+1)\Delta t$ is given by

$$P(0, m+1) = \sum_{j=-n_m}^{j=n_m} Q_{m,j} e^{[-(\alpha_m + j\Delta r)\Delta t]} \tag{38}$$

where n_m is the number of nodes on each side of the central node $m\Delta t$. Then α_m can be written as

$$\alpha_m = \frac{\ln \sum_{j=-n_m}^{j=n_m} Q_{m,j} e^{[-j\Delta r * \Delta t]} - \ln[P(0, m+1)]}{\Delta t} \tag{39}$$

Once α_m has been determined, the $Q_{i,j}$ for $n = m+1$ can be calculated using

$$Q_{m+1,j} = \sum_k Q_{m,k} p(k,j) \exp\left[-(\alpha_m + k\Delta r)\Delta t\right] \tag{40}$$

where $p(k, j)$ is the probability of moving from node (m, k) to node $(m+1, j)$ and the summation is taken for all values of k which is non-zero.

A trinomial tree is also built for the risk-neutral stochastic intensity process λ, which is given as in equation (10)

$$d\lambda(t) = \left[\bar{\theta}(t) - \bar{a}\lambda(t)\right] dt + \bar{\sigma} d\bar{W}(t), \; 0 \leq t \leq T$$

The process of building the intensity rate tree is similar to the process of building the tree for the spot rate. We build the tree specifically for the credit spread λq, which is equivalent to building a tree for λ since one is a multiple of the other. The tree is then fitted to the initial discount curve of credit spreads.

$$\Pr(0, T) = E[e^{(-\int_0^T q\lambda(s)ds)}] = \frac{\bar{P}(0,T)}{P(0,T)} \tag{41}$$

Once the tree for the credit spread is constructed default branches are added to each node of the tree as described in Section 6.2, where default is an absorbing state (i.e. the tree terminates). The trinomial tree for the credit spread will have at each node the probability of default in the next time interval and the probability of survival. The trees are combined after constructing the spot rate and the intensity rate trees. The combined tree can now be used to numerically compute the value of the credit default swap. The independence between the spot rate and the intensity rate processes permitted us to fit the λ-tree separately. In order to combine the trees it is required that both trees should have the same time step size Δt, but they do not necessarily need to have the same space step size. Note that the combined tree is a three-dimensional tree, with two space dimensions (r and λq) and a third time dimension. Table 2 shows the branching probabilities associated with moves in the spot rate and the intensity rate under assumption of independence.

Table 2. Joint branching probabilities for the default-free spot rate and the intensity rate movements for a single period of time given independence between their dynamics

			r move		Marginal
		Down	Middle	Up	
λ move	Up	$p'_u p_d$	$p'_u p_m$	$p'_u p_u$	p'_u
	Middle	$p'_m p_d$	$p'_m p_m$	$p'_m p_u$	p'_m
	Down	$p'_d p_d$	$p'_d p_m$	$p'_d p_u$	p'_d
	Marginal	p_d	p_m	p_u	1

At each time level n, the tree has $(i_{\max} - i_{\min}) * (j_{\max} - j_{\min})$ survival nodes and the same number of default nodes. From node (n, i, j) there are ten different branch possibilities. Both rates r and intensities λ have three possible combinations, and there is a tenth branch to default. The combined tree is constructed to fit the term structure of the default-free bond prices and defaultable bond prices and then used to value the default swap.

6.5 Correlation between Spot and Intensity Rate Processes

In practice the spot rate and the intensity rate of default are correlated with each other. The reason behind this correlation is because both of these processes are influenced by common macroeconomic variables. The procedure for constructing two-factor Hull-White trinomial trees for the case of correlation is slightly different than the case when the two mean reverting processes are independent. First the spot rate tree is built to fit the term structure of default-free discount bond prices. Then the tree for the credit spread is built but not fitted to the term structure of the credit spread. Instead the two trees are combined and the correlation between the two mean reverting processes are incorporated. Default branches are then added to the combined correlated tree. The combined tree is now fitted to the term structure of defaultable bond prices while preserving the fit to default free bond prices. This resulting combined tree which is fitted to the term structure defaultable discount bond prices can now be used to numerically value the default swap given dependence between the spot rate and intensity rate processes.

We use the Hull-White (1994b) procedure that considers correlation in a two-dimensional tree.[28] For a correlation $\rho \neq 0$ the procedure adjusts the transition probabilities given in Table 2 to the probabilities given in Table 3 and Table 4 depending on whether correlation is positive or negative using a correction parameter ϵ,

$$\epsilon = \begin{cases} \frac{1}{36}\rho & \text{for } \rho > 0 \\ -\frac{1}{36}\rho & \text{for } \rho < 0 \end{cases} \tag{42}$$

Table 3 and 4 give the probabilities of the indicated combined movements of r and λ in the combined tree for a given positive or negative correlation $\rho = \pm 36\epsilon$. Default and survival are ignored in these tables, to reach the probabilities for the movements and survival over the next time interval, the probabilities are multiplied by $(1-p)$. The original probabilities are: r: up p_u; middle p_m; down p_d. λ: up p'_u; middle p'_m; down p'_d. Default p. The adjustment for correlation given in Table 3 and 4 only work if ϵ is not too large. Thus there is a maximum value for the correlation that can be implemented for a given time step size Δt. As the refinement is increased ($\Delta t \to 0$) this restriction becomes weaker and the maximum correlation approaches one[29].

Table 3. Joint branching probabilities for the default-free spot rate and the intensity rate movements for a single period give a positive correlation between their dynamics

			r move		Marginal
		Down	Middle	Up	
λ move	Up	$p'_u p_d - \varepsilon$	$p'_u p_m - 4\varepsilon$	$p'_u p_u + 5\varepsilon$	p'_u
	Middle	$p'_m p_d - 4\varepsilon$	$p'_m p_m + 8\varepsilon$	$p'_m p_u - 4\varepsilon$	p'_m
	Down	$p'_d p_d + 5\varepsilon$	$p'_d p_m - 4\varepsilon$	$p'_d p_u - \varepsilon$	p'_d
	Marginal	p_d	p_m	p_u	1

To match the terms structure of the defaultable bond yield given correlation, the tree building procedure is similar to the case when there is independence except the shift of the tree occurs in the λ-dimension without affecting the tree built for the spot rate, thus preserving the spot rate tree.

6.6 Computation Procedure with Stable Innovations

The Hull-White (1994a, 1994b) trinomial tree algorithm described earlier assumed the dynamics of the spot rate have Gaussian innovations. This section

---

[28] Correlation between the spot rate and the intensity rate process is also discussed in Schönbucher (June, 1999). For a more detailed discussion of the procedure used to incorporate correlation, see Hull and White (1994b).

[29] For a detailed discussion see Hull and White (1994b).

Table 4. Joint branching probabilities for the default-free spot rate and the intensity rate movements for a single period, given a negative correlation between their dynamics

			r move		Marginal
		Down	Middle	Up	
λ move	Up	$p'_u p_d + 5\varepsilon$	$p'_u p_m - 4\varepsilon$	$p'_u p_u - \varepsilon$	p'_u
	Middle	$p'_m p_d - 4\varepsilon$	$p'_m p_m + 8\varepsilon$	$p'_m p_u - 4\varepsilon$	p'_m
	Down	$p'_d p_d - \varepsilon$	$p'_d p_m - 4\varepsilon$	$p'_d p_u + 5\varepsilon$	p'_d
	Marginal	p_d	p_m	p_u	

describes the tree construction when spot rate has dynamics driven by stable non-Gaussian innovations. The first step is to estimate the stable parameters for the daily return time series of 1-year Treasury bills. These parameters, shown in Table 1, are estimated using the unconditional maximum likelihood procedure described earlier in Section (5.1).

In Section 5.1 we defined a symmetric α-stable random variable Z, which is constructed as $Z = A^{1/2}G$, where G is a zero mean Gaussian random variable, i.e. $G \sim N(0, \sigma^2)$ and A is an $\alpha/2$-stable random variable totally skewed to the right and independent of G. A is distributed as, $A \sim S_{\alpha/2}(,1,0)$; A is called the $\alpha/2$ stable subordinator.

We model changes in the spot rate as a stable non-Gaussian random variable ($\alpha = 1.314$), however we treat changes in the intensity rate as a Gaussian random variable ($\alpha = 2$) and we can extend the definition for these two stable random variables when they have covariation between them. A (α_1, α_2)-stable vector X is defined by

$$X = (A_1^{1/2}G_1, A_2^{1/2}G_2) \qquad (43)$$

The elements of the vector X are the daily changes in the spot rate, $\Delta r \stackrel{d}{=} A_1^{1/2}G_1$ and the daily changes in the intensity rate, $\Delta \lambda \stackrel{d}{=} A_2^{1/2}G_2$ in one unit of time. The covariation between the changes in the spot rate and the intensity rate of the defaultable bond is as computed described in earlier sections with truncated covariances. This Gaussian dependence ρ, is easily computed (through calibration) and the correlation ρ between the spot and the intensity rates is used to model the joint stable distributions.

In discrete time, when changes in the spot rate Δr are Gaussian, we have $r(t+\Delta t) - r(t) \sim N(0, \sigma^2 \Delta t)$, where $\Delta t = 1$. These daily changes in the spot rate Δr are distributed Gaussian with mean zero and variance σ^2. However, when Δr are modelled employing the stable subordinator we obtain, $r(t+\Delta t) - r(t) \sim A_1^{1/2} N(0, \sigma^2)$. A_1 is a stable random variable totally skewed to the right with stable parameters $\Theta_{A_1} = (\alpha/2, 1, 0)$, where α represents the index of stability, the (right) skewness parameter is equal to 1 and with location parameter equal to zero. Note the stable parameter α was estimated for the daily return series of 1-year treasury bills using the ML procedure. When

$A_1^{1/2}$ is multiplied with the Gaussian random variable with mean zero and variance σ^2 we get a random variable $Z \sim N(0, A_1 * \sigma^2)$. The Gaussian random variable has its variance rescaled to $A_1 * \sigma^2$. This rescaled variance is stochastic since A_1 is a stable random variable. We have perturbed the variance σ^2 of the Gaussian random variable with a stable subordinator, A_1. Since A_1 is a potentially large multiplicative factor we truncate it to take on a maximum value of 10 in absolute value terms. In other words we allow for events that are ten standard deviations from the mean to occur in the economy. Thus we have the change in the spot rate, $r(t + \Delta t) - r(t) \sim N(0, A_1 * \sigma^2)$, as a stable random variable, a truncated Gaussian random variable with a rescaled variance. The subordinator has introduced a stochastic volatility term and increases the probability of outlier events, allowing for extreme default-free spot interest rates to occur. This permits us to represent the empirical data with a greater degree of accuracy. Note that $G = (G_1, G_2)$ are dependent. A_1 and A_2 while independent of G might be dependent. This is achieved by generating A_1 and A_2 by a single uniform random variable or by two independent uniform random variables. The choice of dependence between A_1 and A_2 is an important part of this approach. For simplicity we assume A_1 and A_2 are independent.

In the trinomial tree procedure, employing Gaussian innovations, the size of the spot rate step between nodes at each time step was $\Delta r^* = \sqrt[2]{3V} = \sigma\sqrt[2]{3\Delta t}$. However employing stable innovations the vertical distance between nodes will now be $\Delta r^* = \sqrt[2]{3V_1} = A_1^{1/2}\sigma\sqrt[2]{3\Delta t}$. Each realization of the stable subordinator random variable, A_1, generates a new tree constructed for the spot interest rate.

7 Valuation of the Default Swap on the Tree

Once the tree is constructed and fitted to observed discount bond prices it can be used to value defaultable coupon bonds and credit derivative securities written on these coupon bonds. A credit derivative security is characterized by its payoff given the credit event. In this case the credit event is a default event. We briefly describe the characteristics of a credit default swap whose value we compute on the constructed trinomial tree.

The credit default swap contract is for a fixed period of time, $[0, T]$ and has two counterparties to the contract. Both of the counterparties are assumed to be default-free, i.e. both of them meet their financial obligations. In this contract the protection buyer makes a fixed annual payment s to the protection seller in exchange for a default contingent payment. These payments continue until the maturity of the swap since the bond is allowed to be reorganized and sold at a fraction of its pre-default value, thus permitting multiple defaults.

Let us first establish some notation.[30] Let f_{ij}^n denote the payoff of the default swap if a default event occurs at node (n, i, j). This payoff is the dollar amount the protection seller pays the protection buyer in the event of default. The protection buyer makes an annual payment F_{ij} when the node (n, i, j) is reached which can be interpreted as an insurance premium. Let V_{ij}^N denote the value of the default swap at node (N, i, j), where N is the maturity date of the swap. If default has not occurred at this terminal level of the tree, then the value of the credit default swap is

$$V_{ij}^N = F_{ij}^N \qquad (44)$$

which is simply the final annual payment made by the protection buyer in the event of no default. To compute the value of the default swap at the swap initiation date we use backward induction along the tree. Conditional on survival at time $n\Delta t$, the value of the default swap is the discounted expected value of the swap's value at time $(n+1)\Delta t$, which can be expressed as,

$$V_{ij}^{\prime\prime n} = \sum_{k,l \in Succ(n,i,j)} p_{kl}^n e^{-r_j^n \Delta t} V_{kl}^{n+1} \qquad (45)$$

The value of the swap at time period $n\Delta t$ at the survival node is the discounted expected value of the swap from each of its emanating branches. $Succ(n, i, j)$ represent all the nodes $(n+1, k, l)$ at time period $(n+1)\Delta t$ that are successively connected from node (n, i, j) at time period $n\Delta t$, except for the default node. The value of the default swap at node (n, i, j), at time period n considering both survival and default is given as,

$$V_{ij}^{\prime n} = e^{-\lambda_i^n \Delta t} V_{ij}^{\prime\prime n} + (1 - e^{-\lambda_i^n \Delta t}) f_{ij}^n + F_{ij}^n \qquad (46)$$

The above expression states that the value of the default swap at time period $n\Delta t$ is the expected value of the swap given its payoffs in survival and default together with the annual payment F_{ij}^n made by the protection buyer at time $n\Delta t$.

In the default swap contract the protection buyer pays a periodic fee of $F_{ij}^n = -s$ dollars per annum as insurance premium for the protection. The protection seller pays the par value of the bond less the recovery of the reference coupon bond in the event of default. The protection sellers payment to the buyer can be expressed as

$$f_{ij}^n = 1000 - (1-q)\bar{B}_{ij}^{*n} \qquad (47)$$

where $\bar{B}_{ij}^{*n}$ is the price of the Baa-rated coupon bond at time period $n\Delta t$ at node (n, i, j) just prior to default and $(1-q)$ is the recovery rate of this coupon bond. Since the payoff of the derivative depends on the Baa coupon bond's

---

[30] The notation we employ here is identical to the notation used in Schönbucher (1999).

price, we need its value computed at each node on the tree. The coupon bond's price is obtained using expression (11) by noting that the defaultable coupon bond is a portfolio of defaultable discount bonds or a linear combination of defaultable discount bonds, for which we already have computed values at each node of the tree.

The value of the credit default swap also called the default swap spread is the annual payment $\$s$ the protection buyer makes, and is expressed as a fraction of the par value of the defaultable coupon bond, in our case the par value is \$1000. This value s is found by equating the expected discounted value of the default contingent payments, made by the protection seller, to the present value of the annual payments, made by the protection buyer implying that the value of the default swap contract at time $t = 0$ is zero, i.e. $V_0 = 0$.

8 Data and Computational Results

In this numerical exercise we compute the value of a credit default swap contract which matures in three years where the reference asset is a coupon bond with ten years to maturity. This coupon bond has a coupon rate of 8.5 per cent with coupon payments made annually. The value of the default swap is computed on the trinomial tree for the two cases when the spot rate dynamics have Gaussian and stable non-Gaussian innovations. In both these cases we compute the value of the default swap assuming that the spot and the intensity rate are dependent, for correlation coefficients of 0, 0.5 and 1. For the stable non-Gaussian case we compute a value for the default swap by averaging the results generated from 1000 realizations of the stable subordinator A_1.

For the spot rate we used daily returns of 1-year treasury bills for the period July/15/1959:April/20/2001. To compute the intensity for the Baa rated long term defaultable bond index we used daily returns of an index of Baa long term bonds for the period January/02/1986:April/20/2001 and a recovery rate of 50 per cent.

The estimated parameters for the spot and intensity mean reverting processes are given in Table 5. They were obtained by calibrating the model's bond formulas to market prices using a least squares approach. The Gaussian and stable non-Gaussian parameters estimated for the daily returns of the 1-year treasury bill series and the credit spread for the Baa rated long term bond index are given in Table 1. The stable non-Gaussian parameters have been estimated using the ML procedure outlined in Section 5.1.

In Table 6 we compare the value of the credit default swap, for three different correlation coefficients, assuming the spot rate process is driven by a Gaussian and stable non-Gaussian innovation. For correlation coefficients of 0 and 0.5 the value of the default swap computed under the stable non-Gaussian innovation assumption, is more than double the value of the default swap under the Gaussian innovation assumption. When the correlation coefficient is 1, the default swap value is higher under the stable non-Gaussian case than under the Gaussian case. Modeling defaultable yields with stable non-Gaussian distributions account for extreme losses in the underlying coupon bond's value thus increasing the value of the default swap written on it. One possible reason for much larger default swap spreads under the stable non-Gaussian assumption is because we have assumed that the stable subordinator vector $A = (A_1, A_2)$ is independent of the Gaussian vector $G = (G_1, G_2)$.[31]

Table 5. Estimated parameter values for the default-free spot rate and intensity rate mean reverting processes

a	0.89
σ	0.26
θ	0.055
$\bar{a}$	0.92
$\bar{\sigma}$	0.01
$\bar{\theta}$	0.06
ρ	0.308

Table 6. Default swap spreads (in basis points) for the Gaussian and stable non-Gaussian models given different correlations between the dynamics of the default-free spot and intensity rate processes

	Gaussian (bps)	Stable non-Gaussian (bps)
Rho = 0	232	562
Rho = 0.5	270	565
Rho = 1	306	568

9 Conclusion

In this paper we use a no-arbitrage term structure model for the default-free spot rate and intensity rate process of a Baa rated long term defaultable bond index and their evolution over time. Both the spot rate and the intensity rate of default are modeled using the two-factor extended Vasicek model of Hull and White (1994a, 1994b) and the pricing of defaultable bonds is obtained under the fractional recovery model of Duffie and Singleton (1994) and the multiple default extension of Schönbucher (1996, 1998). We numerically compute the value of a credit default swap using the trinomial tree procedure outlined in Schönbucher (June 1999) which applies the Hull and White (1994a,

---

[31] Research by Bravo Risk Management Group show that A and G are actually dependent.

1994b) trinomial tree algorithm. This algorithm constructs a trinomial tree, which is a three-dimensional tree with two space dimensions representing the stochastic state variables, $r(t)$ and $\lambda(t)$ and the third dimension is time. The trinomial tree approximates the continuous processes of the spot and intensity rate processes in discrete time. Each node on the tree has an absorbing state called default. The fractional recovery model of Duffie and Singleton (1999) employed here assumes that the recovery rate of a defaultable security is a given proportion of its pre-default value. Additionally, the recovery model allows for multiple defaults reflecting the fact that liquidation of the firm rarely occurs in bankruptcies instead debt is reorganized in order to maintain the firm as a going concern avoiding costly liquidation. For sovereign debt liquidation is not an option.

We compute the value of a credit default swap where the reference credit is a Baa-rated coupon bond. This is done numerically on a trinomial tree using the Hull and White (1994a,1994b) algorithm. It is clear that the credit default swap spread is highly sensitive to the assumptions made about the distribution governing the default-free spot rate process. We notice that the default swap spreads are much higher when the spot rate process is modeled with a stable non-Gaussian model than with a Gaussian model. This is because employing stable non-Gaussian distributions result in extreme outcomes for the changes in the default-free spot rate and assigns higher probabilities for these extreme outcomes to occur. It should be noted that modeling the spot rate with a stable subordinator random variable makes our framework a stochastic volatility model. We also notice that the variation in the default swap spreads for different correlation coefficients is much higher in the Gaussian case than the stable non-Gaussian case. A possible reason for the small variation in swap spreads in the stable Paretian model might be because the correlation coefficient is an inadequate measure for the dependence structure between the spot and intensity rate processes when the spot rate is modeled as a stable non-Gaussian process. This is because the correlation coefficient is a measure of central tendency, i.e. accounting for the first two moments of the processes, ignoring higher moment information and extreme co-movements between these two processes.

To make this numerical exercise computationally "simple" we have assumed the intensity process is described by a Gaussian model. However the data unambigiously suggests that this process should be modeled with a stable non-Gaussian model, given that the estimated index of stability for the daily changes in the intensity rate is $\alpha = 1.607$. Future research that models the intensity rate with a stable non-Gaussian model would need to account for extreme co-movements between the spot and intensity rate processes. This would need a more accurate measure for the dependence structure between these processes than the correlation coefficient. In fact, the tail dependence would be a more representative model of dependence accounting for the possibility of extreme co-movements. Copula functions are a popular approach to model dependence structures when the marginal distributions of the spot and intensity rate processes are non-Gaussian.

Another area of future research would be to model the dependence structure between the stable subordinator vector $A = (A_1, A_2)$ and the Gaussian vector $G = (G_1, G_2)$. In other words, it is likely that the stochastic volatility of the spot and intensity processes could be a function of the level of their respective levels. Research in this area, for the stable non-Gaussian class of models is done by Bravo Risk Management Group.[32] Prigent, Renault and Scaillet (2001) find that evidence that Baa spreads display mean-reversion and the volatility of Baa spreads monotonically increase with the level of spreads using nonparametric analysis and assuming that the credit spread dynamics follow a univariate diffusion process.

In this paper we have made the assumption that the recovery rate is constant, for example, an average of historical recovery rates for bonds that are Baa-rated by Moody's Investor Service. However research done in the area of recovery rate uncertainty can be incorporated in our framework. This is important since the recovery rate process is fundamental in determining the credit spread of defaultable instruments which influences the value of credit derivatives.[33]

Fundamental to derivative valuation models are the assumptions made about the stochastic processes that govern the behavior of the underlying risks being traded. Brownian-based Gaussian models are typically used for the analytic tractability they offer but from an empirical standpoint are grossly inadequate. This is because these models fail to capture the empirically observed properties exhibited by the spot rate and credit spread processes. These empirical anomalies include heavy tails, excessive kurtosis, skewness and stochastic volatility. We replace the traditional Brownian-based Gaussian model with a Stable Paretian model to make the model framework consistent with these empirically observed properties. The disadvantage is that we do not have closed-form expressions for the Stable Paretian models. However, we can numerically compute defaultable bond values and credit derivatives using a lattice based model represented by a trinomial tree. Numerical calibration and computation is fast given the Markov term structure model we employ. By generalizing the distribution assumptions made about the default-free spot interest rate process we obtain better market consistent values for the credit default swap since we more accurately model the empirical properties displayed by the spot rate process which influences the value of the credit default swap.

---

[32] For a detailed exposition see Rachev and Mittnik (2000).

[33] We refer the interested reader to Bakshi, Madan and Zhang (2001), Altman et al (2002) and Gupton and Stein (2002) among others for papers on recovery rates and loss given defaults.

References

1. Altman, E.I. (1983/1990). *Corporate Financial Distress.* Wiley, New York.
2. Altman, E.I., Brady, B., Resti, A., Sironi, A. (2002). *The link between default and recovery rates: Implications for credit risk models and procylicality.*
3. Anson, M. J.P. (1999). *Credit Derivatives.* Frank J. Fabozzi Associates, New Hope, Pennsylvania.
4. Artzner, P., and Delbaen, F. (1992). *Credit risk and prepayment option.* 22:81-96.
5. Bakshi, G., Madan, D., Zhang, F. (2001). *Recovery in default risk modeling: Theoretical foundations and empirical applications.* University of Maryland and the Federal Reserve Board.
6. Black, F., and Cox., J. (1976). *Valuing corporate securities: Some effects of bond indenture provisions.* Journal of Finance, 351-367.
7. Black, F., and Scholes, M. (1973). *The pricing of options and corporate liabilities.* Journal of Political Economy, 81 : 637-654.
8. Brémaud, Pierre. (1981). *Point Processes and Queues.* Springer, Berlin, Heidelberg, New York.
9. Clewlow, L., Strickland, Chris. (1998). *Implementing Derivatives Models.* Wiley Series in Financial Engineering.
10. Das, S.R. (1995). *Credit risk derivatives.* Journal of Derivatives, 2(3):7-23.
11. Duffie, D. (1998). *Defaultable term structure models with fractional recovery of par.* Working paper, Graduate School of Business, Stanford University.
12. Duffie, D. (1994). *Forward rate curves with default risk.* Working paper, Graduate School of Business, Stanford University.
13. Duffie, D. (1999). *Credit swap valuation.* Working paper, Graduate School of Business, Stanford University.
14. Duffie, D., and Huang, M. (1996). *Swap rates and credit quality.* Journal of Finance, 51:921-949.
15. Duffie, D., Schroder, M., and Skiadas, C. (1996). *Recursive valuation of defaultable securities and the timing of resolution of uncertainty.* Annals of Probability 6, 1075-1090.
16. Duffie, D., and Singleton. K. (1997). *An econometric model of the term structure of interest rate swap yields.* Journal of Finance, 52(4):1287-1322.
17. Duffie, D., and K. Singleton. (1999). *Modeling term structures of defaultable bonds.* The Review of Financial Studies, 12(4):687-720.
18. Embrechts, P., Klüppelberg, C., Mikosch, T. (1997). *Modelling Extremal Events for Insurance and Finance.* Springer Verlag, Berlin.
19. Flesaker, B., Houghston, L., Schreiber, L., and Sprung, L. *Taking all the credit.* Risk Magazine, 7:105-108.
20. Fons, J.S. (1994). *Using default rates to model the term structures of defaultable bonds.* Financial Analysts Journal, September-October, 25-32
21. Franks, J.R. and Torous, W.N. (1994). *A comparison of financial re-contracting in distressed exchanges and Chapter 11 reorganizations.* Journal of Financial Economics, 35, 349-370
22. Geske, R. *The valuation of corporate liabilities as compound options.*
23. Gupton, G.M. and Stein, R.M. (2002). *LossCalc^{TM} Moody's Model for predicting loss given default (LGD)*, Global Credit Research, Moody's Investor Services.

24. Heath, D., Jarrow, R., and Morton, A. (1992) *Bond pricing and the term structure of interest rates: A new methodology for contingent claims valuation.* Econometrica, 60:77-105
25. Hull, J.C. (2000). Options, Futures, and Other Derivatives. Prentice-Hall International.
26. Hull, J. and A. White. (1990). *Pricing Interest-Rate-Derivative Securities.* Review of Financial Studies, 3, 573-592.
27. Hull, J. and A. White. (1994a). *Numerical procedures for implementing term structure models I : Single-Factor models.* Journal of Derivatives, 2, ,7-16.
28. Hull, J. and A. White. (1994b). *Numerical procedures for implementing term structure models II : Two-Factor models.* Journal of Derivatives, 2, ,37-48.
29. Jarrow, R.A., Lando, D., and Turnbull, S.M. (1993, revised 1994). *A Markov model for the term structure of credit risk spreads.* Working paper, Johnson Graduate School of Management, Cornell University.
30. Jarrow, R.A., and Stuart M. Turnbull. (1995). *Pricing derivatives on financial securities subject to credit risk.* Journal of Finance, 50, 53-85.
31. Jarrow, R.A. and Turnbull, S.M. (2000). *The interaction of market and credit risk.* Journal of Banking and Finance 24, 271-299.
32. Jarrow, R.A. and Yildirim, Y. (2002). *Valuing default swaps under market and credit risk correlation.* The Journal of Fixed Income, 7-19.
33. Lando, D. (1994). *Three essays on contingent claims pricing.* Ph.D. thesis, Graduate School of Management, Cornell University.
34. Lando, D. (1998). *On Cox processes and credit risky securities.* Review of Derivatives Research, 2, (2/3), 99-120.
35. Longstaff, F.A., and Schwartz, E. (June 1995). *The pricing of credit derivatives.* Journal of Fixed Income, 5(1):6-14.
36. Madan, D and H. Unal. (1998). *Pricing the risks of default.* Review of Derivatives Research, 2 (2/3), 121-160.
37. Mandelbrot, B.B., (1963). *The variation of certain speculative prices.* Journal of Business 26, 394-419.
38. Mandelbrot, B.B., (1967). *The variation of some other speculative prices.* Journal of Business 40, 393-413.
39. Martin, B., Rachev, S.T., Schwartz, E.S. (2000). *Stable non-Gaussian models for credit risk management.* Working paper, University of Karlsruhe, Germany.
40. Merton, R.C. (1974). *On the pricing of corporate debt: The risk structure of interest rates.* Journal of Finance, 29:449-470.
41. Mittnik, S., Rachev, S.T., Doganoglu, T. and Chenyao, D. (1996) *Maximum likelihood estimation of stable Paretian models.* Working paper, Christian Albrechts University, Kiel.
42. Mittnik, S., Rachev, S.T., (2000). *Diagnosing and treating the fat tails in financial returns data.* Journal of Empirical Finance 7, 389-416
43. Paulauskas, V. and Rachev, S.T. (1999). *Maximum likelihood estimators in regression models with infinite variance innovation.* Working paper, Vilnius University, Lithuania
44. Prigent, J.L., Renault, O., Scaillet, O. (2001). *An empirical investigation in credit spread indices.*
45. Rachev, S.T., (1991). Probability Metrics and the Stability of Stochastic Models. John Wiley & Sons, Chichester, New York.
46. Rachev, S.T., Mittnik, S. (2000). Stable Paretian Models in Finance. Wiley & Sons, New York.

47. Rachev, S.T., Racheva-Jotova, B., Hristov, B., I. Mandev (1999). Software Package for Market Risk (VaR) Modeling for Stable Distributed Financial Distributed Returns Mercury 1.0 Documentation.
48. Rachev, S.T., Schwartz, E., Khindanova, I., (2000). *Stable modeling of credit risk.* Working paper UCLA.
49. Rachev, S.T., Tokat, Y. (2000) *Asset and liability management: recent advances.* CRC Handbook on Analytic Computational Methods in Applied Mathematics.
50. Samorodnitsky, G., Taqqu, M.S. (1994). Stable Non-Gaussian Random Variables. Chapman and Hall, New York.
51. Schönbucher, P. (August 1996). *The term structure of defaultable bond prices.* Discussion Paper B-384,University of Bonn, SFB 303.
52. Schönbucher, P. (1998). *Term structure modelling of defaultable bonds.* Discussion Paper B-384,University of Bonn, SFB 303. Review of Derivatives Studies, Special Issue : Credit Risk 2 (2/3), 161-192.
53. Schönbucher, P. (1999). Credit Risk Modelling and Credit Derivatives. Ph.D.-thesis, Faculty of Economics, Bonn University.
54. Schönbucher, P. (June 1999). *A tree implementation of a credit spread model for credit derivatives.* Department of Statistics, Bonn University.
55. Tavakoli, Janet M. (1998). Credit Derivatives: A guide to instruments and applications. John Wiley & Sons.
56. Tokat, Y., Rachev, S.T., Schwartz, E.S. (2001). *Asset Liability Management: A review and some new results in the presence of heavy tails.* Ziemba, W.T. (Ed) Handbook of Heavy Tailed Distributions in Finance, Handbook of Finance.
57. Vasicek, O. (1977). *An equilibrium characterization of the term structure.* Journal of Financial Economics, 5, 177-188
58. Wilson, T. (1997a). *Portfolio credit risk (1).* Risk 10 (9), 111-116.
59. Wilson, T. (1997b). *Portfolio credit risk (2).* Risk 10 (10), 56-61.

Basel II in the DaimlerChrysler Bank

Christoph Heidelbach and Werner Kürzinger

Credit Risk Management, DaimlerChrysler Bank AG, Stuttgart, Germany

Summary. The "New Basel Capital Accord" - Basel II – will have far-reaching consequences on banks throughout the world.

We point out the special profile of the DaimlerChrysler Bank as an automotive financial service provider and the special impact of Basel II in this context.

The "New Basel Capital Accord" [1] is a framework developed by the Basel Committee[1] - a panel of the ten largest industry countries - to guarantee transparency and the control of default risk in the financial marked. It will be in force in year 2006[2] world-wide. The Basel II framework has an individual impact on each bank, depending on size, economic environment and business segment.

As a fully licensed bank, the DaimlerChrysler Bank is subject to the "New Basel Capital Accord". It is a key decision to choose the right approach to calculate the capital requirements from the accord which suites the special profile of the DaimlerChrysler Bank.

In the first part of this article we give an introduction to the DaimlerChrysler Bank and work out the product and portfolio specifics. In the second part we point out the crucial role of the risk parameter "Loss given Default" (LGD) for the DaimlerChrysler Bank and set-up a model portfolio to quantify the impact of this risk parameter on the capital needs following Basel II. In the last part we summarise our results.

1 The DaimlerChrysler Bank

The DaimlerChrysler Bank AG is a financial service provider of the DaimlerChrysler AG and one of the leading automotive banks in Germany. It is owned 100% by the DaimlerChrysler Services AG where the Basel II Capital Accord has to be implemented on full consolidated basis. This requires, that each subsidiary financial service provider has to hold its own risk adjusted economic capital.

---

[1] *Bank for International Settlements (BIS)*
[2] *Currently the second consultative document is out*

The DaimlerChrysler Bank serves as a sales promoter for the original equipment manufacturers Mercedes-Benz, smart, Chrysler, Jeep and Setra. The business segments of the DaimlerChrysler Bank are designed to support this task through leasing, financing, insurance, fleetmanagement for passenger cars and commercial vehicles. In the second half of the year 2002, the product spectrum will be extended to investment products and credit card.

Some key numbers of the year 2001 of the DaimlerChrysler Bank are given in Table 1. Among banks in Germany the DaimlerChrysler Bank is considered a medium size institution.

Table 1. Key numbers of the DaimlerChrysler Bank

		Year 2001
Portfolio	In Mio. €	11298
	In units	570078
Balance sheet In Mio. €		10781
Employees		1280

A qualitative comparison of the DaimlerChrysler Bank and an universal bank is given in Table 2. An important characteristic of the DaimlerChrysler Bank is, that almost each contract is secured by a vehicle. Thus, in case of counterparty default, the observed LGD rates are usually very small.

Table 2. Comparison between Universal Bank and DC Bank

	Universal Bank	DC Bank
Exposure Type	Broad spectrum	Private and corporates only
Products	Great variety	Limited number of highly specialised products
Duration	Product Specific	Short duration
Collateral's	Different types of collateral	High degree (>98%) of specific collateral

The role of an automotive sales promoter has some benefits for the DaimlerChrysler Bank. It combines the knowledge of an automobile manufacturer and a financial service provider in one hand. The DaimlerChrysler Bank is able to support the complete value chain management from manufacturing to re-marketing of used vehicles and therefore has a detailed knowledge of the primary and secondary markets. Detailed knowledge of the automotive market enables the DC Bank to estimate future used vehicle prices and thus expected LGD rates at very high precision.

Under Basel II, the securisation by vehicles accompanied by the ability to estimate recovery rates can lead to crucial business advantages. In the IRB-Advanced Approach, the LGD can be estimated by the Bank itself and hence all kinds of collateral can be taken into account indirectly. On the other hand the possibility for a direct consideration of mobile collateral in the Standard Approach is under discussion. We therefore investigate the importance of such collateral in the next section quantitatively.

2 Quantitative Impact of Basel II

To investigate the effects of the New Basel Capital Accord, we set-up a model portfolio (see Figure 1). We assume a normal distribution of ratings and a composite of 70% corporate and 30% private customers and fit the rating distribution to the actual portfolio of the DaimlerChrysler Bank. For the default rates we take german average values from the year 2000 shown in Table 1 [2].

We use the risk wight formula published in November 2001 [3,4,5] and calculate the capital requirements for the model portfolio under the Standard-, the IRB-Foundation and -Advanced Approach. We assume that there is no external rating available, so that in the Standard Approach, credit risk has to be covered by 8% of capital as demanded by Basel I.

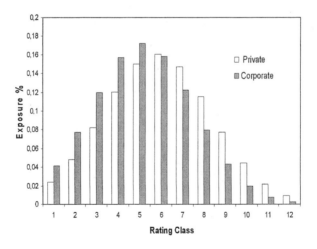

Fig. 1. Rating distribution of corporate and private customers (model portfolio)

Basel II is an evolutionary concept, that rewards increasing complexity of the approaches with reduced demands on economic capital. We investigate the capital requirements and its dependence on the LGD. In the IRB Foundation Approach the LGD is set to 50% for the corporate segment. A bank internal estimate for the LGD is allowed under both IRB Approaches for private (retail) customers and for corporate customers under the IRB Advanced Approach only.

Table 3. Assumed probability of default in rating classes

Rating	PD in percent
1	0.40
2	0.49
3	0.68
4	0.84
5	0.95
6	1.36
7	1.57
8	2.00
9	3.88
10	7.08
11	12.17
12	20.63

In Figure 2, we show the capital needs under the three approaches. Due to the high degree of collateralisation in an automotive bank, a LGD in the range of 20%-40% can be expected. At this range of LGD values the capital requirements due to credit risk are significantly diminished with the growing complexity of the approaches as desired.

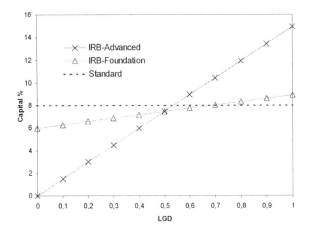

Fig. 2. Capital requirements under different approaches and their dependence on the LGD

In the recent Basel discussions it is under consideration, to extend the class of approved collateral to vehicles. Thus we investigate, how the approval of mobile collateral in credit risk mitigation would affect the capital requirements in the Standard- and the IRB-Foundation Approach. For simplicity we assume the collateral value in terms of the Recovery Rate (RR) to be

$$\text{Collateral} = 80\% \ (1 - \text{LGD}) = 80\% \ \text{RR} \ ,$$

which reflects approximately the conditions in an automotive bank. We follow the framework for an approval of collateral as given in Basel II and allow a haircut to the collateral of 40% which is a conservative estimate, the actual highest haircut is given for stocks at 30%. The IRB-Advanced Approach is not affected by this modification, since here the mobile collateral is completely accounted for in the LGD. In the IRB-Foundation Approach only the corporate part of the portfolio is affected. The results are shown in Figure 3.

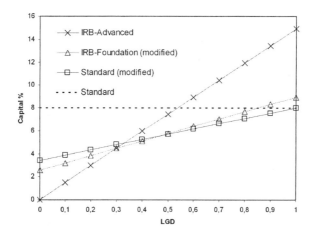

Fig. 3. Impact of approval of vehicles as collateral

In the considered region of LGD from 20%-40%, the advantage of the IRB-Approach is modest and thus the incentive to overcome the technical hurdles of an implementation of the IRB Approach is low.

3 Summary

We give an introduction to the specifics of the DaimlerChrysler Bank and point out its high competence in estimating the LGD.

We compare the different approaches of Basel II to credit risk and the impact of the LGD to the capital needs of the DaimlerChrysler Bank. We point out, that the IRB-Advanced Approach gives clear advantages to an automotive bank, since it allows an indirect approval of vehicles as collateral through a low LGD.

Also we show that a direct approval of vehicles as collateral would give comparable results of capital needs for all 3 approaches, here even the Standard Approach gives the full benefit of high securisation through mobile collateral and hence the competitive advantage of an implementation of a more complex approach is substantially diminished. Thus, current gaps in Basel II and the ongoing consultancy phase impose a high risk for DaimlerChrysler Bank.

References

1. Basel Commitee on Bank Supervision, *The New Basel Capital Accord*, Second Consultative Document, Bank for International Settlements, January 2001
2. *Zeitschrift für das gesamte Kreditwesen*, p. 150, Germany, February 2002
3. Basel Commitee on Bank Supervision, *Potential Modifications to the Commitee's Proposals*, Bank for International Settlements, November 2001
4. Philipp J. Schönbucher, *Factor Modells for Porfolio Credit Risk*, 2000
5. Michael B. Gordy, *A Risk-Factor Model Foundation for Ratings-Based Bank Capital Rules*, 2001

Sovereign Risk in a Structural Approach
Evaluating Sovereign Ability-to-Pay and Probability of Default

Alexander Karmann and Dominik Maltritz

Dresden University of Technology, Chair for Economics, especially Monetary Economics, Dresden, Germany

Summary. We quantify the probability that a sovereign defaults on repayment obligations in foreign currency. Adopting the structural approach as first introduced by Merton, we consider the sovereign's ability-to-pay, characterised by the sum of discounted future payment surpluses, as the underlying process. Its implicit volatility is inferred from market spreads. We demonstrate for the case of Latin America and Russia that our approach indicates default events well in advance of agencies and markets.

1 Introduction

In this paper, we present a model to analyse sovereign risk. Sovereign risk is defined as the risk that the sovereign declares to be unable to fulfil its repayment obligations in a foreign currency which leads to the sovereign's default on liabilities in foreign currency. The model is based on the structural approach to evaluate corporate risk, an approach which dates back to Merton (1974). We will present an adaptation of the Merton model for determining a sovereign's probability of default, PoD.

A default results if the sovereign's ability-to-pay is smaller than the value of the repayment requirements in foreign currency. Thereby, the ability-to-pay is understood as a stochastic variable which follows an Ito process. We assume that the value of repayment obligations at some future time T is publicly known. Then, for any time t prior to T, one can estimate the PoD at T if the parameters of the Ito process as well as of the sovereign's ability-to-pay at time t, A_t, are known.

Modelling a sovereign's ability-to-pay is a challenging task. The approach proposed here relies on what is taken as value of the firm within the theory of the firm. Though there is no directly observable market value of a country available, we can adopt the idea of discounted future net incomes which limit the firm's - here: the sovereign's - capacity to borrow from others. Hence, we will define a sovereign's ability-to-pay to consist of the amount of actual forex reserves and of its potential to attract capital. ¿From economic reasoning, we know that the volume of potential net capital imports today is limited by the volume of discounted future payment surpluses. Estimating these surpluses by a simple time-series approach, we finally derive the levels as well as the drift of the respective ability-to-pay process.

The process' implicit volatility will be inferred from the price spreads in the bond market being interpreted as a risk premium. To do so we adopt the standard Black and Scholes framework. Based on these parameter estimates, it is a straightforward exercise to calculate the sovereign's PoD.

The paper is organized as follows. In Section 2, we modify the structural approach, originally designed to valuate corporate liabilities, to capture the risk of sovereign default. We close with a brief overview over related literature based on option pricing models of sovereign risk. Section 3 deals with the empirical application to calculate PoDs. We demonstrate the steps involved for the case of Argentina 1994 - 2002 and discuss the relationship between input data and PoDs. We extend our investigation to selected countries in Latin America and to Russia. Adopting simple evaluation criteria, we conclude that our model indicates default events well in advance, even when compared to market signals and rating changes.

2 A Structural Approach to Sovereign Risk

2.1 The Structural Approach to Corporate Risk

The structural approach to valuate corporate liabilities relies on the Merton model as the fundamental contribution to this strand of literature. The approach of Merton (1974) is inspired by the seminal paper on option pricing by Black and Scholes (1973).

For the case of a stock option the following assumptions are made:

a) The short term interest rate is known and constant over time.
b) The stock price follows an Ito process. Hence, the distribution of the stock prices is log-normal, the distribution of the returns is normal.
c) There are no dividends.
d) There are no transaction costs.
e) Borrowing any fraction of the underlying is possible, at the short term interest rate.
f) There are no penalties to short selling.
g) The stock option can be exercised at maturity, only.

Black and Scholes (1973, p. 641) argue that for a (delta-) hedged portfolio containing a long position in a stock and a short position in the related call option, the drift rate equals the riskless interest rate, otherwise arbitrage possibilities would result. This argument allows to derive a valuation formula for European stock options.

Merton (1974) shows that this option pricing formula can be used to valuate corporate liabilities as being contingent claims on the value of the firm. The value of the firm represents the underlying of the option contract while the value of the repayment obligations is the strike price. Adopting the assumptions of the Black Scholes model to the topic of corporate liabilities allows to derive valuation formulas for both equity as well as liabilities. The value of the equity is just the value of the call option on the firm's value. The value of the liabilities equals the value of a portfolio in the hand of the creditor consisting of the firm's value and a call option which is short sold.

Empirical applications of the Merton model are described e.g. in Saunders (1999, p. 19 - 37) who also provides an outline of the KMV-Credit-Monitor model well known among practitioners in the financial sector. Another application is given by Delianedis and Geske (1998). Within this type of models, the empirical estimation of the firm's equity and its returns is derived from the market prices of the stock. The volatility of the firm's equity is estimated by historical data on the firm's equity. Solving a system of stochastic differential equations finally leads to the value of the firm and to its volatility. Thus, the value of the liabilities and the probability of default of the firm are determined.

2.2 Adapting the Approach to Sovereign Risk

Macro Fundamentals and the Ability-to-Pay

Sovereign risk is a complex issue not only because of missing 'market' data for a country's net wealth, corresponding to the firm's equity capital, but also because of aggregating economic (solvency, liquidity), institutional (market integration, cooperative enforcement) and political (willingness-to-pay, credibility) aspects (Cantor and Packer 1996). In general terms, it is the repayment prospect (Fischer 1999) of outstanding claims which drives the market perception of sovereign default. In our context, we disregard from 'unwillingness-to-pay', an issue broadly discussed in the 1980ies but believed to be of shrinking relevance due to the increasing integration of individual sovereigns into the world economy (Rogoff 1999, p. 31). Instead, we focus on the sovereign's ability-to-pay at time t, A_t. Thereby, the ability-to-pay is the maximum amount of foreign currency, say US Dollar, the sovereign is able to dispose in order to meet his repayment obligations from borrowing in foreign currency. A_t is composed of already existing forex holdings, FX_t, and the country's potential to attract capital imports, KI_t. From an economic point of view, KI_t will be limited by the sum of discounted future payment surpluses ('net cash flows')NX_t from exports X_t minus imports I_t. We operationalize NX_t by a simple autoregressive process and set KI as the corresponding steady-state capital flow NX^* discounted by an appropriate interest rate ρ^{risk} reflecting the market participants' risk premium in lending to the sovereign. We approximate the risk adjusted discount factor by selecting a sovereign bond of high liquidity and set ρ^{risk} equal to the bond's effective interest rate r_t^{risk} at time t. Formally:

Assumption 1: The sovereign's ability-to-pay A_t is given by

$$A_t = FX_t + KI_t^* \tag{1}$$

where FX_t are the country's foreign exchange reserves and KI_t^* are the potential capital imports equating discounted future income out of net exports NX^*,

$$KI_t^* = \frac{NX^*}{r_t^{risk}}; \tag{2}$$

thereby, NX^* is the steady-state value of the net exports $NX_t := X_t - I_t$ which follow an $AR(1)$ process

$$NX_t = c_0 + c_1 NX_{t-1} \tag{3}$$

and the discount rate r_t^{risk} is the effective interest rate of some highly liquid sovereign bond.

Our procedure to operationalize potential capital import is merely some first-hand approximation for which other alternatives can be formulated. In practitioners' applications, private knowledge may become an important tool to valuate the potential of capital imports. Primarily, capital inflows are provided by investors who outweigh the country's future exports. In addition to these economically based capital inflows, the potential of international aid based on political reasoning may also be taken into account.

We proceed in describing the sovereign's debt structure and the repayment perspectives for a sovereign bond traded in secondary markets to mature at time T. As the decision to default or not hinges on the principle rather than on the coupon payments, we concentrate on zero bonds only. It is often argued that seniorities are involved in serving repayment obligations. Therefore, we assume default to occur if the sum of repayment obligations till time T with priority higher or equal to the sovereign bond considered exceeds the sovereign's ability-to-pay. Or,

Assumption 2. There will be no repayments prior to T. The total amount K of (net) repayment obligations (in international currency) at T with priority higher or equal to the sovereign zero-bond maturing at T is publicly observable. The sovereign will default when holds

$$A_t < K. \tag{4}$$

Relaxing assumption 2 of a single maturity and modelling the debt structure in more detail leads to more complex 'compound' options, as described in Geske (1977). There, for each maturity date, an option contract is specified whose strike price equals the respective repayment obligations. Alternatively, the duration of the liabilities can be taken as a proxy (see Clark (1991, pp. 89) or Delianedis and Geske (1998)). As in general there is no public information on the debt structure, both generalizations would need private information to be implemented.

Market Spreads and PoDs

There are two ways to explain the spread between the price of a risky and a default-free bond. The cost-oriented approach explains the spread as equal to the total costs for holding the risky bond. These costs include a risk premium to account for default (see Edwards (1984) for an early approach). The arbitrage-based approach explains the spread as the implicit price of an insurance against default. This approach can be used after clarification of the economic content of the sovereign's ability-to-pay A_t. According to the Black Scholes mechanism, we characterize the process (A_t) as follows:

Assumption 3: The ability-to-pay process (A_t) is given by

$$dA_t = \mu A_t dt + \sigma A_t dW_t \quad (5)$$

where μ and σ are constant and W is Brownian motion.

To derive a parameter estimation of the process' drift μ we use

Assumption 4: The drift μ is given by the log differences of (A_t) as

$$\mu = log A - log A_{-1} \quad (6)$$

where A is the latest available data for the sovereign's ability-to-pay and A_{-1} is its realization lagged by one year.

To infer for σ, we again invoke the Black Scholes mechanism and assume the existence of a risky and a secure bond with otherwise identical characteristics, more precisely

Assumption 5: There are two types of zero bonds of identical maturity T and identical face value B_t. Both are denominated in one international currency, say US Dollar: risky bonds $B_{t,T}^{risk}$, issued by a foreign sovereign or sovereign institution; and a default-free bond $B_{t,T}^{sec}$, say US-T bonds, with interest rate r.

Non-arbitrage arguments immediately characterize the price a bondholder is willing to pay for an insurance against default. Let us assume that there are no further repayment obligations, i.e. $K = B_T$. In this case, an international investor who holds the entire stock of risky bonds would be willing to pay an insurance against default at a price P which does not exceed the difference $B^{sec} - B^{risk}$ in stock values.

The value of such an insurance equals the value of a hypothetical put option on the sovereign's ability-to-pay where the volume of repayment requirements is the strike price. This holds because the insurance and the put option both have the same pay-off structure. Thereby, we assume that in case of default the sovereign repays as much as he is able to do. In case of non-default, there is no payment from the insurance. In case of default, given our assumption, the insurance pays the difference $K - A_T$ between the repayment requirements and the ability-to-pay at maturity time T. This contingent pay-off structure is identical to the pay-off function of the put. Putting all together, we interpret the price spread between the secure bond and the risky bond as the value of the hypothetical put option introduced above.

Furthermore, if there are other repayment obligations of at least the same priority involved, the bondholders hold only a fraction $\alpha = B_T/K$ of the relevant claims, and the difference in stock values reflects only the part αP of total insurance against the sovereign default or

Assumption 6: Non-arbitrage between holding risky and secure bonds holds. Hence, the price $P_{t,T}$ of a (European) put option to sell the total volume of risky claims with face value K at T is given as follows: let $\alpha := B_T/K$ then

$$B_{t,T}^{sec} = B_{t,T}^{risk} + \alpha P_{t,T}. \tag{7}$$

In practical applications, starting with the price spread per unit of share, expressed as percentage points of the international currency unit, say $b_{t,T}^{sec} - b_{t,T}^{risk}$, these percentage points have to multiplied by the total volume K of repayment obligations to get the option price $P_{t,T}$.

Observing bond spreads in secondary markets and calculating the option price $P_{t,T}$ according to total repayment obligations K, we can use the Black Scholes put price formula

$$P_{t,T} = K exp(-r(T-t))N(-d_2) - A_t N(-d_1)$$

$$\text{where} \quad d_1 = \frac{ln(A_t/K) + (r + \frac{\sigma^2}{2})(T-t)}{\sigma\sqrt{T-t}}$$

$$\text{and} \quad d_2 = d_1 - \sigma\sqrt{T-t} \tag{8}$$

to solve for σ as the market's implicit volatility of the log returns of the ability-to-pay process, by inserting the other input data t, T, r and A_t.

Having identified the parameters μ and σ, the probability of default can be calculated straight forward, using assumption 3, as

$$PoD = P(A_T < K | A_t = A) \tag{9}$$

$$= N_{0,1}\left(\frac{ln(K/A) - (\mu - \frac{\sigma^2}{2})(T-t)}{\sigma\sqrt{T-t}}\right).$$

Equation (9) represents a fundamental as well as market based quantification of sovereign risk which is based on fundamental data and on market information.

2.3 Related Literature on Sovereign Risk

We briefly review some contributions to sovereign risk which also rely on the structural approach. There are three main differences among models covering sovereign risk. First, one has to specify the type of option ('put' or 'call') and the specific option pricing formula used. Second, the underlying has to be determined empirically. This, of course, is the most challenging part in any application and may be based on aspects like e.g. the capital stock of a national economy or the solvency coefficient. Third, the process' characteristics have to be estimated. Especially, volatility estimates can be derived either by using the past realizations of the underlying process or, as implicit volatility, by using the market spreads.

One of the earlier, and remarkably elaborate, contributions is Clark(1991). The focus is in valuating an European call which represents the 'market value of the residents' equity'. Thereby, the underlying is the 'market value of a national economy', V, which is defined as the cash flows of future net exports NX discounted at the economy's rate of return r. Forex reserves and their role are neglected thereby. To estimate r, and finally V, the recursive equation: $\Delta V + NX = rV$ is regressed by rewriting the r.h.s. to consists of a constant (steady-state) return c out of pre-sample period capital stock and of a return rV' out of new capital stock V'. Next, the economy's annual rates of return are calculated as $(\Delta V + NX)/V$ (Clark, 1991, p. 80-81). This is somewhat inconsistent with the process' characteristics implying annual rates of $ln(V/V_{-1})$ (see assumption 3 above). Finally, the volatility is estimated by taking the standard deviation of the annual rate-of-return time series. As an alternative volatility estimate, Clark (1991, p. 101-102) uses the implicit volatility, similarly to our approach. To get somewhat consistent volatility values, he calibrates the model by using different collateralisation levels transforming into different strike prices for the economy.

Klein (1991) and Lichtlen (1997) both adopt a put approach to price new credits or to valuate the fair risk premium for already existing loans. Without employing an empirical application, Klein (1991) proposed a logit approach to estimate the relationship between solvency ratios and fundamentals in order to forecast solvency ratios for new market entrants; thereby, the solvency ratio is the ability-to-pay divided by the repayment requirements. He suggests to approximate the volatility of the underlying by the standard deviation of the bond prices, an assumption which of course does not completely conform with the referred framework. Lichtlen (1997) uses rescheduling events, as documented in the World Bank's 'World Debt Tables'/ resp. 'Global Development Finance', to estimate a logit model of rescheduling probabilities and its macroeconomic determinants. For any tentative σ-value, the put option pricing formula allows to infer from estimated rescheduling resp. default probabilities to the respective solvency ratios. As the overall appropriate volatility value Lichtlen (1991, p. 167) finally choses the one which minimizes on average the distance of volatility and the standard deviation of the respective solvency ratios.

Claessens and Wijnbergen (1993) use a put option formula to price outstanding debt and apply their approach to the question whether Mexico or the international lending community did win from a Brady bond deal. Their 'net amount of financing to serve foreign debt' consists of three elements: expected non-oil current account; adjustments to serve senior debt, FDI and reserve accumulation; and expected oil earnings (Claessens and Wijnbergen (1993, p. 971)). They approximate the volatility as the standard deviation of the forward prices for oil.

Karmann and Plate (2000) differ from our approach mainly in the definition of the underlying as forex reserves available to the debtor nation which consist of actual reserves and the amount of net exports expected till the expiration date (one year ahead). In another recent contribution, Leerbass (1999) bases his notion of the underlying on a stock index of the national economy, arguing that this would closely reflect discounted future GDP and thereby the economy's ability-to-pay.

3 Evaluating the Model

3.1 Input Data Generation

In our model, some data are available on an annual base, like forex reserves or repayment requirements, others are given on a daily base, like bond prices or interest rates (all data are taken from DATA STREAM ; the sample period is begin of 1980 till end of January 2002). This comination deserves some closer description on the generation of the ability-to-pay values.

For each year t^y, the respective steady-state forecast NX^* of net exports will be estimated by regressing equation (3) with monthly data for exports X and imports I for the sample period 1980.1 till end of the year $t^y - 1$. Using the coefficient estimates $\hat{c}_0$ and $\hat{c}_1$ we get

$$NX^* \equiv \hat{c}_0/(1 - \hat{c}_1) \qquad (3')$$

as the steady-state forecast for the year t^y.

For any day $t^d(t^y)$ of the particular year t^y we valuate the PoDs for exactly one year ahead. Implicitly, this asks for considering risky and secure bonds maturing at day $t^d(t^y + 1)$ and their spread. But, typically there are few, or just one, risky bonds issued by the sovereign. We therefore meet the assumption that the term structure of the risk premium is flat. The option value $P_{t^d(t^y),t^d(t^y+1)}$ is now given by the relation

$$\alpha P_{t^d(t^y),t^d(t^y+1)} = exp\left(-ln(1 + r^{sec}_{t^d(t^y),t^d(t^y+1)})\right) - \qquad (10)$$
$$exp\left(-ln(1 + r^{risk}_{t^d(t^y),t^d(t^y+1)})\right).$$

3.2 Evaluation Criteria

Evaluating the performance of a method proposed to quantify risk needs a set of explicitely defined criteria and standards, as the Mathieson and Schinasi (1999, Annex V, p. 192) point out. A straight forward condition is that the approach taken has to clearly indicate default events by signalling them well in advance. The second criterion is the performance of the approach in comparison to markets and agencies. I.e. we have to relate our results on PoDs with the movement of market spreads and with changes of ratings from agencies, like S&P's or Moody's, or from market analysts, as represented by Institutional Investor or Euromoney.

While we will take up this first set of criteria in evaluating our approach, there are some other criteria proposed in the literature which would deserve a more detailed discussion about their appropriateness. PoDs could be transformed into ones ('default') and zeros ('non default') by some threshold value, say $p = 0,5$, to count the hits and misses. This is a criterion well known in evaluating 'early warning' systems. But one has to assign appropriate relative weights for type-I and type-II errors, a task which heavily depends on the intended use of such a model: to maximize profits or to minimize losses from international portfolio investments. Another criterion is the degree of correlation between sovereign PoDs and corporate defaults of the respective countries. There should be a high correlation between the creditworthiness of the sovereign and the one of the corporate sector (see Mathieson and Schinasi (1999, p. 193)). Finally, the question arises how durable the PoD signals are because frequent 'large' jumps would affect the predictive power of the values derived from the model.

3.3 The Case of Latin America and Russia

We start with the case of Argentina, finally rated as SD ('selected default') by S&P's in Nov. 6, 2001, due to nonpayment on debt obligations. In fig. $A.ARG.I$, the PoDs are shown as calculated for the observation period from begin of 1995 till early 2002. The relevant input data are presented in fig. $A.ARG.II$, containing the sovereign's ability-to-pay A_t at date t and the repayment obligations K for the respective year, and in fig. $A.ARG.III$, containing the market spread between the risky and the secure asset.

To analyse the PoDs of Argentina, we distinguish 5 different periods of time:

1. 1995   where PoDs are at a high level of 40 %.
2. 1996/97 where PoDs are at a low level of 10 %.
3. 1998/99 where PoDs increase again up to 30 - 40 %.
4. 2000   where PoDs increase strongly to 70 %.
5. 2001   where PoDs are at a 100 % level.

In 1995, our macro fundamentals suggest a 'non critical' economic situation. The high PoD levels mainly reflect the high risk premia prevalent in many bond markets just after the Mexican crisis. Indead, the fear of contagion is a phenomenon not limited to the Tequila crisis but observed also during the Asian crisis (see e.g. Karmann, Gressmann and Hott (2002) quantifying contagion for Asia). In 1996/97, the former tension in bond markets calmed down lowering the spreads for Argentinian dollar-denominated bonds. On the other side, increasing repayment obligations were fully matched by the sovereign's ability-to-pay which improved within this period of time. The resulting PoD levels of around 10% increased in the second half year of 1997. This was a consequence of rising risk premia during the Asian crisis reflecting the market's fear that the Asian crisis may affect Argentina.

The period 1998/99 is characterized by two countervailing effects. The trend of increasing repayment obligations remained valid but the sovereign's ability-to-pay shrunk due to deteriorating net exports perspectives NX. Hence, PoD levels rose to reach around 40%. The Russia crisis of mid 1998 drove PoDs up to some 60%, a consequence mainly of the considerable increase in market spreads.

In 2000, news on macro-fundamentals let jump the PoD levels to 80% when repayment obligations now nearly coincided with the sovereign's ability-to-pay and the drift became negative. In contrast, the market spreads did not react significantly. In 2001, the fundamentals, as calculated, worsened even more bringing the sovereign's solvency ratio down below one. This, together with the negative drift, led to a second jump of the PoDs up to a 100% level in the beginning of 2001, long before market spreads started to rise in reflection of growing expectations that Argentina would fail to fulfil repayment requirements.

To evaluate the explanatory power of our model for Argentina, we concentrate on the co-movements of PoD figures and rating changes. Starting in 1995, Argentina, rated as $BB-$, improved considerably in terms of calculated PoDs well in advance of the subsequent upgrading in April 2nd, 1997. Three quarters later, our PoDs indicate growing concern on repayment perspectives. News on fundamentals at the end of the year 2000 may have led S&P's to downgrade Argentina, by a slight change to $BB-$ in November 14th, 2000, to be followed by further downgradings during 2001 till SD. Our PoDs, when updated with the news available end of the year 2000, already suggest that a default will be almost sure within a one-year period. While our PoDs react well in advance of rating changes they also lead w.r.t. market spreads as the period from December 1998 till end of the year 2000 shows when markets did not react.

We complete our exercise by investigating PoDs for two sovereigns, Ecuador and Russia, having defaulted and for two countries, Chile and Venezuela, not having defaulted in the entire period 1995 - 2002.

The case of Ecuador is somewhat between fundamental crisis and market herding. Ecuador defaulted on Brady bonds in October 1st, 1999 (see http://www.east-west.be/news4.html#F) when our model predicted a PoD of around 70 % after starting at a close to 15 % level in April 1998 (earlier data were not available). As there is no sovereign rating by S&P's we concentrate on comparing PoDs and market spreads. The increase of calculated PoDs is, at first, a direct consequence of increasing spreads during the Russia crisis. It is enforced by fundamentals deteriorating since early 1999, in terms of decreasing expected flows NX, a negative drift - expressing on-average negative returns - and a higher discounting. As the latter expresses market expectations, we conclude that our model and the market both are signaling an increasingly risky situation during the nine-month period prior to default.

The case of Russia seems to be more a matter of perceived credibility. Though macro fundamentals remained unchanged and forex reserves accumulation (12,2 bill. USD) nearly covered repayment obligations (13 bill. USD), the expected net exports (10 bill. USD) were discounted by the market at extremely high rates. This implicitly means that market belief strongly limited the sovereign's potential to borrow from international markets, regardless of solvency aspects. In contrast to Argentina and Ecuador, there was a rapid and remarkable increase of PoDs before default was announced by August 17th, 1998.

Does this mean that markets have not foreseen default and/or fundamentals did not react properly? A closer analysis of the PoDs and the input data reveals that during the year 1997, the PoD was around 5%. This corresponds to the observed lending boom during that period (see Semenkov (2000, p. 27)). At the end of 1997, the POD increased to a level of about 20% at the end of the year, together with a slight increase in the market spread which lowered the sovereign's ability-to-pay. News on macro fundamentals at the beginning of 1998 led to lower solvency ratios while spreads remained unchanged till mid of the year. The following increase of the PoDs, starting with a 30% levels during the first six months to increase to 40% by August 1st and finally to 80% till the date of default, was merely driven by the increase of spreads from a 10% level to a 65% level.

Putting together, news on fundamentals of Russia at the beginning of 1998 directed PoDs to react somewhat in advance before half-a-year later market participants started to panic. Like in a second generation crisis model (see e.g. Obstfeld (1984)), expectations of market participants switched to drive the economy towards the bad equilibrium. W.r.t. rating changes we see that the process of downgrading began in June 9th, 1998 when our PoDs were already back at some 25% level. Even one week before the announcement of default, Russia was still related as $B+$ while our PoD levels had already reached some 70 % (August 11th, 1998).

Chile represents the case of a sovereign without any default and without any change of rating during the entire period since bond data are available. Macro fundamentals indicate a solvency ratio of as large as 3. As expected, net exports are negative here, in this case the ability-to-pay just consists of the country's foreign exchange reserves. The calculated PoDs seem to be in a reasonable range of 10 - 15% according to low spreads (1,5 - 2,0%) which had been unaffected even by the Argentina crisis. Venezuela is a similar case.

There is no sovereign default in terms of nonpayments on governmental bonds though, admittedly, there are considerable arrears over a long period of time resulting from non-repayments by some governmentally owned steel manufacturers. Venezuela's solvency ratio is at least as large as the corresponding one of Chile but market spreads are higher and also reflect contagion fears during the Tequila crisis and the Russia crisis. Consistently, the PoDs of Venezuela are higher than the ones of Chile, well in line with a sovereign rating below the grade of $A-$ for Chile. But, comparing the co movement between PoDs and rating grades the decision of S&P's to downgrade Venezuela by Februar 23rd, 1996, seems to be questionable in the light of our data indicating considerably reduced PoDs of less than 10%. But also the process thereafter of successive upgrading by S&P's is not supported directly by our PoDs which remain within a range of 15 - 20% since 1996.

4 Summary

We presented a model to compute the probability of default of sovereigns on their foreign exchange liabilities. The ability-to-pay of a sovereign results from the existing foreign exchange reserves and the potential of possible capital imports which are approximated using a time series model. For the development of the ability-to-pay, a stochastic process of the Ito type is assumed. We estimate the volatility of the process by applying the Black Scholes formula for put options to the price-spreads on the bond markets. Thereby, the price-spread between a bond regarded as free of default risk and a risky bond issued by the sovereign of a developing country is interpreted as risk premium for the risk of default.

The computed probability of default depends on macro economic fundamentals which primarily determine the ability-to-pay. In addition, data from the international capital markets are used. These market data reflect the risk assessment of the market participants so that they are well capable for estimating the volatility as the risk parameter of the process.

Our approach is applied to some countries. The probabilities of default determined so far are discussed in detail. It is shown that a strong rise of the computed probabilities of default precedes the occurrence of default in each case. Also related events, like financial crises in other countries, are reflected by the model. Furthermore, we see that the computed probabilities of default run clearly ahead of rating migrations.

Thus, we conclude that our model seems to be well convenient to determine the sovereign probability of default, whereby areas for future research remain, like the prediction of future payment surpluses.

5 Appendix

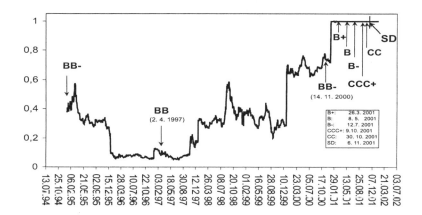

Fig. 1. A.ARG.I: Probability of default for Argentina

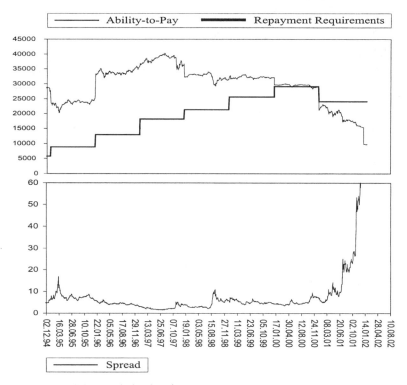

Fig. 2. A.ARG.II/III: Input data for Argentina

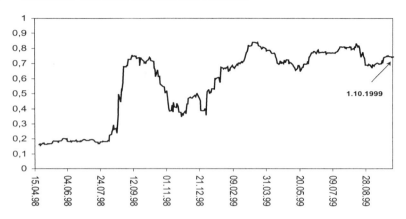

Fig. 3. A.ECU.I: Probability of default for Ecuador

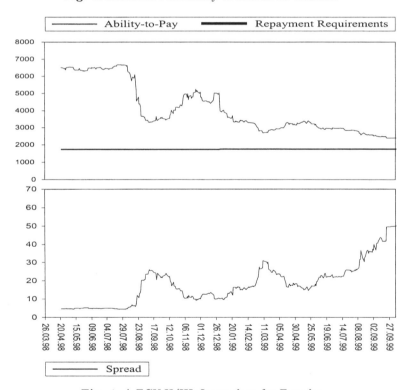

Fig. 4. A.ECU.II/III: Input data for Ecuador

Sovereign Risk in a Structural Approach   105

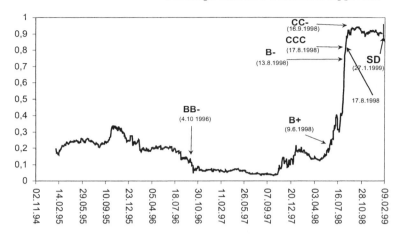

Fig. 5. A.RUS.I-a: Probability of default for Russia

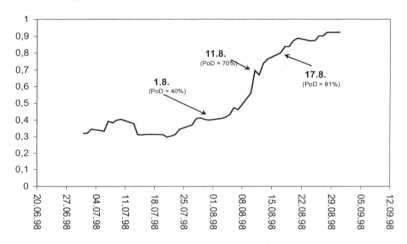

Fig. 6. A.RUS.I-b: Probability of default for Russia - summer 1998

106 Alexander Karmann and Dominik Maltritz

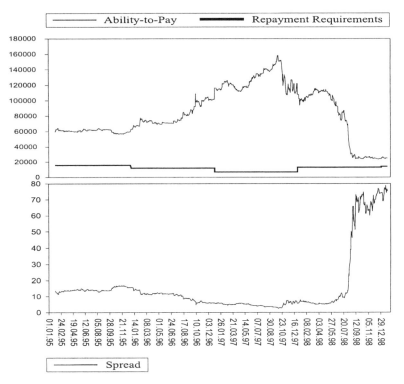

Fig. 7. A.RUS.II/III: Input data for Russia

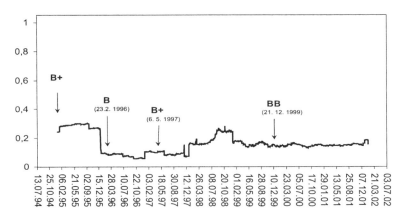

Fig. 8. A.VEN.I: Probability of default for Venezuela

Sovereign Risk in a Structural Approach 107

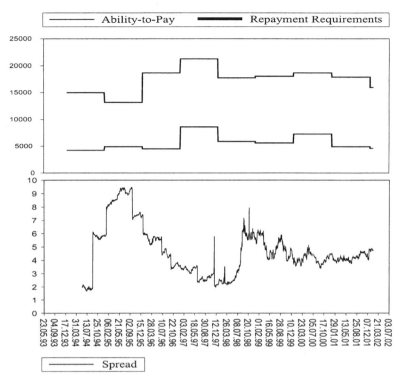

Fig. 9. A.VEN.II/III: Input data for Venezuela

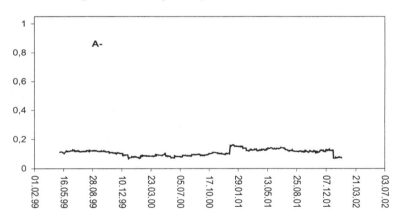

Fig. 10. A.CHI.I: Probability of default for Chile

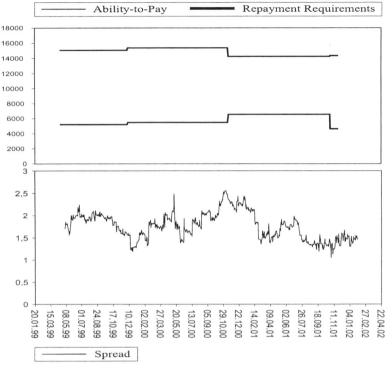

Fig. 11. A.CHI.II/III: Input data for Chile

References

1. Black, F. and M. Scholes, The Pricing of Options and Corporate Liabilities, *Journal of Political Economy*, **81** (1973), 637 – 659.
2. Cantor, R. and F. Packer , Determinants and Impact of Sovereign Credit Ratings, *FRBNY Economic Policy Review*, **2** (1996), 37 – 54.
3. Claessens, S. and S. van Wijnbergen, Secondary Market Prices and Mexico's Brady Deal, *Quarterly Journal of Economics*, **108** (1993), 965 – 982.
4. Clark, E. A., *Cross-Border Investment Risk*, London: Euromoney Publications, 1991.
5. Delianedis, G. and R. Geske, Credit Risk and Risk Neutral Default Probabilities: Information about Rating Migrations and Default, unpublished working paper, 1998.
6. Fischer, S., On the Need for an International Lender of Last Resort, *Journal of Economic Perspectives*, **13** (1999), 85 – 104.
7. Geske, R., The Valuation of Corporate Liabilities as Compound Options, *Journal of Financial and Quantitative Ananlysis*, **12** (1977), 541 – 552.
8. Karmann, A., Gressmann, O. and Ch. Hott, Contagion of Currency Crises - Some Theoretical and Empirical Analysis, *Research Notes in Economics & Statistics - Deutsche Bank Research*, **02-2** (2002).

9. Karmann, A. and M. Plate, Country Risk-Indicator. An Option Based Evalution. Implicit Default Probabilities of USD Bonds., in: Bol, G. and G. Nakhaeizadeh (eds.), *Datamining and Computational Finance*, Heidelberg; New York: Physica, 2000, 43 – 50.
10. Karmann, A., Sovereign Risk, Reserves, and Implicit Default Probabilities: An Option Based Spread Analysis, in: Karmann, A. (ed.), *Financial Structure and Stability*, Heidelberg; New York: Physica, 2000, 232 – 244.
11. Klein, M., Bewertung von Länderrisiken durch Optionspreismethoden, *Kredit und Kapital*, **24** (1991), 484 – 507.
12. Leerbass, F. B., A Simple Approach to Country Risk, unpublished working paper, 1999.
13. Lichtlen, M. F., *Management von Länderrisiken*, Bern: Paul Haupt, 1997.
14. Mathieson, D. and G. Schinasi, *International Capital Markets. Developments, Prospects and Key Political Issues*, Washington D. C.: International Monetary Fund, 1999.
15. Merton, R. C., On the Pricing of Corporate Debt: The Risk Structure of Interest Rates, *Journal of Finance*, **29** (1974), 449 – 470.
16. Obstfeld, M., The Logic of Currency Crises, *Cahiers Economiques et Monétaires*, **43** (1984), 189-213.
17. Rogoff, K., International Institutions for Reducing Global Financial Instability, *Journal of Economic Perspectives*, **13** (1999), 21 – 41.
18. Saunders, A., *Credit Risk Measurement*, New York: John Wiley and Sons, 1999.
19. Semenkov, V., August '98 as a Turning Point of Russian Reforms, in: Hoelscher, J. (ed.), *Financial Turbulence and Capital Markets in Transition Countries*, Basingstoke, Hampshire: Macmillan, 2000, 27 – 38.
20. Standard & Poor's, Sovereign Rating History, see: http://www.standardandpoors.com/RatingsActions/RatingsLists/Sovereigns/index.html
21. Vasicek, O. A., Credit Valuation, working paper, San Francisco: KMV, 1984.

An Extreme Analysis of VaRs for Emerging Market Benchmark Bonds *

Rüdiger Kiesel[1], William Perraudin[2], and Alex Taylor[3]

[1] Universität Ulm, Germany, and London School of Economics, Great Britain
[2] Birkbeck College, Bank of England, and CEPR, Great Britain
[3] Judge Institute of Management, University of Cambridge, Great Britain

Summary. This paper examines the practical usefulness of Extreme Value Theory (EVT) techniques for estimating Value-at-Risk (VaR). Unlike most past studies, the performance of EVT estimators is explicitly compared with that of naive VaR estimators based on quantiles of empirical return distributions. We show that for confidence levels similar to those commonly used in market risk calculations, EVT and naive estimators yield almost identical results when applied to one-day emerging market bond returns. For extremely small confidence levels, the EVT and the naive estimators yield different results on actual data but the differences disappear in a Monte Carlo exercise assuming t-distributed return innovations.

1 Introduction

The Value at Risk (VaR) approach has become the industry standard technique for assessing the riskiness of a portfolio. Defined as the loss which will be exceeded on a given fraction of occasions if the portfolio in question is held for a particular length of time, VaR has the major advantage of simplicity and transparency. The importance of VaR calculations has been reinforced by the role VaRs have been given by the Market Risk Amendment of the Basel Accord. Under the Basel rules (see [2]), banks are permitted to base regulatory capital for trading book risks on VaRs supplied by their own internal risk management models.

Though simple to define, the VaR of a portfolio may be estimated in a wide range of different ways.[4] [20] and [22] examine various straightforward approaches to estimating VaRs and assess their performance on an out-of-sample basis. Employing simulated foreign exchange portfolios, [20] argues

---

* We thank Vincenzo Zinni for advice on data. Correspondence should be addressed to authors at Birkbeck College, 7-15, Gresse Street, London, W1P 2LL. The views expressed in the paper are those of the authors and not necessarily those of the Bank of England.
[4] [12] surveys several approaches to VaR calculation.

that it is important to allow for the predictability of volatility. In contrast, using returns on actual portfolios held by a large bank with an active trading operation, [22] suggest that, at least for well-diversified portfolios comprising government bond and foreign exchange exposures, changes in the likelihood of major losses are hard to forecast and hence modeling the conditional volatility of returns is not helpful.

Recently, several authors have suggested that statistical techniques for analyzing extreme realizations of random variables, termed Extreme Value Theory (EVT), may have useful applications in risk management. Early studies include [23] [14]. Recent contributions which stress financial applications include [9], [24], [21], [27] and [26].

Although EVT has aroused much interest among risk managers, it has been slow to catch on. One reason is that the techniques involved are complex compared to more standard VaR modeling. A second is that there are several potential pitfalls in applying EVT analysis. (Some problems concerning the uncritical use of EVT are mentioned in [10].) In particular, the theory generally presumes independent observations and estimation may require difficult judgments (as we shall see below, the choice of tail thresholds can be tricky).

A third reason is that proponents of EVT modeling often also advocate non-standard risk measures. See, for example, [24] and [14]. For a discussion of some of the problematic aspects of VaR and an overview of alternative risk measures [1]). A fourth reason is that there have been few comparative studies which have used EVT for VaR estimation and compared the results with simpler approaches. The current paper aims to fill this gap by applying a range of up-to-date EVT methods and comparing the results with those obtained using simpler techniques.

This paper examines how EVT techniques perform in a representative market risk application to emerging market bonds. When the confidence level and sample size are similar to those generally applied in market risk VaR calculations, the EVT approaches yield results almost identical to those obtained using naive VaR estimators based on quantiles of the empirical distribution of returns. Only when the confidence level is decreased substantially do differences in the emerge between the EVT and the naive estimates when actual return data is employed. In a Monte Carlo exercise with t-distributed return innovations, differences between EVT and naive estimators are even slower to appear as the confidence level is decreased. A general finding of our research (in line with the conclusions of [10] and [27]) is that, before applying EVT methods, it is important to pre-whiten the data using GARCH or other models capable of filtering out conditional heteroskedasticity.

The data used in this study are daily holding returns on emerging market bonds. Our bonds are selected on the basis that they are the main benchmark instruments for banks involved in emerging market, fixed income trading. The events of the last two years including the Asian Crisis, the Russian default and the market turbulence associated with the collapse of LTCM, serve to emphasize how important it is to understand the risks in emerging market exposures. Our study aims to contribute to this understanding. Lastly, we supplement our VaR estimates based on actual return data with Monte Carlo comparison of EVT techniques.

The structure of the paper is as follows. Section 2 briefly surveys relevant results in Extreme Value Theory and discusses different approaches to VaR estimation. Section 3 discusses our implementation of these different approaches on emerging market bond return data. Section 4 reports the results of our estimations and Monte Carlo comparisons of different techniques. Section 5 concludes. Appendix A describes our improvement of the Weighted Least Squares tail estimator of [21]. Appendix B provides more details of other EVT tail estimators which we implement.

2 Extreme Value Theory

2.1 Outline of Relevant Results

In this section, we introduce EVT techniques which are useful for estimating VaRs. Further background and examples of applications may be found in the books [13] and [29] and the articles [27] and [26]. We shall be concerned with a family of stationary, independent random variables, $\{X_n\}_{n \in \mathbb{N}}$ with the common distribution function, F. We are interested in estimating probabilities of extreme events in a given data set which we assume to be generated by $\{X_n\}$.

For return distributions typically encountered in finance one assumes that the tail behaviour of F is asymptotically given by

$$1 - F(x) \equiv \bar{F}(x) \sim x^{-\alpha} L(x) \qquad for \qquad x \to \infty, \qquad (1)$$

with $\alpha > 0$ and a slowly varying[5] function, L. $\xi = 1/\alpha$ is often called the tail index.

From a theoretical point of view (1) corresponds to F being in the maximum domain of attraction of the Fréchet distribution, one of the so-called standard extreme-value distributions.

[24] has suggested that one might base a risk measure for portfolios on the capital required to cover maximum losses over some time period up to a given confidence level. Such a measure would directly employ the result in equation 1 and just requires that one estimate the parameter, ξ. However, the presence of the slowly varying function $L(.)$, which leads to a semi-parametric estimation problem, implies serious difficulties for the appropriateness of such a - purely parametric - measure.

Under assumption (1)) the distribution of excesses

$$F_u(x) \equiv \mathbb{P}(X - u < x | X > u),$$

---

[5] A slowly varying function is defined as some function, $L(x)$ such that $\lim_{x \to \infty} L(\lambda x)/L(x) = 1$ for all real λ.

i.e. the conditional probability of losses, given that a threshold u is exceeded, can be approximated for high thresholds by a scaled generalized Pareto distribution $G_{\beta,\xi}$. Recall that

$$G_{\beta,\xi}(x) = 1 - \left(1 + \xi \frac{x}{\beta}\right)^{-\frac{1}{\xi}}, \qquad (2)$$

defined with the appropriate domains, and β a positive parameter which depends on the threshold u.

To estimate the tail of F, we define $\bar{F}_u(y) \equiv 1 - F_u(y)$ and then employ the identity:

$$\bar{F}(u+y) = \bar{F}(u)\bar{F}_u(y). \qquad (3)$$

Thus, the problem of estimating (high) quantiles of F, reduces to one of estimating the parameters ξ and β in the approximating generalized Pareto distribution, which can be done by Maximum Likelihood (ML) estimation, and to estimate $\bar{F}(u)$ for which we can use the empirical tail, i.e.

$$\hat{\bar{F}}(u) = \bar{F}_n(u) = \frac{1}{n}\sum_{i=1}^{n} \mathbf{1}_{\{X_i > u\}} = \frac{N_u}{n}, \qquad (4)$$

with N_u be the number of excesses over the threshold level, u, in the sample. Combining the estimates, we find

$$\hat{\bar{F}}(u+y) = \frac{N_u}{n}\left(1 + \hat{\xi}\frac{y}{\hat{\beta}}\right)^{-\frac{1}{\hat{\xi}}}. \qquad (5)$$

From this, we can estimate p-quantiles as

$$\hat{x}_p = u + \frac{\hat{\beta}}{\hat{\xi}}\left(\left(\frac{n}{N_u}(1-p)\right)^{-\hat{\xi}} - 1\right). \qquad (6)$$

The only remaining problem is to choose the threshold, u. This should be done so that the empirical estimator for $\bar{F}$ is reasonably accurate, but without allowing 'average' observations from the distribution to dilute the excess tail estimator $\bar{F}_u$. Effectively, there is a trade off between bias and variance. Graphical methods are commonly employed to find an 'optimal' choice of u. One such method relies on the mean excess function, defined as $e(u) \equiv \mathbb{E}(X - u | X > u)$. Recall that for X with distribution function given as $G_{\xi,\beta}$ we have

$$e(u) = \mathbb{E}(X - u | X > u) = \frac{\beta + \xi u}{1 - \xi}, \qquad (7)$$

with u and ξ from appropriate domains. So, the mean excess function is linear and we can use the empirical mean-excess function

$$e_n(u) = \frac{1}{N_u}\sum_{i \in \Delta_n(u)}(X_i - u), \quad u > 0 \qquad (8)$$

with $\Delta_n(u) = \{i : X_i > u\}$, to investigate the choice of u. From the linearity of the mean excess function one would choose u such that $e_n(x)$ is approximately linear for $x \geq u$.

A second approach to estimating VaRs is based on the Hill estimator of the tail parameter, α. If we assume that $\bar{F}(x) = Cx^{-\alpha}$, $x \geq u > 0$ and $C = u^\alpha$, i.e. $\bar{F}$ comes from a fully specified Pareto distribution, we obtain as MLE for α

$$\hat{\alpha} = \left(\frac{1}{n} \sum_{j=1}^{n} \log X_{n-j+1:n} - \log u \right)^{-1}.$$

In the general case we only know that under (1) $\bar{F}$ behaves like a Pareto distribution above a certain threshold u. Then, conditionally on the number k of exceedances of the threshold we get the conditional MLE

$$\hat{\alpha}^{(H)} = \hat{\alpha}^{(H)}_{k,n} = \left(\frac{1}{k} \sum_{j=1}^{k} \log X_{n-j+1:n} - \log X_{n-k:n} \right)^{-1}. \tag{9}$$

This is the Hill estimator.[6] The MLE for the (unknown) constant is: $\hat{C}_{k,n} = X_{n-k:n}^{\hat{\alpha}^{(H)}_{k,n}} k/n$.

Note here the important fact that the number of order statistics, k, depends on the threshold u. Given estimates of $\hat{\alpha}^{(H)}_{k,n}$ and $\hat{C}_{k,n}$ (conditional on a choice of k), the estimated tail is

$$\hat{\bar{F}}(x) = \frac{k}{n} \left(\frac{x}{X_{n-k:n}} \right)^{-\hat{\alpha}^{(H)}_{k,n}} \tag{10}$$

and an estimator for the p-quantile is

$$\hat{x}_p = \left(\frac{n}{k}(1-p) \right)^{-\frac{1}{\hat{\alpha}^{(H)}_{k,n}}} X_{n-k:n}. \tag{11}$$

It can be shown that the Hill estimator is weakly consistent under fairly general conditions (including some dependency structures) as $k = k(n)$ tends to infinity at an appropriate rate. Typically, the asympotic variance for $\hat{\alpha}^{(H)}_{k,n}$ decreases with increasing k, however for large k a bias may enter, so we face a bias-variance tradeoff situation. [13] (see their §6.4) provide a detailed discussion of the statistical properties of the Hill estimator together with various alternative approaches, such as Pickand's Estimator and the Deckers-Einmahl-de Haan Estimator.

As just emphasized, a crucial choice to make in computing the Hill estimator is the threshold u and the value k which this implies. The simplest

---

[6] Incidentally, if F is in the domain of attraction of the Frechet distribution, the Hill estimator equals the empirical mean excess function of $\log X$ calculated at the threshold, u.

procedure is to examine so-called Hill plots, i.e., plots of $\hat{\alpha}_{k,n}^{(H)}$ against k, and to select the 'optimal' k from a region of the graph at which the relationship appears to have settled down. Other graphical approaches would be to use the mean excess function, as discussed above, or to compare the empirical tail distribution with a Pareto distribution by means of Quantile-Quantile plots using different threshold values.

2.2 Regression Hill-type Estimators

Recently, new tail index estimators have been proposed which generalize the Hill estimator in a natural way, avoiding the problem of selecting k. To simplify the exposition, we suppose that F satisfies the so-called Hall condition:

$$\bar{F}(x) := 1 - F(x) = Cx^{-1/\xi}\left(1 + Dx^{-\rho/\xi} + o\left(x^{-\rho/\xi}\right)\right). \tag{12}$$

This amounts to allowing only a special type of slowly varying function, L.

The statistical properties and the performance of the Hill estimator (under assumption (12)) depend crucially on the optimal choice of the sample fraction $k = k_n$ used in the estimation of ξ (which is determined by the second order parameter ρ). If k_n is too large, then the estimators have a large bias. On the other hand, if k_n is small, their variance is large. In practice, the selection of an optimal $k_n = k_n^{opt}$ (or equivalently the estimation of ρ) is very difficult. Several approaches have recently been proposed.

To avoid having to select a single cut-off number, k_n, (based on Hill plots or some other method), [21] suggest that one use a regression approach in which information from estimates based on a range of different cut-off points is pooled. More precisely, [21] estimate α by taking the intercept of a regression of Hill-type estimates $\hat{\xi}_{k,n}^{(H)}(= 1/\hat{\alpha}_{k,n}^{(H)})$ on k. They employ a weighted regression to take account of the heteroskedasticity in the regression errors.

The justification may be sketched as follows. Under the condition in equation (12), one may asymptotically evaluate the mean and variance of Hill's estimator (compare [16] and [8])

$$\mathbb{E}(\xi_{k,n}) = \xi - \frac{D\xi\rho}{1+\rho}C^{-\rho}\left(\frac{k}{n}\right)^{\rho}\{1 + O(1/k) + O(k/n) + o(1)\} \tag{13}$$

$$Var(\xi_{k,n}) = \frac{\xi^2}{k}\{1 + O(1/k) + o((k/n)^{\rho})\} \tag{14}$$

(where $O(.), o(.)$ are the Landau Symbols). If ρ equals unity, $\xi_{k,n}$ and k are linearly related. Supposing that ρ is close to unity may be motivated, first, on the grounds that it is related to the common procedure of choosing $k(n)$ at a point where ξ estimates behave almost linearly in k. Second, one may note that ρ is indeed approximately unity for many relevant distributions, e.g., for symmetric stable distributions with index $\alpha = 0.5$ we have $\xi = 1/\alpha = 2$ and $\rho = 1$, Students t-distribution with 2 degrees of freedom, Frechet distributions. A third justification for treating ρ as approximately one is that previous studies (e.g., [5] and [11]) suggest small-sample estimates of ρ are rather inaccurate and that it is preferable to choose a fixed value of ρ rather than to estimate it freely.

Thus, the [21] linear approximation appears reasonable:

$$\xi_{k,n} = \beta_0 + \beta_1 k + \epsilon_k, \quad k = 1,\ldots,\kappa. \tag{15}$$

However, the correlated nature of the $\xi_{k,n}$ $k = 1,2,\ldots,n$ implies that the Gauss-Markov conditions which ensure the optimality of Ordinary Least Squares (OLS) do not hold. More efficient estimates may be obtained by allowing for the covariance structure of the regression errors. [21] employ Weighted Least Squares (WLS). In Appendix A, we show how to improve on their technique using a full Generalized Least Squares (GLS) method. In brief, one may write (15) in matrix notation as

$$\boldsymbol{\xi} = \boldsymbol{T}\boldsymbol{\beta} + \boldsymbol{\epsilon} \tag{16}$$

with $I\!\!E(\epsilon) = 0, \boldsymbol{Cov}(\epsilon) = \boldsymbol{C}$. To perform GLS, one must, therefore, find a matrix $\boldsymbol{W}$ such that

$$\boldsymbol{Cov}(\boldsymbol{W}\epsilon) = \boldsymbol{W}\boldsymbol{C}\boldsymbol{W}' = \boldsymbol{I}. \tag{17}$$

Utilizing the structure of the Hill estimators as linear transformations of the log order statistics we outline in the Appendix how to find such a matrix, $\boldsymbol{W}$.

2.3 Other Approaches

In our empirical application below, we implement GPD fits, and regression-based generalizations of the Hill estimator. We also implement another method to estimate ξ recently proposed by [11], which obtains $k(n)$ from an iterative selection procedure. [11] base their choice on the law of the iterated logarithm for the Hill estimator. Finally, we implement an estimation strategy outlined by [5], who use a regression approach based on a Pareto quantile plot. Details of both approaches are provided in Appendix B.

Finally, another approach would be to select the sample fraction by a re-sampling procedure. A simple bootstrap approach fails because it underestimates the bias of the Hill estimator. However, [9] have recently proposed a method based on a combination of sub-sample bootstrap estimates to obtain a consistent estimator of the optimal number of order statistics needed for the Hill estimator. Since this method requires a large sample size (they use a sample size of at least 5000), we did not implement their approach for our current problem.

3 Empirical Investigation

3.1 Data and Preliminary Tests

We implement the various VaR estimation techniques described above on a data set of daily returns on emerging market bonds. The bond issues included in our sample are selected on the basis that they are the main benchmarks for investors trading in emerging market fixed income exposures.[7] The bonds included (which are all US dollar-denominated) are all Brady bonds except the Moroccan bond (which was pre-Brady) and the Indonesian bond (which is a Eurobond). The bonds may be identified by their maturity dates and coupon rates given in Table 1. The length of the data series ranges from 500 to 2000 observations. The data was kindly supplied to us by an investment bank. The time period covered varies from 2/2/91 to 13/4/99 in the case of Mexico and Venezuela, to from 2/7/96 to 13/4/99 in the case of Peru.[8] The returns data employed in the study consisted of daily log changes in (dirty) bond prices.

We perform unit root tests on the bond returns, namely Augmented Dickey-Fuller (ADF) and Phillips-Perron (PP) tests. These generally reject the presence of a unit root in the log return process, y_t, at a 1% significance level. Furthermore, there is no evidence of a drift or a time trend in the log price process x_t. Stationarity of the log returns y_t cannot be rejected at the 10% significance level using a Kwiatkowski, Phillips, Schmidt, and Shin test (KPSS). Together with the results from the ADF and PP tests, this allows us to conclude that the log returns should be modeled as stationary processes.[9]

Comparison of the unconditional distribution of the y_t with a normal distribution suggests they are heavy-tailed. QQ-plots for the data curve down on the left and up on the right, indicating heavier than normal tails. Table 1 reports the first two moments of the daily bond returns and their skewness and kurtosis coefficients. The kurtosis coefficients clearly exceed the value 3 which one expects for a Gaussian distribution. The skewness parameter is significantly different from zero, which suggests an asymmetric distribution. The standard Bera-Jarque test (also reported in Table 1) rejects the hypothesis of normality.

The autocorrelation function of y_t suggests no dependence structure. However, the raw time series of y_t point to volatility clustering, in that days with large price movements tend to occur together. The correlogram of y_t^2 and the Box-Ljung statistic (also reported in Table 1) convincingly confirm this.[10]

---

[7] We excluded benchmark issues for some additional countries because the length of the time series available was too short.

[8] The numbers of observations were: Mexico 2,089, Venezuela 2,089, Equador 1,178, Brazil 1,235, Argentina 1,695, Peru 692, Morocco 1,595, Nigeria 1,593, Russia 697, Poland 1,222, Indonesia 667.

[9] For details and discussion of these tests see e.g. [15] or [25].

[10] The Portmanteau Ljung-Box Test involves the following steps: 1. Estimate y_t with the "best-fitting" ARMA model and obtain the sequence of fitted errors $\hat{\epsilon}_t^2$.

Both show that significant correlation exists at quite long lags. This suggests that y_t^2 may follow a process well approximated by an ARMA(1,1), since it is hard to reproduce persistence in shocks with low correlation using simple AR processes. We also use a more formal Lagrange Multiplier test for ARCH disturbances generated by a range of orders for regression models for y_t and the corresponding residuals.[11] All these tests reject the null hypothesis that there are no ARCH effects in the errors on at least a 99% level of significance.

3.2 Pre-Whitening

The above time series analysis suggests that the log returns follow a GARCH-type process. However, in modeling the tail behavior of the returns, we wish to avoid specifying the distribution of innovations in the GARCH model. This is possible if we implement a GARCH model with normal errors as a Pseudo-Maximum Likelihood approach[12] and then analyze the residuals from this preliminary model using the range of EVT techniques described in the last section. We shall refer to the residuals from the initial GARCH model as the pre-whitened returns data.

We try several models from the GARCH-family (e.g., M-GARCH, Exponential GARCH) on the raw returns data. It appears that a parsimonious model of the GARCH(1,1)-type without drift fits the data satisfactorily. Our basic model for the raw data may, therefore, be summarized as:

$$y_t = \sigma_t \epsilon_t, \quad \epsilon_t \sim F,$$
$$\sigma_t^2 = a + a_1 \epsilon_{t-1}^2 + b_1 \sigma_{t-1}^2. \tag{18}$$

The process in equation (18) is stationarity if and only if $a_1 + b_1 < 1$. This condition cannot be rejected for any of the fits to our raw return series at a 99% level. ACF plots for the squared residuals from the GARCH model together with formal BL and LM tests suggest that the ARCH-effects have been removed by the pre-whitening and we may proceed under the i.i.d. assumption for the residuals. This is important, since most of the EVT results are only

---

Calculate $\hat{\sigma}^2 = \frac{1}{T}\sum_{t=1}^{T} \hat{\epsilon}_t^2$. 2. Calculate (and plot) the sample autocorrelations of the squared residuals $\rho(i) = \sum(\hat{\epsilon}_t^2 - \hat{\sigma}^2)(\hat{\epsilon}_{t-i}^2 - \hat{\sigma}^2)/\sum(\hat{\epsilon}_t^2 - \hat{\sigma}^2)^2$. 3. $Q = T(T+2)\sum_{i=1}^{n}\rho(i)/(T-i) \sim \chi^2(n)$ if $\hat{\epsilon}_t^2$ are uncorrelated. Rejecting the hypothesis the null that the $\hat{\epsilon}_t^2$ are uncorrelated is equivalent to rejecting the null of no ARCH or GARCH errors.

[11] The LM test consists of the steps: 1. Use OLS to estimate the most appropriate AR(n) (or regression model) $y_t = a_0 + a_1 y_{t-1} + \ldots a_n y_{t-n} + \epsilon_t$. 2. Regress the squared residuals on a constant and q lagged values $\hat{\epsilon}_t^2 = \alpha_0 + \alpha_1 \hat{\epsilon}_{t-1}^2 + \ldots + \alpha_q \hat{\epsilon}_{t-q}^2$. If there are no ARCH or GARCH effects, the estimated values of $\alpha_1, \ldots, \alpha_q$ should be zero. So the usual R^2 statistic will be low. Under the null hypothesis $TR^2 \sim \chi^2(q)$.

[12] Compare [17], §4.1.

established for i.i.d. random variables (or special dependence structures). QQ-plots still give evidence of somewhat heavy tails for the pre-whitened returns, however. The fact that the pre-whitened returns continue to exhibit fat tails justifies our use of non-parametric EVT approaches to fit the tail behavior.

We illustrate some of the above findings using the Mexican return data. Figure 1 exhibits ACF plots for returns and squared returns respectively of the actual and pre-whitened Mexican bond returns. The ACF plots for actual and pre-whitened returns show no indication of dependence. On the other hand, the ACF plots for squared raw returns indicates significant autocorrelation and at quite long lags. Once the series has been pre-whitened, the autocorrelation of squared returns is largely eliminated.

To compute quantiles, $\hat{y}_p$, for the one-day predictive distribution using the GARCH model described in equation (18), we use

$$\hat{y}_p = \hat{\sigma}_{t+1} \hat{x}_p \tag{19}$$

where $\hat{x}_p$ is the corresponding quantile from the distribution of the standardized residuals, estimated by EVT methods, and

$$\hat{\sigma}_{t+1}^2 = \hat{a} + \hat{a}_1 \hat{\epsilon}_t^2 + \hat{b}_1 \hat{\sigma}_t^2$$

with $(\hat{a}, \hat{a}_1, \hat{b}_1)$ the estimated vector of coefficients of the GARCH model, and $\hat{\epsilon}_t = y_t - \hat{y}_t$ the residuals. This approach of pre-whitening has also been used by [27].

4 Results from the EVT Analysis

4.1 Tail Index Estimates

Estimates of tail index estimates provide an indication of how many moments exist in that $\xi < 1/n$ implies that the nth moment is finite. This may be useful since, for example, if fourth moments do not exist, confidence intervals for estimates of covariances are likely to be unreliable. Hill-type estimators provide semi-parametric estimates of the tail index, ξ, and hence may be preferred to the fully parametric GPD-fit methods if one's primary interest is in ξ. However, since the excess distribution is fitted more accurately in the GPD-fit approach, if one's main aim is quantile estimation, the latter approach might be preferred.

In Table 2, we present tail index estimates for the pre-whitened return series.[13] The estimates we report come from GPD fits and modified GLS Hill estimates. We also report results based on estimations of a parametric GARCH model with Student's t-distributed residuals. The estimated number

---

[13] We carried out similar estimates using the raw returns but these are not reported due to space constraints.

of degrees of freedom in a Student's t-distribution is the inverse of the tail index. Thus, in Table 2, we report the inverse of the number of degrees of freedom that we estimated.

In general, the t-fit tail index is larger than the modified Hill tail index estimate. The reason for this is that the t-fit uses the entire dataset (not just the tails) and thus is also affected by the peakedness of the distribution. It is difficult to compare tail index estimates obtained from GPD fits with those obtained from the Modified Hill approach since the GPD fit involves a second parameter. The substantial variation in the GPD tail index estimates across different return series and the fact that, in one case, we even obtain a negative tail index tends to confirm our argument made above that the modified GLS Hill approach is more to be relied on for estimates of ξ.

The tail index estimates we obtain are in the ranges 0.139 to 0.324 and 0.2432 to 0.452 for the upper and lower tails respectively. This suggests the innovation distribution is skewed to the left (consistent with the skewness coefficients for the raw series in Table 1). These results imply that the second moment exists but that the fourth moment is unlikely to exist. The tail index estimation results for the raw series (which we do not report) similarly suggest negative skewness and are in the range 0.3012 to 0.3972 for the upper tail (except for Russia which has an index of 0.6136), and 0.2939 to 0.4542 for the lower tail.

4.2 Quantile Estimates

Table 3 reports quantile estimates for the pre-whitened returns. For each return series, we provide estimates based on the modified Hill approach (i.e., with GLS pooling of information), GPD-fits and Student's t-fits. We also include "empirical estimates", i.e., the corresponding quantiles of the empirical distribution of the pre-whitened returns. Since the data series has been pre-whitened, the quantiles are in units of standard deviations. The fact that the 1% quantiles are typically greater than 2.33 shows that the distributions are more fat-tailed than normal distributions, therefore.

Noticeable in Table 3 is the fact that the quantile estimates based on GDP fits and the modified Hill approach are close to the empirical quantile. Only the t-fit quantiles are clearly different. Generally, the latter overstate the absolute value of the absolute quantile, suggesting that the distribution is never as fat-tailed as the t-fit suggests. The relative over-estimate of the absolute quantile is greater for upper than for lower tails since the data is negatively skewed and the t-distribution is symmetric.

Table 5 reports quantile estimates for pre-whitened Mexican bond returns. These were reasonably representative of results we obtained for the other bond returns which we do not present owing to space constraints. From the results already reported, it is clear that quantile estimates do not vary substantially for different estimation approaches unless the quantile is far out in the tails.

We therefore focus, in Table 5, on quantiles at or below 1% and at or above 99%

The estimates based on the regression approaches and the GPD fit are reasonably close in the negative tail, although the GPD fit quantiles are smaller in the case of the positive tail. For the more extreme quantiles, the empirical quantile estimator is consistently closer to the mean than the EVT estimates. Even excluding the empirical quantile estimate, variation in the estimates is quite considerable. The proportional discrepancy between the highest and the lowest estimates is about 50% for the quantiles furthest from the mean.

This variation is to be expected since our datasets contain fewer than 2,000 observations and we focus on quantiles corresponding to the lowest or highest 1, 2, 10, or 20 observations. However, the GPD estimates show the smallest variation as measured by their standard errors and might on these grounds be preferred.[14]

Once again, we illustrate our findings using the Mexican bond data. Figure 2 shows the basic return series without any pre-whitening.[15] Volatility clustering is apparent in the plot. The figure also shows 1% VaR quantiles calculated using the modified Hills EVT estimator (i) on the raw returns and (ii) on the pre-whitened returns. The former performs poorly in that the exceptions[16] cluster in periods when the volatility is also clustering. In the case of pre-whitened returns, this phenomenon is not apparent.

4.3 Monte Carlo Comparison of the Different Estimators

In Tables 6 and 7, we report results on a small simulation study for the VaR estimation methods discussed above. The Monte Carlo consists of 1,000 samples, each of which contains 1,000 observations. This was roughly the average number of observations in our data series. The data are drawn from a Student's t_ν-distribution, with $\nu = 2, 3, 4$ degrees of freedom. Recall that for the t_ν-distribution the νth moment exists, but no higher moment, and that $\xi = 1/\nu$ and $\rho = 2/\nu$. We only report results on estimators for the tail index and upper quantiles since the t_ν-distributions are symmetric.

Rather than performing a full mean-squared error analysis, we use the mean of the empirical tail-index estimator distribution and as standard errors, the standard deviation of this distribution

The Monte Carlo tail index estimates (see Table 6) are all reasonably close to the true values. We include estimates of Monte Carlo standard errors which suggest the deviations from true values are statistically significant but they

---

[14] This is consistent with results communicated to us privately by A. McNeil.
[15] Note we plot *minus* the change in the log prices, so losses correspond to *positive* movements.
[16] Exceptions in the Basel terminology are defined as observations for which losses exceed the VaR (see [2]) and hence returns crosses the VaR quantiles shown in the plot.

are too small to be economically or financially significant. The two regression-based approaches and the BVT technique seem to perform better for the case of 4 degrees of freedom than the Drees methods.

Table 7 gives Monte Carlo results for quantile estimates derived from the 1,000 t-distributed samples. The results are not very conclusive since none of the EVT methods yields estimates which on average differ substantially from the average empirical quantile. One should note here that the t-distribution is quite well-behaved and EVT estimates are likely to perform well in this case. It is not surprising that we find more variation across estimates in the last subsection where we were employing actual data.

4.4 VaR and Risk Management

Another way to assess VaR techniques is to calculate VaRs using a rolling window of observations and then to see how many realized losses exceed the estimated VaRs. Table 8 reports the results of such calculations. In line with the Basel terminology[17], we refer to cases in which losses exceed VaRs as "exceptions". The rolling data window employed in the calculations contains 500 observations. The VaRs were calculated for the raw series and for pre-whitened returns.

The results show that after pre-whitening using a GARCH model, either EVT methods (and here we employ our modified, GLS, Hill estimator) or empirical quantiles yield similar, quite accurate results. The GARCH-t and the GARCH-normal parametric models yields too few and too many exceptions respectively. Using EVT on non-pre-whitened returns yields far too may exceptions. These results are similar to those obtained by [27] using major stock market indices.

To illustrate these results, in Figure 2, we show scatter diagrams of the raw and pre-whitened daily returns for Mexican bonds together with plots of the estimated VaRs, day by day. The problem that is evident with the raw data is that the VaRs do not increase when volatility rises despite the fact that there is clear evidence of volatility clustering. This leads to an excessive number of exceptions.

5 Conclusion

In this paper, we exposit and implement a range of EVT techniques for estimating quantiles and hence for calculating VaRs for financial returns. Amongst the methods we employ is an improved version of the regression-based Hill estimator of [21]. The improvement consists of using a GLS approach which is likely to be more efficient than the WLS technique described by [21].

Our conclusions are as follows.

---
[17] See [2]

1. Simple empirical quantiles yield similar results to the more sophisticated EVT techniques as long as the quantile is not too far out in the tails. For datasets of a size often employed in market risk VaR calculations, say 500 observations, VaRs with a 1% confidence level are broadly the same if one uses EVT methods or simple estimates based on empirical quantiles. This conclusion is important since much of the recent literature on applications of EVT to risk management problems fails to make explicit comparisons with simpler estimation approaches.
2. The primary usefulness of EVT methods in financial applications is, therefore, in the context of VaRs with extremely small confidence levels. Some banks have recently started to estimate credit risk VaRs at confidence levels of as little as 3 basis points. These are generally estimated for long holding periods such as a year or more so they are not directly comparable with standard market risk VaR calculations based on high frequency data of the kind discussed in this paper. But, EVT may have valuable applications in such contexts.
3. Using actual data, we find that for very small confidence levels, EVT-based quantiles appear to be closer to the mean (implying smaller VaRs) than those based on the empirical quantiles of the pre-whitened returns. However, this finding is not conclusive since when we perform Monte Carlo estimates of quantiles for Student's t-distributed random variables, we find that the empirical quantiles are not very different from what appear to be the more dependable EVT based estimates, namely the GPD fits and the modified Hill estimates, even for quantiles far out in the tails.
4. Our results suggest that pre-whitening of returns using GARCH-type models which remove conditional heteroskedasticity is very important for the emerging market bond data we employ. This conclusion contrasts with the results of [22] who examined VaRs for returns on well-diversified portfolios mainly comprising interest rate exposures from five different G7 government bond markets. It is to be expected, however, that results of this kind will be sensitive to the statistical nature of the return series one is studying.

A Appendix

A.1 Generalized Least-Squares Estimator

Recall that we denote the order statistic based on n observations $X_1, \ldots, X_n$ by
$$X_{1:n} \leq \ldots \leq X_{n:n},$$
then Hill's estimator based on the k largest observations is given by

$$\hat{\xi} = \hat{\xi}_{k,n}^{(H)} = \frac{1}{k} \sum_{j=1}^{k} \log X_{n-j+1:n} - \log X_{n-k:n}. \quad (9)$$

As argued in §2.4 we can view the sequence of Hill's estimators as a linear model
$$\hat{\xi}_{k,n} = \beta_0 + \beta_1 k + \epsilon_k, \quad k = 1, \ldots, \kappa. \quad (15)$$
Writing (15) in matrix notation
$$\hat{\xi} = T\beta + \epsilon \quad (16)$$
with
$$T' = \begin{bmatrix} 1 & 1 & \ldots & 1 \\ 1 & 2 & \ldots & \kappa \end{bmatrix}$$
we get
$$\mathbb{E}(\hat{\xi}) = T\beta \quad \boldsymbol{C}ov(\hat{\xi}) = \boldsymbol{C}ov(\epsilon).$$
So in order to transform the regression problem to the standard Gauss-Markov setting we have to find the covariance matrix of the Hill estimators. Setting
$$\boldsymbol{Y} = (\log(X_{n-\kappa:n}), \log(X_{n-\kappa+1:n}), \ldots, \log(X_{n:n})) \quad (20)$$
we find
$$\hat{\xi} = \boldsymbol{A}\boldsymbol{Y}, \quad (21)$$
where the transformation matrix $\boldsymbol{A}$ is given as
$$\boldsymbol{A} = \begin{bmatrix} 0 & 0 & \ldots & 0 & 0 & -1 & 1 \\ 0 & 0 & \ldots & 0 & -1 & \frac{1}{2} & \frac{1}{2} \\ \ldots & \ldots & \ldots & \ldots & \ldots & \ldots & \ldots \\ -1 & \frac{1}{\kappa} & \ldots & \frac{1}{\kappa} & \frac{1}{\kappa} & \frac{1}{\kappa} & \frac{1}{\kappa} \end{bmatrix} \quad (22)$$

From assumption (12)
$$U(x) := \inf\{y : F(y) \geq 1 - 1/x\} \sim ax^\xi(1 + bx^{-\rho} + o(x^{-\rho})) \quad (23)$$
follows (see [18]). Since we already strengthened this assumption to the case $\rho = 1$ (23) leads to
$$U(x) \sim ax^\xi(1 + bx^{-1} + o(x^{-1})) \sim ax^\xi$$
in case $\xi \geq 1$, which is the case of heavy tails we are interested in. So we continue under the assumption
$$U(x) \sim ax^\xi \quad (24)$$
which is equivalent (again see [18]) to
$$\bar{F}(x) := 1 - F(x) = Cx^{-\frac{1}{\xi}}. \quad (25)$$
Using the quantile transformation lemma (see [13], Lemma 4.1.9) together with (24) we get jointly in $j = 1, \ldots, \kappa + 1$

$$\log X_{n-j+1:n} \stackrel{d}{=} \log U(U_{j:n}^{-1})$$

$$= \log(aU_{j:n}^{-\xi}) = -\xi \log U_{j:n} + \log a$$

where $U_{1:n} \leq U_{2:n} \leq \ldots \leq U_{n:n}$ denote the order statistic of a uniform $(0,1)$ sample of size n. Using this transformation we only have to find the covariance structure of $\boldsymbol{U} = (\log U_{j:n}, 1 \leq j \leq n)'$ in order to find the covariance structure of $\boldsymbol{Y}$. By using a standard Taylor approximation we can compute the covariance structure of $\boldsymbol{U}$ from the well-known covariances of the order statistic of the uniform variables. Indeed, it is well known (see e.g. [28], p.129) that

$$\mathbb{E}(U_{i:n}) = \frac{i}{n+1} \tag{26}$$

$$\mathbb{V}ar(U_{i:n}) = \frac{1}{n+1}\frac{1}{n+2}\left(1 - \frac{i}{n+1}\right)$$

$$\boldsymbol{C}ov(U_{i:n}, U_{j:n}) = \frac{1}{n+1}\frac{1}{n+2}\left(1 - \frac{j}{n+1}\right) \quad i \leq j. \tag{27}$$

By means of a first order Taylor approximation we get

$$\boldsymbol{C}ov(\log U_{i:n}, \log U_{j:n}) \approx \frac{(n+1)}{i}\frac{(n+1)}{j}\mathbb{E}\left(\left(U_{i:n} - \frac{i}{n+1}\right)\left(U_{j:n} - \frac{j}{n+1}\right)\right)$$

$$= \frac{(n+1)^2}{ij}\boldsymbol{C}ov(U_{i:n}, U_{j:n})$$

$$= \frac{(n+1)}{(n+2)ij}\left(1 - \frac{j}{n+1}\right).$$

Defining

$$\boldsymbol{C} = (\boldsymbol{C}ov(\log U_{i:n}, \log U_{j:n}))_{1 \leq i,j \leq \kappa+1} \tag{28}$$

we get

$$\boldsymbol{C}ov(\hat{\boldsymbol{\xi}}) = \boldsymbol{C}ov(\boldsymbol{AY}) = \boldsymbol{AC}ov(\boldsymbol{Y})\boldsymbol{A}' = \xi^2 \boldsymbol{ACA}' := \xi^2 \boldsymbol{\Omega}. \tag{29}$$

Since $\boldsymbol{\Omega}$ is symmetric and positive definite (being a covariance matrix), we can write it as $\boldsymbol{\Omega} = \boldsymbol{\Gamma D \Gamma}'$, where $\boldsymbol{\Gamma}$ is orthogonal and $\boldsymbol{D}$ is a diagonal matrix with positive diagonal elements. We now set $\boldsymbol{W} = \left(\boldsymbol{\Gamma D}^{-\frac{1}{2}}\right)^{-1}$. Pre-multiplying both sides of (16) we get

$$\boldsymbol{W}\hat{\boldsymbol{\xi}} = \boldsymbol{WT}\boldsymbol{\beta} + \boldsymbol{W}\boldsymbol{\epsilon}.$$

Setting $\boldsymbol{\xi}^* = \boldsymbol{W}\hat{\boldsymbol{\xi}}, \boldsymbol{T}^* = \boldsymbol{WT}, \boldsymbol{\epsilon}^* = \boldsymbol{W}\boldsymbol{\epsilon}$ we get

$$\boldsymbol{\xi}^* = \boldsymbol{T}^*\boldsymbol{\beta} + \boldsymbol{\epsilon}^*, \tag{30}$$

where $\mathbb{E}(\epsilon^*) = 0$ and

$$\boldsymbol{Cov}(\epsilon^*) = \boldsymbol{Cov}(\boldsymbol{W}\epsilon) = \boldsymbol{W}\boldsymbol{Cov}(\epsilon)\boldsymbol{W}'$$

$$= \boldsymbol{W}\boldsymbol{Cov}(\hat{\xi})\boldsymbol{W}' = \xi^2\boldsymbol{W}\boldsymbol{\Omega}\boldsymbol{W}' = \xi^2\boldsymbol{I}$$

Therefore the Gauss-Markov conditions hold for model (30) and we can apply the usual theory to obtain the ordinary least-squares estimator for β (which will be the best linear unbiased estimator (BLUE)). We get

$$\hat{\beta}_{MLS} = (\boldsymbol{T}^{*'}\boldsymbol{T}^*)^{-1}\boldsymbol{T}^{*'}\xi^* \tag{31}$$

$$= (\boldsymbol{T}'\boldsymbol{W}'\boldsymbol{W}\boldsymbol{T})^{-1}\boldsymbol{T}'\boldsymbol{W}'\boldsymbol{W}\hat{\xi} \tag{32}$$

$$= (\boldsymbol{T}'\boldsymbol{\Omega}^{-1}\boldsymbol{T})^{-1}\boldsymbol{T}'\boldsymbol{\Omega}^{-1}\hat{\xi}. \tag{33}$$

The first coordinate of $\hat{\beta}_{MLS}$ is then our covariance adjusted modified least squares estimator for the tail index ξ.

A.2 Further Tail Index Estimation Methods

Threshold Selection

Under condition (12) it is possible to show that the optimal value (in terms of minimizing the mean squared error of the Hill estimator) of the threshold value has the asymptotics

$$k_n^{opt} \sim \left(\frac{C^{2\rho}(\rho+1)^2}{2D^2\rho^3}\right)^{1/(2\rho+1)} n^{2\rho/(2\rho+1)} \tag{34}$$

(compare [19], a textbook exposition is [13], §6.4 and in particular Theorem 6.4.9). Unfortunately, the parameter ρ is unknown (besides the fact that it is not clear a priori that (12) holds). So efforts to improve the selection of the threshold k centre around estimating ρ and/or the optimal order.

Drees-Kaufmann Approach

The basic idea is to construct a sequential approach to find a consistent estimator of the optimal threshold k_n^{opt}. The construction is based on the observation that the maximum random fluctuation of $i^{\frac{1}{2}}(\hat{\xi}_{n,i} - \xi)$, $2 \leq i \leq k_n$ is of order $(\log\log n)^{\frac{1}{2}}$, i.e.

$$\max_{2\leq i\leq k_n} \left|\hat{\xi}_{n,i} - \xi - b_{n,i}\right| = O\left((\log\log n)^{\frac{1}{2}}\right) \tag{35}$$

for all intermediate sequences (k_n), where $b_{n,i} \in \mathbb{R}$, $2 \leq i \leq k_n$, denotes some bias term of the Hill estimator. They define a 'stopping time' for the sequence of Hill estimators by

$$\bar{k}_n(r_n) := \min\left\{ k \in \{2,\ldots,n\} \,\Big|\, \max_{2\leq i \leq k_n} i^{\frac{1}{2}} \left|\hat{\xi}_{n,i} - \hat{\xi}_{n,k}\right| > r_n \right\}, \qquad (36)$$

where the thresholds r_n constitute a sequence that is of larger order than $(\log\log n)^{\frac{1}{2}}$ but of smaller order than $n^{\frac{1}{2}}$. Similarly, one can define a deterministic sequence $\tilde{k}_n$ by

$$\tilde{k}_n(r_n) := \min\left\{ k \in \{2,\ldots,n\} \,\Big|\, \max_{2\leq i \leq k_n} i^{\frac{1}{2}} |b_{n,i} - b_{n,k}| > r_n \right\}. \qquad (37)$$

In view of (35) $\bar{k}_n(r_n)$ and $\tilde{k}_n(r_n)$ are asymptotically equivalent.

From (13) we know that $b_{n,i} \sim (i/n)^\rho$ and so $\bar{k}_n(r_n) \sim (r_n n^\rho)^{2/(2\rho+1)}$, so that

$$\left(\frac{\bar{k}_n(r_n^\zeta)}{(\bar{k}_n(r_n))^\zeta} \right)^{1/(1-\zeta)} \sim n^{2\rho/(2\rho+1)}$$

with $\zeta \in (0,1)$ has optimal order and hence can serve as a basis for an estimator of k_n^{opt}. Indeed, the final estimator is obtained by adjusting the constants according to (34). However, this involves again estimating the second order parameter ρ, which in turn is based on all Hill estimators up to the index $\bar{k}_n(r_n)$. We refer to Drees and Kaufmann [11] for details. Drees and Kaufmann point out that in Monte-Carlo experiments with moderate-sized samples the naive approach of using $\rho = 1$ performs better than using an estimator for ρ, even if ρ is mis-specified. We therefore implemented two variants of their estimator, one using an estimated value for ρ and the other with ρ set equal to one. Following their recommendations we used $\zeta = 0.7$ and for the threshold $r_n = 2.5\tilde{\xi}n^{0.25}$, with $\tilde{\xi}$ the standard Hill estimator from a sample size $2\sqrt{n^+}$, where n^+ denotes the number of positive observations in the given sample.

Beirlant-Vynckier-Teugels Approach

This approach is based on the Pareto quantile plot. As log-transformed Pareto distributed random variables are (strictly) exponentially distributed, one can visually check the hypothesis of strict Pareto behaviour from a sample $X_1, \ldots, X_n$ by inspecting the scatterplot with coordinates

$$\left(-\log\left(\frac{j}{n+1}\right), \log X^*_{n-j+1} \right) \quad (j = 1,\ldots,n).$$

Here $\log X^*_{n-j+1} := \log U_n[n/(j-1)]$ are used as non-parametric estimators of the theoretical quantiles $\log U[n/(j-1)]$, which must stand in linear relationship to the (approximations of the) corresponding theoretical quantiles

$-\log[j/(n+1)]$ of the standard exponential distribution. If the Pareto quantile plot is linear, then the slope of a fitted line provides an estimate for $\xi = 1/\alpha$. Under the general model (12), one has $\log L(x)/\log x \to 0$ as $x \to 0$. From this it follows that the Pareto quantile plot eventually will be linear for smaller j values. The faster $\log L(x)/\log x \to 0$ tends to zero, the clearer this phenomenon appears. An estimate of the tail index α can thus be obtained from estimating the slope of the Pareto quantile plot from a point $(-\log[(k+1)/(n+1)], \log X^*_{n-k})$ to the right of which the linearity starts to appear. The equation defining a line through $(-\log[(k+1)/(n+1)], \log X^*_{n-k})$ with slope a is given by

$$y = \log X^*_{n-k} + a\left(x + \log\left[\frac{k+1}{n+1}\right]\right)$$

from which an estimator $\hat{a}_{n,k}$ for $\xi = 1/\alpha$ is obtained with a least-squares algorithm for the fit of the line through the points with coordinates $(-\log[j/(n+1)], \log X^*_{n-j+1})$, $j = 1,\ldots,k$. Beirlant, Vynckier and Teugels show that the estimator can be improved (in the sense of minimizing the mean squared error) by introducing optimal weights in the above regression problem and applying a right-sided weighted least squares method. These weights again involve the second order parameter ρ

In this approach the problem of determining the number of extremes k_n^{opt} to use in the extreme value analysis can be regarded as a diagnostic regression problem, deciding on the point from which an 'optimal' linear fit is obtained through the quantile plot. Beirlant, Vynckier and Teugels solve this problem using an iterative procedure. Details of this procedure and a further discussion can be found in [3], [4], and [5]. Related results are [6] and [7].

Table 1. Descriptive statistics for bond returns

Bond	Maturity Date	Coupon (%)	Mean*	Variance*	Skewness	Kurtosis	Bera-Jarque†	Box-Ljung
Mexico	31/12/19	6.25	2.72	1.25	−0.17	13.83	1.02	1893
Venezuela	18/12/07	5.94	0.77	3.30	−1.72	31.90	7.37	1371
Ecuador	27/02/15	6.00	−0.88	10.02	−1.04	18.16	1.14	667
Brazil	15/04/14	3.50	4.07	4.58	−0.43	13.92	0.62	73.3
Argentina	31/03/05	6.19	1.93	2.97	−0.34	23.74	3.04	956
Peru	07/03/17	3.25	5.40	3.42	−0.34	12.33	2.52	300
Morocco	01/01/09	6.56	3.48	1.66	−0.17	16.74	1.26	790
Nigeria	15/11/20	6.25	3.37	2.30	−0.34	10.68	0.39	421
Russia	15/12/15	5.97	−27.60	22.40	0.45	18.23	0.68	504
Poland	27/10/14	4.00	5.52	1.58	−0.66	29.30	3.53	237
Indonesia	01/08/06	7.75	−5.03	4.13	−1.02	26.93	1.60	102

* Means and variances are for daily log returns and are multiplied by 10,000.
† The Bera-Jarque statistic is multiplied by 10,000 (the 1% level of significance is 201).
The Box-Ljung statistic is computed using the squares with 30 lags (the 1% level of significance is 50.89)

Table 2. Tail index estimates for pre-whitened residuals

	For Upper Tails of Residual Distribution					
	Modified Hill		GPD-fit		t-fit	
Bond	ξ	std	ξ	std	ξ	std
Mexico	0.232	0.031	0.180	0.503	0.271	0.025
Venezuela	0.255	0.033	0.077	0.623	0.379	0.030
Ecuador	0.192	0.032	0.082	0.440	0.279	0.030
Argentina	0.221	0.034	0.084	0.493	0.305	0.028
Brazil	0.182	0.031	−0.092	0.592	0.185	0.026
Peru	0.273	0.057	0.313	0.493	0.346	0.051
Morocco	0.139	0.023	−0.174	0.752	0.338	0.037
Nigeria	0.232	0.035	0.118	0.562	0.330	0.035
Russia	0.250	0.055	0.118	0.482	0.236	0.035
Poland	0.324	0.059	0.282	0.514	0.445	0.024
Indonesia	0.398	0.123	0.314	0.777	0.500	0.022

	For Lower Tails of Residual Distribution			
	Modified Hill		GPD-fit	
Bond	ξ	std	ξ	std
Mexico	0.281	0.038	0.220	0.595
Venezuela	0.288	0.041	0.119	0.807
Ecuador	0.371	0.068	0.227	0.741
Argentina	0.277	0.039	0.179	0.703
Brazil	0.243	0.033	0.179	0.587
Peru	0.265	0.064	0.110	0.714
Morocco	0.301	0.051	0.119	0.722
Nigeria	0.276	0.051	−0.038	0.914
Russia	0.304	0.052	0.288	0.539
Poland	0.342	0.061	0.179	0.747
Indonesia	0.452	0.150	0.190	1.301

Table 3. Quantiles for residuals

Bond	Method	\multicolumn{6}{c}{Quantile}					
		0.005	0.01	0.05	0.95	0.99	0.995
Mexico	emp	−3.800	−3.019	−1.466	1.524	2.766	3.287
	WLS	−3.789	−2.724	−1.579	1.562	2.797	3.559
	GPD	−3.542	−2.833	−1.507	1.562	2.614	3.169
	t-fit	−4.883	−3.930	−2.185	2.185	3.930	4.883
Venezuela	emp	−3.808	−3.091	−1.592	1.463	2.458	2.805
	WLS	−3.826	−2.977	−1.666	1.487	2.535	3.205
	GPD	−3.895	−3.138	−1.590	1.444	2.563	3.090
	t-fit	−6.701	−5.073	−2.490	2.490	5.073	6.701
Ecuador	emp	−4.366	−3.212	−1.563	1.364	2.175	2.526
	WLS	−4.190	−3.287	−1.591	1.357	2.462	3.057
	GPD	−4.165	−3.237	−1.581	1.384	2.180	2.556
	t-fit	−4.992	−4.000	−2.205	2.205	4.000	4.992
Argentina	emp	−3.670	−3.304	−1.598	1.419	2.376	2.961
	WLS	−4.100	−2.929	−1.684	1.419	2.570	3.269
	GPD	−3.888	−3.068	−1.618	1.418	2.318	2.744
	t-fit	−5.376	−4.247	−2.274	2.274	4.247	5.376
Brazil	emp	−3.317	−2.961	−1.585	1.538	2.420	2.677
	WLS	−3.553	−2.751	−1.661	1.544	2.661	3.253
	GPD	−3.523	−2.886	−1.617	1.545	2.374	2.695
	t-fit	−3.881	−3.262	−1.982	1.982	3.261	3.881
Peru	emp	−3.674	−3.020	−1.523	1.454	2.787	3.502
	WLS	−4.100	−3.117	−1.636	1.465	2.820	3.732
	GPD	−3.524	−2.938	−1.548	1.479	2.711	3.465
	t-fit	−6.067	−4.684	−2.391	2.391	4.684	6.067

Table 4. Contd. quantiles for residuals

Bond	Method	\multicolumn{6}{c}{Quantile}					
		0.005	0.01	0.05	0.95	0.99	0.995
Morocco	emp	−3.757	−3.022	−1.495	1.698	2.552	2.956
	WLS	−3.744	−2.850	−1.574	1.753	2.845	3.215
	GPD	−3.577	−2.910	−1.529	1.705	2.662	2.999
	t-fit	−5.928	−4.595	−2.368	2.368	4.595	5.928
Nigeria	emp	−3.551	−3.004	−1.610	1.546	2.660	3.286
	WLS	−3.761	−2.912	−1.713	1.560	2.594	3.338
	GPD	−3.598	−3.057	−1.587	1.540	2.602	3.125
	t-fit	−5.781	−4.503	−2.343	2.343	4.503	5.781
Russia	emp	−3.539	−2.608	−1.513	1.373	2.448	2.790
	WLS	−3.334	−2.622	−1.574	1.341	3.017	4.073
	GPD	−3.599	−2.828	−1.531	1.415	2.328	2.778
	t-fit	−4.432	−3.633	−2.098	2.098	3.633	4.432
Poland	emp	−3.210	−2.860	−1.415	1.360	2.523	3.200
	WLS	−3.738	−2.816	−1.531	1.393	2.642	3.274
	GPD	−3.733	−2.924	−1.409	1.369	2.572	3.284
	t-fit	−8.284	−6.018	−2.714	2.714	6.018	8.284
Indonesia	emp	−5.717	−3.501	−1.068	1.084	2.776	3.698
	WLS	−6.725	−3.494	−1.518	1.296	2.906	3.462
	GPD	−4.957	−3.625	−1.154	1.091	2.756	3.777
	t-fit	−9.924	−6.964	−2.920	2.920	6.964	9.924

Table 5. Comparison of quantiles for pre-whitened Mexican bond returns

Method	\multicolumn{8}{c}{Quantile}							
	0.0005	0.001	0.005	0.01	0.99	0.995	0.999	0.9995
emp	−4.9936	−4.8063	−3.8001	−3.019	2.7661	3.2865	3.9903	4.1099
WLS-s	−6.2117	−5.1294	−2.9858	−2.4655	2.7224	3.459	6.2493	7.9404
WLS-f	−6.3135	−5.1975	−3.0133	−2.4807	2.6814	3.3839	6.0589	7.6463
Drees-s	−7.898	−6.2345	−3.4192	−2.699	3.0157	4.0116	7.7009	10.2442
Drees-f	−10.1775	−7.6606	−3.9452	−2.9695	2.9114	3.8123	7.1673	9.3852
Teugels	−9.3438	−7.1468	−3.7594	−2.8754	3.0051	3.9912	7.6458	10.1549
GPD	−6.6985	−5.5988	−3.5419	−2.8328	2.6136	3.1693	4.7617	5.6031
t-fit	−9.5202	−7.828	−4.8831	−3.9295	3.9295	4.8831	7.828	9.5202

Table 6. Monte Carlo comparison of tail index estimators

df	Method				
	DS-1	DS-2	BVT	WLS-1	WLS-2
2	0.5281	0.5454	0.5326	0.4743	0.4880
	(0.0060)	(0.0082)	(0.0062)	(0.0067)	(0.0102)
3	0.3859	0.3976	0.3556	0.3374	0.3398
	(0.0049)	(0.0060)	(0.0022)	0.0040	0.0053
4	0.3154	0.3256	0.2477	0.2791	0.2728
	(0.0035)	(0.0051)	(0.0015)	0.0024	0.0030

Table 7. Monte Carlo comparison of quantile estimators

df	α	q_α	Method						
			Emp	GPD	DS-1	DS-2	BVT	WLS-1	WLS-2
2	0.95	2.920	2.908	2.930	2.911	2.947	2.934	2.941	2.941
			(0.057)	(0.053)	(0.058)	(0.065)	(0.063)	(0.060)	(0.061)
2	0.99	6.965	6.825	7.004	7.011	7.208	6.998	6.501	6.653
			(1.310)	(0.978)	(1.160)	(1.472)	(1.160)	(1.140)	(1.572)
2	0.995	9.925	9.636	9.910	10.249	10.663	10.265	9.270	9.589
			(5.200)	(3.627)	(4.370)	(5.584)	(3.953)	(4.207)	(6.326)
3	0.95	2.353	2.350	2.353	2.365	2.365	2.368	2.356	2.356
			(0.021)	(0.021)	(0.023)	(0.023)	(0.022)	(0.024)	(0.024)
3	0.99	4.499	4.541	4.575	4.476	4.557	4.273	(4.171)	(4.190)
			(0.271)	(0.158)	(0.271)	(0.321)	(0.141)	(0.203)	(0.251)
3	0.995	5.705	5.841	5.857	5.932	6.087	5.549	5.391	5.430
			(0.819)	(0.515)	(0.849)	(1.050)	(0.354)	(0.653)	(0.863)
4	0.95	2.132	2.126	2.119	2.138	2.138	2.141	2.134	2.134
			(0.016)	(0.011)	(0.014)	(0.014)	(0.016)	(0.015)	(0.015)
4	0.99	3.747	3.706	3.743	3.616	3.675	3.274	3.422	3.393
			(0.124)	(0.090)	(0.118)	(0.161)	(0.061)	(0.098)	(0.109)
4	0.995	4.604	4.512	4.561	4.558	4.667	3.942	4.205	4.156
			(0.336)	(0.239)	(0.332)	(0.483)	(0.139)	(0.250)	(0.286)

Table 8. Number of VaR-exceptions

Quantile	Method					
	Empirical	EVT	Garch-Emp	Garch-EVT	Garch-Normal	Garch-t
99.5%	1.01	1.01	0.69	0.76	1.13	0.44
99%	1.76	1.76	1.13	1.13	1.32	0.94

The number of observations was 2,089 with a window length of 500, so the number of VaR calculations was 1,589.

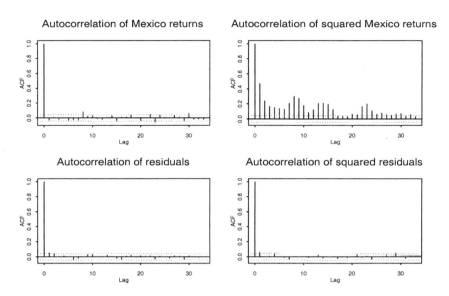

Fig. 1. Autocorrelation functions for Mexican bond returns

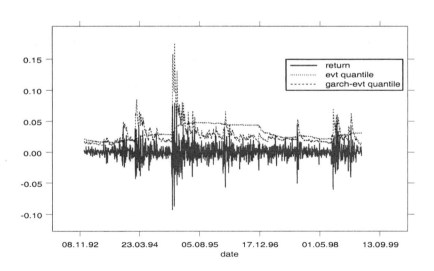

Fig. 2. Mexican bond returns and VaRs

Notes to Figures

Figure 1
The figure shows autocorrelation functions (ACFs) for raw and pre-whitened holding returns and squared holding returns on the Mexican benchmark bond. The pre-whitened returns are residuals from a GARCH(1,1) model. The horizontal dotted line provides an approximate 95 % confidence interval for the autocorrelation estimate at each lag.

Figure 2
Returns and 1% VaR quantiles plotted against time are shown for the Mexican bond. The returns and quantiles are multiplied by minus 1 (i.e., are in units of losses). The "evt quantile" results are calculated by applying the Modified GLS EVT technique to the raw holding returns. The "garch-evt quantile" results are calculated by using the Modified GLS method on pre-whitened returns and then adjusting for the fitted GARCH volatilities to get the quantile of actual returns.

References

1. Artzner, P., F. Delbaen, J. Eber, and D. Heath, 1997, Thinking Coherently, *Risk* 10.
2. Basel Committee on Banking Supervision, 1996, Overview of the Amendment to the Capital Accord to Incorporate Market Risk, unpublished mimeo Bank for International Settlements Basel.
3. Beirlant, J., J. Teugels, and P. Vynckier, 1996a, Excess Functions and Estimation of the Extreme-Value Index, *Bernoulli* 2, 293–318.
4. Beirlant, J., J. Teugels, and P. Vynckier, 1996b, *Practical Analysis of Extreme Values*. (Leuven University Press Leuven).
5. Beirlant, J., P. Vynckier, and J. Teugels, 1996, Tail Index Estimation, Pareto Quantile Plots, and Regression Diagnostics, *Journal of the American Statistical Association* 91, 1659–1667.
6. Beurlant, J., G. Dierckx, Y. Goegebeur, and G. Matthys, 1999, Tail Estimation and an Exponential Regression Model, *Extremes* forthcoming.
7. Caers, J., J. Beirlant, and P. Vynckier, 1998, Bootstrap Confidence Intervals for Tail Indices, *Computational Statistics and Data Analysis* 26, 259–277.
8. Dacarogna, M.M., U.A. Muller, and O.V. Pictet, 1996, Hill, Bootstrap and Jackknife Estimators for Heavy Tails, unpublished mimeo Olsen Associates Zurich.
9. Danielsson, J., and C. de Vries, 1997, Beyond the Sample: Extreme Quantile and Probability Estimation, Working paper LSE http://www.hag.hi.is/jond/research.
10. Diebold, F., T. Schuermann, and J. Stroughair, 1987, Pitfalls and Oppportunities in the Use of Extreme Value Theory in Risk Management, in J.D. Moody Refenes, A.-P.N., and A.N. Burgess, eds.: *Advances in Computational Finance* (Kluwer Academic Publishers, Amsterdam).
11. Drees, H., and E. Kaufmann, 1998, Selecting the Optimal Sample Fraction in Univariate Extreme Value Estimation, *Stochastic Process Applications* 75, 149–172.
12. Duffie, Darrell, and Jun Pan, 1997, An Overview of Value at Risk, *Journal of Derivatives* 4, 7–49.

13. Embrechts, P., C. Klüppelberg, and P. Mikosch, 1997, *Modelling Extremal Events*. (Springer Berlin).
14. Embrechts, Paul, Sidney Resnick, and Gennady Samorodnitsky, 1998, Living on the Edge, *Risk* 11, 96–100.
15. Enders, W., 1995, *Applied Econometric Time Series*. (Wiley and Sons New York).
16. Goldie, C.M., and R.L. Schmith, 1987, Slow Variation with Reminder: Theory and Applications, *Quarterly Journal of Mathematics Oxford (2)* 38, 38–45.
17. Gourieroux, C., 1997, *ARCH Models and Financial Applications*. (Springer Berlin).
18. Hall, P., 1982, On Some Simple Estimates of an Exponent of Regular Variation, *Journal of the Royal Statistical Society, Series B* 44, 37–42.
19. Hall, P., and A.H. Welsh, 1985, Adaptive Estimation of Parameters of Regular Variation, *Annals of Statistics* 13, 331–341.
20. Hendricks, D., 1996, Evaluation of Value-at-Risk Models Using Historical Data, *Federal Reserve Bank of New York Economic Policy Review* 2, 39–69.
21. Huisman, R., K. Koedijk, C. Kool, and F. Palm, 1997, Fat Tails in Small Samples, Working paper Limburg Institute of Financial Economics Maastricht University.
22. Jackson, Patricia D., David Maude, and William R.M. Perraudin, 1997, Bank Capital and Value at Risk, *Journal of Derivatives* 4, 73–89.
23. Longin, Francois M., 1997a, The Asymptotic Distribution of Extreme Stock Markert Returns, *Journal of Business* 69, 383–407.
24. Longin, Francois M., 1997b, From Value at Risk to Stress Testing: The Extreme Value Approach, Working Paper 97004 CERESSEC Paris.
25. Maddala, G.S., and I.-M. Kim, 1998, *Unit Root, Cointegration, and Structural Change*. (Cambridge University Press Cambridge UK).
26. McNeil, A., 1998, Calculating Quantile Risk Measures for Financial Return Series Using Extreme Value Theory, unpublished mimeo ETH Zurich.
27. McNeil, Alexander, and Rudiger Frey, 2000, Estimation of tail-related risk measures for heteroskedastic financial time series: an extreme value approach, *Journal of Empirical Finance* 7, 271–300.
28. Reiss, R.D., 1989, *Approximate Distributions of Order Statistics*. (Springer Berlin).
29. Reiss, R.D., and M. Thomas, 1997, *Statistical Analysis of Extreme Values*. (Birkhauser Verlag Basel Boston Berlin).

Default Probabilities in Structured Commodity Finance

Daniel Kluge and Frank Lehrbass

Deutsche Genossenschafts-Hypothekenbank AG, Hamburg, Germany

1 Introduction

Looking at any credit risk model the assignment of a default probability to a certain net exposure is necessary to make the model work. In the case of corporate loans, for instance, the bank may either use published default statistics by the rating agencies, so-called "Distance to Default" Scores provided by KMV, or default statistics from an internal rating procedure as described in [3]. Besides there is the possibility to extract default related figures from market prices (such as Default Digital or Default Swap Spreads) as has been done with respect to country risk for example in [4].

In specialized lending situations such as commodity related project finance things are more complex. In practice banks base their investment decision in such situations on figures obtained from cash flow modeling.

A detailed overview on cash flow models can be found in [7]. To motivate our approach we simply observe that any cash flow model will produce forecasts of revenues and costs associated to the considered project and as a consequence predictions of cash available for debt service. Based on these figures bank debt can be adequately structured, i.e. it will be structured in such a way that the predicted cash available at the debt coupon dates suffices to pay the liabilities.

In the following we will focus on the project revenues forecasted by the cash flow model. We may assume that revenues are determined by the commodity spot price and production level. In cash flow modeling certain levels for this figures are assumed at certain points in time.

The default risk of the project itself stems from possible deviations of the commodity spot price and/or production level from their asserted values. It should be noted that spot price variation is a risk that in the context of project finance cannot be hedged appropriately, in particular in the presence of production level uncertainty.

Assessing the credit risk associated to a specific investment project, it is therefore necessary to measure the "combined downside risk" of spot price and production level movements. Modeling these risk factors in practice, there has to occur a trade off between the sophistication and the practicability of the employed model. In the sequel a workable approach to the measurement of the combined downside risk mentioned above will be presented.

We will start with the simple case, where there is only uncertainty about future spot prices. Then a method how to incorporate production level uncertainty will be introduced. This method is merely developed along the lines of [2]. Finally, for an application, we consider a special case of the general model, particularly well suited to situations where catastrophe risk, e.g. the risk of a political turmoil, is the key driver of production level uncertainty. In fact this is no minor issue in the context of project finance: "There is an element of political risk associated with most project financings by virtue of the fact that they are often related to capital-intensive infrastructure development, in which governments obviously take a keen interest" [6], p.5.

It should be noted that our approach is not specific to commodity related project finance. It is applicable whenever sufficient market data are available for the key drivers of project revenues and/or expenses.

2 Spot Price Variation Only

The cash flow model will determine revenue levels and consequently spot price levels at certain points in time, below which available cash will be insufficient to service bank debt. The project's default probability at these points in time is then given by the probability that the spot price will fall below the corresponding critical level.

To obtain these probabilities we need some idea about the spot price dynamics. A model where parameter estimation is practicable and model complexity is adequate is the so-called model 3 in [8]. It is a three factor model incorporating a stochastic convenience yield and a stochastic short rate. The short rate process itself only affects the spot price dynamics under the risk neutral measure. However, for the purpose of credit risk quantification, only the subjective probability measure is relevant. Under this probability measure the dynamics of the model are as follows:

$$\begin{cases} dS_t = (\mu - \delta_t)S_t dt + \sigma_1 S_t dZ_1(t) \\ d\delta_t = k(\alpha - \delta_t)dt + \sigma_2 dZ_2(t) \\ dr_t = a(m - r_t)dt + \sigma_3 dZ_3(t) \\ dZ_1(t)dZ_2(t) = \rho_1 dt \\ dZ_2(t)dZ_3(t) = \rho_2 dt \\ dZ_1(t)dZ_3(t) = \rho_3 dt \end{cases} \quad (1)$$

S denotes the commodity spot price, δ the convenience yield of the commodity and r the short rate.

This model is able to capture the mean reverting behavior often observed for commodity spot prices in reality. The economic intuition behind this is as follows. If the spot price is high, high cost producers enter the market putting downward pressure on the spot price. Conversely, if the spot price is low, high cost producer will go out of the market putting upward pressure on the spot price. For a detailed discussion of this model and its advantages over simpler models the reader is referred to [8].

When deriving a formula for the default probabilities, it is obvious from (1) that we only need the dynamics S and δ. In particular, there exists a 2-dim standard Brownian motion $W_t = (W_1(t), W_2(t))$ with

$$Z_1(t) = W_1(t)$$
$$Z_2(t) = \rho_1 W_1(t) + \bar{\rho}_1 W_2(t) \quad (\bar{\rho}_1 = \sqrt{1 - \rho_1^2}) \quad (2)$$

Writing out the processes for S, δ yields

$$S_t = S_0 \exp\left(\int_0^t (\mu - \delta_u - \tfrac{1}{2}\sigma_1^2) du + \int_0^t \sigma_1 dZ_1(u) \right) \quad (3)$$

$$\delta_t = \alpha + e^{-kt}(\delta_0 - \alpha) + \sigma_2 e^{-kt} \int_0^t e^{ku} dZ_2(u) \quad (4)$$

With $\sigma_S = \sigma_1 \begin{pmatrix} 1 \\ 0 \end{pmatrix}$ and $\sigma_\delta(u,t) = \sigma_2 e^{-k(t-u)} \begin{pmatrix} \rho_1 \\ \bar{\rho}_1 \end{pmatrix}$ we obtain

$$S_t = S_0 \exp\left((\mu - \tfrac{1}{2}\sigma_1^2)t - \int_0^t \delta_v dv + \int_0^t \sigma_S \cdot dW_u \right) \quad (5)$$

$$\delta_t = \alpha + e^{-kt}(\delta_0 - \alpha) + \int_0^t \sigma_\delta(u,t) \cdot dW_u \quad (6)$$

Substituting the solution for the convenience yield (2.6) into the equation for the spot price (2.5) and using Corollary 2 in [1], p.99, we get

$$S_t = D_t \exp(\xi_t) \quad (7)$$

with

$$D_t = S_0 \exp\left((\mu - \tfrac{1}{2}\sigma_1^2 - \alpha)t - \tfrac{1}{k}(1 - e^{-kt})(\delta_0 - \alpha) \right) \text{ nonstochastic} \quad (8)$$

$$\xi_t = \int_0^t \left(\sigma_S - \int_u^t \sigma_\delta(u,v) dv \right) \cdot dW_u \sim N(0, \sigma_\xi(t)^2) \quad (9)$$

$$\sigma_\xi(t)^2 = \int_0^t \left\| \sigma_S - \int_u^t \sigma_\delta(u,v) dv \right\|^2 du \quad (10)$$

Solving the right hand side of (2.10) yields

$$\sigma_\xi(t)^2 = \left(\sigma_1^2 + \frac{2\rho_1\sigma_1\sigma_2}{k} + \frac{\sigma_2^2}{k^2}\right) t - 2\left(\frac{\rho_1\sigma_1\sigma_2}{k^2} + \frac{\sigma_2^2}{k^3}\right)(1 - e^{-kt})$$
$$+ \frac{\sigma_2^2}{2k^3}(1 - e^{-2kt}) \qquad (11)$$

Let $\Phi(\ )$ be the standard normal distribution function, then we get

$$P(S_T < L) = P(D_T \exp(\xi_T) < L) = P(\xi_T < \ln(L/D_T))$$
$$= P\left(\frac{\xi_T}{\sigma_\xi(T)} < \frac{\ln(L/D_t)}{\sigma_\xi(T)}\right)$$

Hence

$$P(S_T < L) = \Phi\left(\frac{\ln(L/D_T)}{\sigma_\xi(T)}\right) \qquad (12)$$

3 Adding Production Level Variation

Apart from the development of spot prices the revenues of the project are influenced by movements in the production level. For a debt coupon date T we assume that the actual production level is the product of the level $\bar{m}_T$ asserted in the cash flow model and the realization of a jump process Δ_T taking the form

$$\Delta_T = \prod_{i=1}^{N_T}(1 + \kappa_i) \qquad (13)$$

where N is a Poisson process with intensity λ and κ_i is the relative jump amplitude of the i-th jump. The jump process Δ is assumed to be independent of the spot price dynamics, which is reasonable for the case where changes in the production level due to production outages are considered.

Revenues are given by the product of spot price and production level, i.e. by $S_T \Delta_T \bar{m}_T$. Hence default triggering revenue levels lead to default triggering levels for $\hat{S}_T := S_T \Delta_T$. The dynamics of $\hat{S}$ turn out to be

$$\frac{d\hat{S}_t}{\hat{S}_t} = (\mu - \delta_t)dt + \sigma_1 dZ_1(t) + \kappa_t dN_t \qquad (14)$$

Drift compensation is not an issue at this point, since the jump component does not represent a source of uncertainty associated to the actual spot price, but an outside source of risk only appearing in the context of the specific investment project.

Following standard jump diffusion approaches (compare to [5], [2]), we will assume that the jump amplitudes κ_i are i.i.d. with

$$\ln(1+\kappa_i) \sim N(\ln(1+\bar{\kappa}) - \tfrac{1}{2}\omega^2, \omega^2) \qquad (15)$$

where each of the jump amplitudes are uncorrelated with both the Poisson process and the spot price dynamics. If L is the default triggering level for $\hat{S}_T$, then the default probability is given by

$$\begin{aligned} P(\hat{S}_T < L) &= \sum_{j=0}^{\infty} P(\{\hat{S}_T < L\} \cap \{N_T = j\}) \\ &= \sum_{j=0}^{\infty} P(N_T = j) \cdot P(\hat{S}_T < L | N_T = j) \end{aligned} \qquad (16)$$

According to (3.14), we obtain

$$\begin{aligned} \hat{S}_T|_{N_T=j} &= S_0 \left(\prod_{i=1}^{j}(1+\kappa_i)\right) \exp\left((\mu - \tfrac{1}{2}\sigma_1^2)T - \int_0^T \delta_u du + \int_0^T \sigma_S \cdot dW_u\right) \\ &= D_T \exp(\xi_T^{(j)}) \end{aligned} \qquad (17)$$

with D_T as in (2.8) and

$$\begin{aligned} \xi_T^{(j)} &= \sum_{i=1}^{j} \ln(1+\kappa_i) + \int_0^T \left(\sigma_S - \int_0^T \sigma_\delta(u,v) dv\right) \cdot dW_u \\ &\sim N\left(\mu_\xi^{(j)}(T), \sigma_\xi^{(j)}(T)^2\right) \end{aligned} \qquad (18)$$

where

$$\mu_\xi^{(j)}(T) = j \cdot (\ln(1+\bar{\kappa}) - \tfrac{1}{2}\omega^2) \qquad (19)$$

$$\sigma_\xi^{(j)}(T)^2 = j\omega^2 + \sigma_\xi(T)^2 \qquad (20)$$

In our situation we therefore find

$$P\left(\hat{S}_T < L | N_T = j\right) = P(D_T \exp(\xi_T^{(j)}) < L) = \Phi\left(\frac{\ln(L/D_T) - \mu_\xi^{(j)}(T)}{\sigma_\xi^{(j)}(T)}\right) \qquad (21)$$

Using (3.16) we end up with

$$P\left(\hat{S}_T < L\right) = \sum_{j=0}^{\infty} e^{-\lambda T} \cdot \frac{(\lambda T)^j}{j!} \cdot \Phi\left(\frac{\ln(L/D_T) - \mu_\xi^{(j)}(T)}{\sigma_\xi^{(j)}(T)}\right) \qquad (22)$$

4 Application – Catastrophe Risk

In our application we assumed that the only risk that may cause the production level to deviate from the asserted level is the risk of a political turmoil or expropriation risk. If such an event occurs, revenues from the project jump down to zero and remain at zero from there on. The adequate jump process must therefore have constant relative jump amplitudes $\kappa_i \equiv -1$. In this case we get

$$P(\hat{S}_T < L | N_T = 0) = P(S_T < L)$$
$$P(\hat{S}_T < L | N_T > 0) = 1$$
(23)

Consequently (3.22) becomes

$$P\left(\hat{S}_T < L\right) = e^{-\lambda T} \Phi\left(\frac{\ln(L/D_T)}{\sigma_\xi(T)}\right) + (1 - e^{-\lambda T})$$
(24)

with D_T, $\sigma_\xi(T)$ as in (6). With time measured in years, λ is the expected number of political turmoils per year.

In order to estimate the relevant parameters we chose a slightly different approach as can be found [8]. In any case, since the convenience yield of the commodity is an unobservable variable, Kalman filtering is the appropriate estimation technique. In [8] the commodity spot price is also treated as an unobservable variable, which is contrary to our approach.

The measurement equation of the corresponding state space form is given by the formula for log futures prices of the commodity, which depend linearly on the log spot price, the convenience yield as well as the short rate. The state equation is given by a discretisized version of the convenience yield dynamics (see also [8]). The parameters to be estimated in this state space form are

k, α, ν (market price of convenience yield risk)
a, m, φ (market price of interest rate risk)
$\sigma_1, \sigma_2, \sigma_3, \rho_1, \rho_2, \rho_3$

The drift parameter μ of the spot price dynamics must be estimated separately after the Kalman filtering.

Ideally, the estimation of the listed parameters should be carried out simultaneously with Kalman filtering, which was not practicable. As in [S97], we therefore estimated the parameters of the short rate process (i.e. a, m,σ_3, φ) in advance. A daily 1-month USD zero-yield supplied by DATASTREAM was used as a proxy for the short rate. In order to capture mean reversion, the sample period was larger than the sample period for the Kalman filtering, extending from 01/01/1989 to 08/31/2000. No shorter maturity yield was chosen due to the well known end-of-month effects. The parameters a, m,gσ_3 were estimated by discretizing the short rate dynamics and using maximum likelihood. The market price φ of interest rate risk was determined such that the 3-month USD zero-yields were fitted best.

To carry out the Kalman filtering, we used daily data on futures and spot prices for copper and aluminum from the London Metal Exchange (LME) ranging from 01/01/1998 to 08/31/2000. Although commodity contracts are also traded at the New York Mercantile Exchange/Comex (COMEX) and the Shanghai Metal Exchange (SHME), LME was our first choice, since it is the most important commodity exchange, predominantly due to its ability to use worldwide delivery points.

However, as in [8], further reduction of the number of parameters to be estimated in the Kalman filtering was necessary. Since we treated the spot price as observable, we chose to first approximate $\sigma_1^2 \Delta t$ by the sample variance of the first differences of the log spot prices and then to approximate $\rho_3 \sigma_1 \sigma_3 \Delta t$ by the sample covariance between the first differences of log spot prices and short rates.

Finally the parameters to be estimated with Kalman filtering were k, $\alpha, \nu, \upsilon_2, \rho_1, \rho_2$. For these we obtained maximum likelihood estimates.

After the Kalman filtering, the drift parameter μ of the spot price dynamics under the subjective probability measure had to be determined. Discretisizing the dynamics of the spot we chose the approximation

$$\mu \approx \frac{\overline{\Delta \ln S}}{\Delta t} + \tfrac{1}{2}\sigma_1^2 + \alpha \qquad (25)$$

The parameter estimates are given below in the order they were estimated with standard errors in brackets, where available:

	Copper Grade A		High Grade Primary Aluminum	
a	0.43675	(0.21525)		
m	0.05192	(0.00869)		
σ_3	0.01260			
γ	-0.75987			
σ_1	0.20736		0.16583	
ρ_3	0.00493		0.02859	
k	1.10023	(0.04121)	1.69881	(0.09447)
α	0.01409	(0.00830)	-0.00486	(0.09447)
ν	-0.44170	(0.9744)	-0.49703	(0.12383)
σ_2	0.12031	(0.01038)	0.19002	(0.00932)
ρ_1	0.61787	(0.03294)	0.67681	(0.01125)
ρ_2	0.14343	(0.04266)	0.10755	(0.00760)
μ	0.04648		0.01870	

Consider now the investment of a bank into a copper mine with yearly debt coupon dates. Assume that the current copper spot price is $S_0 = \$$

1.987,00 and the current estimate for the convenience yield is $\delta_0 = 1,5\%$. Let $L = \$ 1.200,00$ be the critical spot price level at the end of each year $T = 1,\ldots,5$, below which revenues are insufficient for debt service. Using the above parameter estimates for copper, we get the default probabilities at the debt coupon dates as shown below.

(i) without catastrophe risk:

T	1	2	3	4	5
$\mathbf{P}(S_T <L)$	1,51%	7,20%	12,04%	15,51%	18,06%

(ii) with catastrophe risk ($\lambda = 0,01$, i.e. on average one catastrophe per 100 years)

T	1	2	3	4	5
$\mathbf{P}(S_T <L)$	2,49%	9,04%	14,64%	18,83%	22,05%

5 Concluding Remarks

The above modeling of the combined downside risk of spot price and production level uncertainty works quite well in the case where production level uncertainty is foremost due to catastrophe risk. Also, the determination of the jump intensity parameter λ in this case can be handled easily using historical market data.

In the case, where production level uncertainty must be modeled by allowing for variable jump amplitudes, there is need for model refinement. In particular, correlated jump amplitudes should be admitted, since e.g. a downward jump (due to a production outage) is likely to be followed by an upward jump (production level returning to normal during the next coupon period). In this case, there may also be accumulation of cash reserves available for later debt service, if the production level and/or spot price level where higher than scheduled. Also, the determination of the jump process parameters will become more complex.

References

1. D.Heath, R.Jarrow, A.Morton, *Bond Pricing and the Term Structure of Interest Rates: A new Methodology for Contingent Claims Valuation*, Econometrica, Vol.60, No.1, January 1992, pp.77-105.
2. J.Hilliard, J.Reis, *Valuation of Commodity Futures and Options under Stochastic Convenience Yields, Interest Rates, and Jump Diffusion in the Spot*, Journal of Financial and Quantitative Analysis, Vol.33, No.1, March 1998, pp.61-86.
3. F.B.Lehrbass, *Rethinking risk-adjusted returns*, RISK, Vol.12, No.4, April 1999, pp.35-40 (Credit Risk Special Report).

4. F.B.Lehrbass, *A Simple Approach to Country Risk* in: J.Franke, W.Härdle, G.Stahl, *Measuring Risk in Complex Stochastic Systems*, Berlin, Springer 2000.
5. R.C.Merton, *Option Pricing when Underlying Stock Returns are Discontinious*, Journal of Financial Economics, Vol.3, 1976, pp.125-144.
6. Moody's, *Key Credit Risks of Project Finance*, Special Comment, April 1998, New York: Moody's Investors Service.
7. P.E.Nevitt, *Project Financing*, Euromoney Publications, London, 1979.
8. E.Schwartz, *The Stochastic Behavior of Commodity Prices: Implications for Valuation and Hedging*, Journal of Finance, Vol.52, No.3, July 1997, pp.922-973.

Kendall's Tau for Elliptical Distributions *

Filip Lindskog[1], Alexander McNeil[2], and Uwe Schmock[1]

[1] RiskLab, Departement of Mathematics, ETH Zentrum, Zürich, Switzerland
[2] Departement of Mathematics, ETH Zentrum, Zürich, Switzerland

Summary. By using well known properties of elliptical distributions we show that the relation between Kendall's tau and the linear correlation coefficient for bivariate normal distributions holds more generally (subject to only slight modifications) for the class of elliptical distributions. We mention applications of this result to calibrating elliptical distributions and their copulas in the context of multivariate financial time series models and portfolio credit risk models in particular.

1 Introduction

It is well known, and easily demonstrated, that for the two-dimensional normal distribution with linear correlation coefficient ρ the relation

$$\tau = \frac{2}{\pi} \arcsin \varrho, \qquad (1)$$

between Kendall's tau and the linear correlation coefficient holds (cf. [2, p. 290], where the calculations are traced back to publications of T. J. Stieltjes from 1889 and W. F. Sheppard from 1898). However, it does not seem to be at all well known that the elegant relationship (1) also holds (subject to only slight modifications) for all non-degenerate elliptical distributions, and this is the main result (Theorem 2) of this short communication.

The result is not only of theoretical interest; it is also extremely useful for statistical purposes. For example, it can be used to build a robust estimator of linear correlation for elliptically distributed data.

Many multivariate datasets encountered in practice, such as financial time series data for market and credit risk management, are not multivariate normally distributed but may plausibly be modelled by another member of the elliptical family with heavier tailed margins. In this situation it is well known

---

* Research supported by Credit Suisse, Swiss Re and UBS through RiskLab, Switzerland.

that the standard estimator of correlation, based on normal assumptions and maximum-likelihood theory, is both inefficient and lacks robustness; many alternative covariance and correlation estimators have been proposed including M-estimators, estimators based on multivariate trimming and estimators based on variances of sums and differences of standardized variables (cf. [3] for an overview). Formula (1) provides an appealing bivariate method; we simply estimate Kendall's tau using the standard textbook estimator and invert the relationship to get the Kendall's tau transform estimate of ρ. Simulation studies suggest that this simple method performs better than most of its competitors, see Figure 1 and [8].

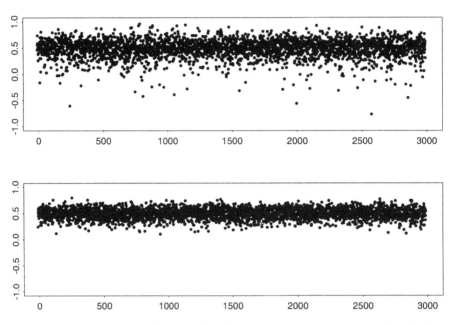

Fig. 1. For 3000 independent samples of size 90 from a bivariate t_3-distribution with linear correlation 0.5, the upper graph shows the standard estimator and the lower graph Kendall's tau transform estimator for the linear correlation

Note that, unlike almost all other methods of correlation estimation, the Kendall's tau transform method directly exploits the geometry of elliptical distributions and does not require us to estimate variances and covariances. This is advantageous when interest focusses explicitly on correlations, as it often does in financial derivative pricing applications. More generally the relationship can be used to calibrate the correlation matrices of higher dimensional elliptical distributions, although in some cases the matrix of pairwise correlations must be adjusted to ensure that the resulting matrix is positive definite; see [8] and [10] for details.

In the context of portfolio credit risk modelling, elliptical distributions—in particular t-distributions—and their copulas have been suggested to describe the dependence structure of latent variables representing asset values (or their proxies) in so-called structural models of defaults [7]. In these models, pioneered by Black and Scholes (1973) and Merton (1974) in the univariate case, defaults are caused by firms' asset values falling below their liabilities. Standard industry models (KMV, CreditMetrics) implicitly assume multivariate normal dependence, and the difference between such models and models based on latent variables with other elliptical distributions can be profound, see the simulation studies in [6] and [7]. When we implement latent variable models based on multivariate t- or more general elliptical distributions, we require statistically robust and computationally feasible methods for calibrating them to asset value data, or other suitable proxy data such as equity returns; see [9] for practical applications of this approach.

In Section 2 of this note we review the definition and some properties of elliptical distributions. The new result is stated in Section 3 and proved in Section 4.

2 Definitions and Basic Properties

All random variables mentioned in this paper are real or $\mathbb{R}^n$-valued and all matrices have real entries. Furthermore, all random variables mentioned are assumed to be defined on a common probability space $(\Omega, \mathcal{A}, \mathbb{P})$.

Definition 1. *If X is a p-dimensional random (column) vector and, for some vector $\mu \in \mathbb{R}^p$, some $p \times p$ nonnegative definite symmetric matrix Σ and some function $\phi : [0, \infty) \to \mathbb{R}$, the characteristic function $\varphi_{X-\mu}$ of $X - \mu$ is of the form $\varphi_{X-\mu}(t) = \phi(t^T \Sigma t)$, we say that X has an elliptical distribution with parameters μ, Σ and ϕ, and we write $X \sim E_p(\mu, \Sigma, \phi)$.*

When $p = 1$, the class of elliptical distributions coincides with the class of one-dimensional symmetric distributions.

Theorem 1. *$X \sim E_p(\mu, \Sigma, \phi)$ with $\mathrm{rank}(\Sigma) = k$ if and only if there exist a random variable $R \geq 0$ independent of U, a k-dimensional random vector uniformly distributed on the unit hypersphere $\{z \in \mathbb{R}^k \mid z^T z = 1\}$, and a $p \times k$ matrix A with $AA^T = \Sigma$, such that*

$$X \stackrel{d}{=} \mu + RAU. \qquad (2)$$

For the proof of Theorem 1 and details about the relation between R and ϕ, see Fang, Kotz and Ng (1987) [5] or Cambanis, Huang and Simons (1981) [1].

Remark 1. (a) Note that the representation (2) is not unique: if $\mathcal{O}$ is an orthogonal $k \times k$ matrix, then (2) also holds with $A' \triangleq A\mathcal{O}$ and $U' \triangleq \mathcal{O}^\mathrm{T} U$.

(b) Note that elliptical distributions with different parameters can be equal: if $X \sim E_p(\mu, \Sigma, \phi)$, then $X \sim E_p(\mu, c\Sigma, \phi_c)$ for every $c > 0$, where $\phi_c(s) \triangleq \phi(s/c)$ for all $s \geq 0$.

For $X = (X_1, \ldots, X_p)^\mathrm{T} \sim E_p(\mu, \Sigma, \phi)$ with $\mathbb{P}\{X_i = \mu_i\} < 1$ and $\mathbb{P}\{X_j = \mu_j\} < 1$, we call $\varrho_{ij} \triangleq \Sigma_{ij}/\sqrt{\Sigma_{ii}\Sigma_{jj}}$ the *linear correlation coefficient* for X_i and X_j. If $\mathrm{Var}(X_i)$ and $\mathrm{Var}(X_j)$ are finite, then $\varrho_{ij} = \mathrm{Cov}(X_i, X_j)/\sqrt{\mathrm{Var}(X_i)\mathrm{Var}(X_j)}$.

Definition 2. Kendall's tau *for the random variables* X_1, X_2 *is defined as*

$$\tau(X_1, X_2) \triangleq \mathbb{P}\{(X_1 - \tilde{X}_1)(X_2 - \tilde{X}_2) > 0\} - \mathbb{P}\{(X_1 - \tilde{X}_1)(X_2 - \tilde{X}_2) < 0\},$$

where $(\tilde{X}_1, \tilde{X}_2)$ *is an independent copy of* (X_1, X_2).

3 Main Results

The main result of this note is the following theorem; its proof is a combination of Lemmas 2 and 7 below. For the case of normal distributions, see also Lemma 6.

Theorem 2. *Let* $X \sim E_p(\mu, \Sigma, \phi)$. *If* $i, j \in \{1, \ldots, p\}$ *satisfy* $\mathbb{P}\{X_i = \mu_i\} < 1$ *and* $\mathbb{P}\{X_j = \mu_j\} < 1$, *then*

$$\tau(X_i, X_j) = \left(1 - \sum_{x \in \mathbb{R}} (\mathbb{P}\{X_i = x\})^2\right) \frac{2}{\pi} \arcsin \varrho_{ij}, \quad (3)$$

where the sum extends over all atoms of the distribution of X_i. *If in addition* $\mathrm{rank}(\Sigma) \geq 2$, *then* (3) *simplifies to*

$$\tau(X_i, X_j) = \left(1 - (\mathbb{P}\{X_i = \mu_i\})^2\right) \frac{2}{\pi} \arcsin \varrho_{ij}, \quad (4)$$

which further simplifies to (1) *if* $\mathbb{P}\{X_i = \mu_i\} = 0$.

The following lemma states that linear combinations of independent elliptically distributed random vectors with the same dispersion matrix Σ (up to a positive constant, see Remark 1) remains elliptical. This lemma is of independent interest.

Lemma 1. *Let* $X \sim E_p(\mu, \Sigma, \phi)$ *and* $\tilde{X} \sim E_p(\tilde{\mu}, c\Sigma, \tilde{\phi})$ *for* $c > 0$ *be independent. Then for* $a, b \in \mathbb{R}$, $aX + b\tilde{X} \sim E_p(a\mu + b\tilde{\mu}, \Sigma, \phi^*)$ *with* $\phi^*(u) \triangleq \phi(a^2 u)\, \tilde{\phi}(b^2 cu)$.

Proof. For all $t \in \mathbb{R}^p$,

$$\begin{aligned}
\varphi_{aX+b\tilde{X}-a\mu-b\tilde{\mu}}(t) &= \varphi_{a(X-\mu)}(t)\, \varphi_{b(\tilde{X}-\tilde{\mu})}(t) \\
&= \phi\bigl((at)^\mathrm{T} \Sigma (at)\bigr)\, \tilde{\phi}\bigl((bt)^\mathrm{T} (c\Sigma)(bt)\bigr) \\
&= \phi(a^2 t^\mathrm{T} \Sigma t)\, \tilde{\phi}(b^2 c t^\mathrm{T} \Sigma t).
\end{aligned}$$

4 Proof of Theorem 2

The following lemma gives the relation between Kendall's tau and the linear correlation coefficient for elliptical random vectors of pairwise comonotonic or countermonotonic components. It proves Theorem 2 for the case $\text{rank}(\Sigma) = 1$.

Lemma 2. *Let $X \sim E_p(\mu, \Sigma, \phi)$ with $\text{rank}(\Sigma) = 1$. If $\mathbb{P}\{X_i = \mu_i\} < 1$, and $\mathbb{P}\{X_j = \mu_j\} < 1$, then*

$$\tau(X_i, X_j) = \left(1 - \sum_{x \in \mathbb{R}} (\mathbb{P}\{X_i = x\})^2\right) \frac{2}{\pi} \arcsin \varrho_{ij}. \tag{5}$$

Proof. Let $\tilde{X}$ be an independent copy of X. Let $X \stackrel{d}{=} \mu + RAU$ and $\tilde{X} \stackrel{d}{=} \mu + \tilde{R}A\tilde{U}$ be stochastic representations according to Theorem 1, where $(\tilde{R}, \tilde{U})$ denotes an independent copy of (R, U). In particular, A is a $p \times 1$ matrix and U is symmetric $\{1, -1\}$-valued. Furthermore, $\mathbb{P}\{X_i = \mu_i\} < 1$ and $\mathbb{P}\{X_j = \mu_j\} < 1$ imply $A_{i1} \neq 0$ and $A_{j1} \neq 0$. Therefore,

$$\varrho_{ij} = A_{i1}A_{j1}/\sqrt{A_{i1}^2 A_{j1}^2} = \text{sign}(A_{i1}A_{j1}) = \frac{2}{\pi}\arcsin \varrho_{ij}, \tag{6}$$

$(X_i - \tilde{X}_i)(X_j - \tilde{X}_j) \stackrel{d}{=} A_{i1}A_{j1}(RU - \tilde{R}\tilde{U})^2$ and

$$\mathbb{P}\{RU = \tilde{R}\tilde{U}\} = \sum_{x \in \mathbb{R}} (\mathbb{P}\{RU = x\})^2 = \sum_{x \in \mathbb{R}} (\mathbb{P}\{X_i = x\})^2. \tag{7}$$

If $A_{i1}A_{j1} > 0$, then by Definition 2

$$\tau(X_i, X_j) = \mathbb{P}\{(RU - \tilde{R}\tilde{U})^2 > 0\} = 1 - \mathbb{P}\{RU = \tilde{R}\tilde{U}\}$$

Using (6) and (7), the result (5) follows. If $A_{i1}A_{j1} < 0$, then

$$\tau(X_i, X_j) = -\mathbb{P}\{(RU - \tilde{R}\tilde{U})^2 > 0\}$$

and the result (5) follows in the same way.

Lemma 3. *Let $X \sim E_p(\mu, \Sigma, \phi)$ with $\text{rank}(\Sigma) = k \geq 2$ and let $\tilde{X}$ be an independent copy of X. If $\mathbb{P}\{X_i = \mu_i\} < 1$, then $\mathbb{P}\{X_i = \tilde{X}_i\} = (\mathbb{P}\{X_i = \mu_i\})^2$.*

Proof. Let $X \stackrel{d}{=} \mu + RAU$ be a stochastic representation according to Theorem 1. Define $A_i \triangleq (A_{i1}, \ldots, A_{ik})$ and $a \triangleq A_i A_i^T$. Since $\mathbb{P}\{X_i = \mu_i\} < 1$, the case $a = 0$ is excluded. By choosing an orthogonal $k \times k$ matrix $\mathcal{O}$ whose first column is A_i^T/a and using Remark 1(a) if necessary, we may assume that $A_i = (a, 0, \ldots, 0)$, hence $X_i \stackrel{d}{=} \mu_i + aRU_1$. Note that U_1 is a continuous random variable because $k \geq 2$. Hence $\mathbb{P}\{aRU_1 = x\} = 0$ for all $x \in \mathbb{R} \setminus \{0\}$, and it follows that

$$\mathbb{P}\{X_i = \tilde{X}_i\} = \sum_{x \in \mathbb{R}} (\mathbb{P}\{X_i = x\})^2 = \sum_{x \in \mathbb{R}} (\mathbb{P}\{aRU_1 = x\})^2 = (\mathbb{P}\{X_i = \mu_i\})^2.$$

Lemma 4. Let $X \sim E_p(\mu, \Sigma, \phi)$ with $\text{rank}(\Sigma) = k \geq 2$, and let $\tilde{X}$ be an independent copy of X. If $\mathbb{P}\{X_i = \mu_i\} < 1$ and $\mathbb{P}\{X_j = \mu_j\} < 1$, then

$$\tau(X_i, X_j) = 2\,\mathbb{P}\{(X_i - \tilde{X}_i)(X_j - \tilde{X}_j) > 0\} - 1 + (\mathbb{P}\{X_i = \mu_i\})^2. \qquad (8)$$

Proof. Since $Y \triangleq X - \tilde{X} \sim E_p(0, \Sigma, \phi^2)$ by Lemma 1, there exists a stochastic representation $Y = RAU$ according to Theorem 1. By Lemma 3, $\mathbb{P}\{Y_i = 0\} = (\mathbb{P}\{X_i = \mu_i\})^2 < 1$ and similarly $\mathbb{P}\{Y_j = 0\} < 1$. Define $A_i \triangleq (A_{i1}, \ldots, A_{ik})$ and $A_j \triangleq (A_{j1}, \ldots, A_{jk})$. With the same arguments as in the proof of Lemma 3, it follows that $A_i U$ and $A_j U$ are continuous random variables, which implies that $\mathbb{P}\{A_i U = 0\} = 0$ and $\mathbb{P}\{A_j U = 0\} = 0$. Therefore,

$$\mathbb{P}\{Y_i Y_j = 0\} = \mathbb{P}\{R = 0\} = \mathbb{P}\{Y_i = 0\} = (\mathbb{P}\{X_i = \mu_i\})^2.$$

Since $\tau(X_i, X_j) = 2\,\mathbb{P}\{Y_i Y_j > 0\} - 1 + \mathbb{P}\{Y_i Y_j = 0\}$, the conclusion follows.

Lemma 5. Let $X \sim E_p(0, \Sigma, \phi)$ and $\tilde{X} \sim E_p(0, c\Sigma, \tilde{\phi})$ with $\text{rank}(\Sigma) \geq 2$ and $c > 0$. If $\mathbb{P}\{X_i = 0\} < 1$ and $\mathbb{P}\{\tilde{X}_i = 0\} < 1$, then

$$\mathbb{P}\{X_i X_j > 0\}(1 - \mathbb{P}\{\tilde{X}_i = 0\}) = \mathbb{P}\{\tilde{X}_i \tilde{X}_j > 0\}(1 - \mathbb{P}\{X_i = 0\}).$$

Proof. Take $X \stackrel{d}{=} RAU$ according to Theorem 1 and set $W \triangleq AU$. Then

$$\begin{aligned}
\mathbb{P}\{X_i X_j > 0\} &= \mathbb{P}\{RW_i RW_j > 0\} \\
&= \mathbb{P}\{RW_i RW_j > 0 \mid R > 0\}\,\mathbb{P}\{R > 0\} \\
&= \mathbb{P}\{W_i W_j > 0\}\,\mathbb{P}\{R > 0\}.
\end{aligned}$$

Furthermore, $\tilde{X} \stackrel{d}{=} \sqrt{c}\tilde{R}W$ according to Theorem 2, and a similar calculation shows

$$\begin{aligned}
\mathbb{P}\{\tilde{X}_i \tilde{X}_j > 0\} &= \mathbb{P}\{c\tilde{R}^2 W_i W_j > 0 \mid \tilde{R} > 0\}\,\mathbb{P}\{\tilde{R} > 0\} \\
&= \mathbb{P}\{W_i W_j > 0\}\,\mathbb{P}\{\tilde{R} > 0\}.
\end{aligned}$$

As in the proof of Lemma 3, it follows that W_i has a continuous distribution. Therefore, $\mathbb{P}\{R > 0\} = 1 - \mathbb{P}\{X_i = 0\}$ and $\mathbb{P}\{\tilde{R} > 0\} = 1 - \mathbb{P}\{\tilde{X}_i = 0\}$, and Lemma 5 follows.

Although the next result for normal distributions is well known, we give a proof for completeness of the exposition and for showing where the arsine comes from.

Lemma 6. Let $X \sim \mathcal{N}_p(\mu, \Sigma)$. If $\mathbb{P}\{X_i = \mu_i\} < 1$ and $\mathbb{P}\{X_j = \mu_j\} < 1$, then

$$\tau(X_i, X_j) = 2\,\mathbb{P}\{(X_i - \tilde{X}_i)(X_j - \tilde{X}_j) > 0\} - 1 = \frac{2}{\pi}\arcsin \varrho_{ij},$$

where $\tilde{X}$ is an independent copy of X.

Proof. Using $\sigma_i \triangleq \sqrt{\Sigma_{ii}} > 0$, $\sigma_j \triangleq \sqrt{\Sigma_{jj}} > 0$ and $\varrho_{ij} \triangleq \Sigma_{ij}/\sigma_i\sigma_j$, we have

$$\Sigma^{ij} = \begin{pmatrix} \Sigma_{ii} & \Sigma_{ij} \\ \Sigma_{ji} & \Sigma_{jj} \end{pmatrix} = \begin{pmatrix} \sigma_i^2 & \sigma_i\sigma_j\varrho_{ij} \\ \sigma_i\sigma_j\varrho_{ij} & \sigma_j^2 \end{pmatrix}.$$

Define $Y \triangleq X - \tilde{X}$ and note that $(Y_i, Y_j) \sim \mathcal{N}_2(0, 2\Sigma^{ij})$. Furthermore, $(Y_i, Y_j) \stackrel{d}{=} \sqrt{2}(\sigma_i V \cos\varphi_{ij} + \sigma_i W \sin\varphi_{ij}, \sigma_j W)$, where $\varphi_{ij} \triangleq \arcsin\varrho_{ij} \in [-\pi/2, \pi/2]$ and (V, W) is standard normally distributed. By the radial symmetry of (Y_i, Y_j),

$$\tau(X_i, X_j) = 2\mathbb{P}\{Y_i Y_j > 0\} - 1 = 4\mathbb{P}\{Y_i > 0, Y_j > 0\} - 1$$
$$= 4\mathbb{P}\{V \cos\varphi_{ij} + W \sin\varphi_{ij} > 0, W > 0\} - 1.$$

If Φ is uniformly distributed on $[-\pi, \pi)$, independent of $R \triangleq \sqrt{V^2 + W^2}$, then $(V, W) \stackrel{d}{=} R(\cos\Phi, \sin\Phi)$ and

$$\tau(X_i, X_j) = 4\mathbb{P}\{\cos\Phi\cos\varphi_{ij} + \sin\Phi\sin\varphi_{ij} > 0, \sin\Phi > 0\} - 1$$
$$= 4\mathbb{P}\{\Phi \in (\varphi_{ij} - \pi/2, \varphi_{ij} + \pi/2) \cap (0, \pi)\} - 1 = 4\frac{\varphi_{ij} + \pi/2}{2\pi} - 1,$$

which simplifies to $(2/\pi)\arcsin\varrho_{ij}$.

Lemma 7. *Let $X \sim E_p(\mu, \Sigma, \phi)$ with $\mathrm{rank}(\Sigma) = k \geq 2$. If $\mathbb{P}\{X_i = \mu_i\} < 1$ and $\mathbb{P}\{X_j = \mu_j\} < 1$, then*

$$\tau(X_i, X_j) = \left(1 - (\mathbb{P}\{X_i = \mu_i\})^2\right)\frac{2}{\pi}\arcsin\varrho_{ij}. \tag{9}$$

Proof. Let $\tilde{X}$ be an independent copy of X. By Lemma 4, we can use (8). By Lemmas 1 and 3, $X - \tilde{X} \sim E_p(0, \Sigma, \phi^*)$ with $\mathbb{P}\{X_i = \tilde{X}_i\} = (\mathbb{P}\{X_i = \mu_i\})^2 < 1$ and $\mathbb{P}\{X_j = \tilde{X}_j\} = (\mathbb{P}\{X_j = \mu_j\})^2 < 1$. If $Z, \tilde{Z} \sim \mathcal{N}_p(\mu, \Sigma/2)$ are independent, then $Z - \tilde{Z} \sim \mathcal{N}_p(0, \Sigma)$. By Lemma 5,

$$\mathbb{P}\{(X_i - \tilde{X}_i)(X_j - \tilde{X}_j) > 0\} = \mathbb{P}\{(Z_i - \tilde{Z}_i)(Z_j - \tilde{Z}_j) > 0\}(1 - (\mathbb{P}\{X_i = \mu_i\})^2).$$

Substituting this into (8) and using Lemma 6, the result (9) follows.

Remark 2. After completion of this note we found that relation (1) was already proved by a different method in [4, Theorem 3.1] for elliptical distributions having a density. The extensions (3) and (4) are not given in [4].

References

1. S. Cambanis, S. Huang, and G. Simons, *On the theory of elliptically contoured distributions*, J. Multivariate Anal. **11** (1981), 368–385.
2. H. Cramér, *Mathematical Methods of Statistics*, Princeton University Press, Princeton, 1946.

3. S. J. Devlin, R. Gnanadesikan, and J. R. Kettenring, *Robust estimation of dispersion matrices and principal components*, J. Amer. Statist. Assoc. **76** (1981), 354–362.
4. H.-B. Fang, K.T. Fang, and S. Kotz, *The meta-elliptical distributions with given marginals*, J. Multivariate Anal. **82** (2002), no. 1, 1–16.
5. K.-T. Fang, S. Kotz, and K.-W. Hg, *Symmetric Multivariate and Related Distributions*, Chapman & Hall, London, 1987.
6. R. Frey and A. McNeil, *VaR and expected shortfall in portfolios of dependent credit risks: conceptual and practical insights*, Journal of Banking & Finance **26** (2002), 1317–1334.
7. R. Frey, A. McNeil, and M. Nyfeler, *Copulas and credit models*, Risk **10** (2001), 111–114.
8. F. Lindskog, *Linear correlation estimation*, Research report, RiskLab Switzerland, August 2000, http://www.risklab.ch/Papers.html#LCELindskog.
9. R. Mashal and A. Zeevi, *Beyond correlation: extreme co-movements between financial assets*, Preprint, Columbia Graduate School of Business, 2002, http://www.columbia.edu/~rm586.
10. P. J. Rousseeuw and G. Molenberghs, *Transformation of non positive semidefinite correlation matrices*, Comm. Statist. A—Theory Methods **22** (1993), no. 4, 965–984.

Exploring Credit Data *

Marlene Müller and Wolfgang Härdle

Institut für Statistik und Ökonometrie, Humboldt-Universität zu Berlin, Germany

Summary. Credit scoring methods aim to assess the default risk of a potential borrower. This involves typically the calculation of a credit score and the estimation of the probability of default.

One of the standard approaches is logistic discriminant analysis, also referred to as logit model. This model maps explanatory variables for the default risk to a credit score using a linear function. Nonlinearity can be included by using polynomial terms or piecewise linear functions. This may give however only a limited reflection of a truly nonlinear relationship. Moreover, an additional modeling step may be necessary to determine the optimal polynomial order or the optimal interval classification.

This paper presents semiparametric extensions of the logit model which directly allow for nonlinear relationships to be part of the explanatory variables. The technique is based on the theory generalized partial linear models. We illustrate the advantages of this approach using a consumer retail banking data set.

1 Introduction

Credit scoring methods aim to estimate the default risk of potential borrowers and to classify them into groups according to their default risk. This involves typically the calculation of a credit score and the estimation of the probability of default (PD).

From a statistical point of view, classification between risky and non-risky borrowers is first of all a discriminant analysis problem. Classical solutions to this are linear or quadratic discriminant analysis and — on a more advanced level — logistic discriminant analysis. All these methods are based on a score depending on explanatory variables. Typically, the score summarizes the explanatory variables in a predefined form (linear or quadratic). More complex

---

* The research for this paper was supported by Sonderforschungsbereich 373 "Quantifikation und Simulation Ökonomischer Prozesse", Humboldt-Universität zu Berlin (Germany).We are grateful to Bernd Rönz, Humboldt-Universität zu Berlin, for his cooperation at an earlier stage of this project.

nonlinear mappings can be considered by using polynomial terms or piecewise linear functions. This gives, however, only an imprecise reflection of a truly nonlinear relationship. Moreover, an additional modeling step is necessary to determine the optimal polynomial order or the optimal interval classification.

Recently developed methods allow for a flexible modeling via neural networks and classification trees, for applications see [1] and [12]. Overviews on these methods for consumer credit risk can be found in [9] and [8]. These nonparametric approaches do not restrict the possible nonlinear impact of explanatory variables. However, it is often hard to interpret the resulting relationships between the explanatory variables and the classification rule. This motivates our semiparametric approach.

We consider a modification of logistic discriminant analysis that allows for a more flexible handling of a subset of the explanatory variables. Our approach is based on generalized partial linear models which extend the "easy to interpret" structure of the logistic model by nonparametric components. A particularly interesting feature of logistic discriminant analysis (equivalently: fitting a logit model) is that simultaneously credit scores and PDs are estimated. This leads to a growing interest in the logit model for redesigning credit rating systems according to the requirements of the New Basel Capital Accord ("Basel II", cf. [2]). The paper is organized as follows: Section 2 explains the data structure for cross-sectional credit samples and provides the notation of the data that we use throughout the paper. Section 3 recalls the important terms for logistic discriminant analysis (the logit model) and presents the results for our specific sample. Section 4 introduces the semiparametric extension of the logit model. We estimate here several specifications of this semiparametric model and compare the resulting fits to the estimated logit model. Finally, Section 5 discusses the estimated models with respect to performance criteria.

2 Data Structure

Before we describe the data that we use in the following, let us consider a typical example for a cross-sectional credit data set. Suppose we have a sample of customers that apply for a loan to buy a car. Assume further, that we have information if these customers paid their installments without problems ($Y = 0$) or not ($Y = 1$). For the sake of simplicity, we will call these two categories non-default and default in the following. Obviously, we have now a default indicator Y and explanatory variables $X = (X_1, \ldots, X_p)$ for each member of the sample. Table 1 shows for illustration descriptive statistics on a subsample of the credit data used in [4] and [5].

Note that Table 1 reflects the usual structure of the explanatory variables in credit data sets: The variables may be of discrete (binary, categorical) or of continuous form. For the discrete data, a sufficiently complex representa-

Table 1. Example data: Sample on loans for cars

		Yes	No	(in %)	
Y	default	26.4	73.6		
X_1	previous loans OK	66.2	33.8		
X_2	employed	73.2	26.8		
		Min	Max	Mean	S.E.
X_3	duration (in months)	4	54	21.8	10.6
X_4	amount (in DM)	428	14179	3902.3	2621.9
X_5	age (in years)	19	75	34.2	10.8

tion is possible by using dummy variables. For the continuous variables, an appropriate way of including them into the score has to be found.

The data that we explore and analyze in the rest of this paper have been provided by the French bank Compagnie Bancaire. The used estimation sample consists of 6180 cases (clients) and 24 variables:

- Response variable Y (credit worthiness, binary, 1 denotes default). The number of faulty clients is relatively small (6%) which is typical for credit data.
- Metric explanatory variables X2 to X9. All of them have (right) skewed distributions. Variable X6 is discrete with only five different realizations. X8 and X9 in particular have one realization which covers a majority of observations.
- Categorical explanatory variables X10 to X24. Six of them are binary. The others have three to eleven categories (not ordered).

In addition to the estimation sample, the bank provided us with a validation data set of 1998 cases. Table 2 gives the number of non-defaults and defaults in the estimation and validation data sets. We refer to [17] for additional details.

Table 2. Defaults and non-defaults in the French bank sample

	Estimation data set		Validation data set	
0 (non-defaults)	5808	(94%)	1891	(94.6%)
1 (defaults)	372	(6%)	107	(5.4%)
total	6180		1998	

We now describe the variables in the estimation sample in more detail. The validation sample will be only used to evaluate the semiparametric models.

We plot first the estimated probability density functions for the metric variables using histograms and kernel density estimates. For more statistical

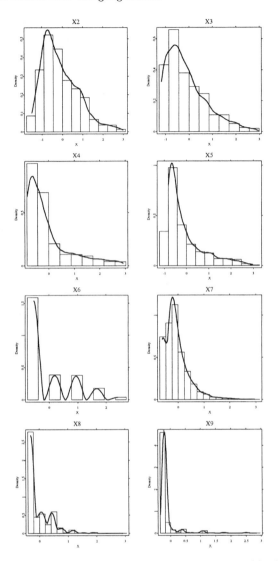

Fig. 1. Histograms and kernel density estimates, variables X2 to X9

and numerical details on density estimation we refer to the monographs of [19], [10], or [18]. Figure 1 shows the density estimates for the variables X2 to X9. For the kernel estimators we employed a rule-of-thumb bandwidths as smoothing parameter. From the figure we can conclude that the variables X6, X8 to X9 are of quasi-discrete structure. Since nonparametric components require continuous variation of the relevant variables, we will therefore concentrate on variables X2 to X5 and X7 for a nonparametric analysis.

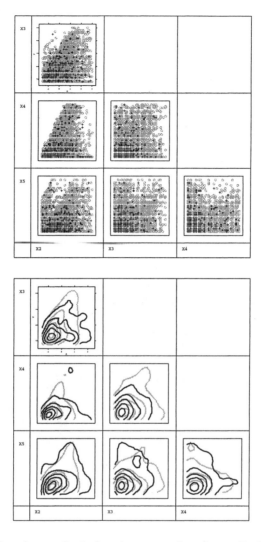

Fig. 2. Scatterplots (upper display) and contour-plots (lower display), variables X2 to X5. Default observations and contours for default are emphasized in black.

As a second step in exploration we display bivariate plots for the variables X2 to X5. Figure 2 shows all bivariate scatterplots of X2 to X5. Due to the large number of non-defaults, it is difficult to capture their bivariate distribution. We therefore show bivariate contours of the estimated densities for defaults and non-defaults. These density estimates are again kernel estimates using a rule-of-thumb bandwidth [11, 18]. The figure shows that the

3 Logistic Credit Scoring

Logistic discriminant analysis assumes that the probability of belonging to the group of faulty clients is given by

$$P(Y = 1|X) = F\left(\sum_{j=2}^{24} \beta_j^\top X_j + \beta_0\right), \tag{1}$$

where

$$F(u) = \frac{1}{1 + \exp(-u)}$$

is the logistic (cumulative) distribution function. X_j denotes the j-th variable itself if it is metric ($j \in \{2,\ldots,9\}$) or a vector of dummies if it is categorical ($j \in \{10,\ldots,24\}$). For all categorical variables we used the first category as reference.

Model (1) can be motivated as follows: Suppose that we know the true (negative) credit score which has the form

$$Y^* = v(X) - u$$

with $v(\bullet)$ denoting a "regression" (or index) function and u an error term. We observe a default if the score Y^* is positive. (For practical purposes we consider higher score values to indicate higher risk of default.) Thus, our model is

$$Y = \begin{cases} 1 \text{ if } Y^* = v(X) - u > 0, \\ 0 \text{ otherwise.} \end{cases}$$

This is equivalent to (1) if u has a (standard) logistic distribution and

$$v(X) = \sum_{j=2}^{24} \beta_j^\top X_j + \beta_0 = \beta^\top X \tag{2}$$

holds. Modifications of the logit model usually concern the distributional assumptions (a Gaussian distribution of u leads to the probit model) or the assumptions on the index function $v(\bullet)$.

The logit model is estimated by maximum–likelihood (cf. [14]). Table 3 shows the estimated coefficients β_j. We find that most of the variables contribute to the explanation of the response Y. As mentioned above, the modeling for the categorical variables is sufficiently complex due to their representation by dummy variables. For the metric variables we achieve different levels of significance. Variables X2, X3, X6, X8, and X9 have coefficients significantly

Table 3. Results of the logit estimation. Bold coefficients are significant at 5%

Variable	Coefficient	S.E.	t-value	Variable	Coefficient	S.E.	t-value
constant	**-2.605280**	0.5890	-4.42	X19#2	-0.086954	0.3082	-0.28
X2	**0.246641**	0.1047	2.35	X19#3	0.272517	0.2506	1.09
X3	**-0.417068**	0.0817	-5.10	X19#4	-0.253440	0.4244	-0.60
X4	-0.062019	0.0849	-0.73	X19#5	0.178965	0.3461	0.52
X5	-0.038428	0.0816	-0.47	X19#6	-0.174914	0.3619	-0.48
X6	**0.187872**	0.0907	2.07	X19#7	0.462114	0.3419	1.35
X7	-0.137850	0.1567	-0.88	X19#8	**-1.674337**	0.6378	-2.63
X8	**-0.789690**	0.1800	-4.39	X19#9	0.259195	0.4478	0.58
X9	**-1.214998**	0.3977	-3.06	X19#10	-0.051598	0.2812	-0.18
X10#2	-0.259297	0.1402	-1.85	X20#2	-0.224498	0.3093	-0.73
X11#2	**-0.811723**	0.1277	-6.36	X20#3	-0.147150	0.2269	-0.65
X12#2	-0.272002	0.1606	-1.69	X20#4	0.049020	0.1481	0.33
X13#2	0.239844	0.1332	1.80	X21#2	0.132399	0.3518	0.38
X14#2	-0.336682	0.2334	-1.44	X21#3	**0.397020**	0.1879	2.11
X15#2	**0.389509**	0.1935	2.01	X22#2	-0.338244	0.3170	-1.07
X15#3	0.332026	0.2362	1.41	X22#3	-0.211537	0.2760	-0.77
X15#4	**0.721355**	0.2580	2.80	X22#4	-0.026275	0.3479	-0.08
X15#5	0.492159	0.3305	1.49	X22#5	-0.230338	0.3462	-0.67
X15#6	**0.785610**	0.2258	3.48	X22#6	-0.244894	0.4859	-0.50
X16#2	**0.494780**	0.2480	2.00	X22#7	-0.021972	0.2959	-0.07
X16#3	-0.004237	0.2463	-0.02	X22#8	-0.009831	0.2802	-0.04
X16#4	0.315296	0.3006	1.05	X22#9	0.380940	0.2497	1.53
X16#5	-0.017512	0.2461	-0.07	X22#10	-1.699287	1.0450	-1.63
X16#6	0.198915	0.2575	0.77	X22#11	0.075720	0.2767	0.27
X17#2	-0.144418	0.2125	-0.68	X23#2	-0.000030	0.1727	-0.00
X17#3	**-1.070450**	0.2684	-3.99	X23#3	-0.255106	0.1989	-1.28
X17#4	-0.393934	0.2358	-1.67	X24#2	0.390693	0.2527	1.55
X17#5	**0.921013**	0.3223	2.86				
X17#6	**-1.027829**	0.1424	-7.22				
X18#2	0.165786	0.2715	0.61				
X18#3	0.415539	0.2193	1.89				
X18#4	**0.788624**	0.2145	3.68				
X18#5	**0.565867**	0.1944	2.91	df			6118
X18#6	0.463575	0.2399	1.93	Log-Lik.			-1199.6278
X18#7	**0.568302**	0.2579	2.20	Deviance			2399.2556

different from zero. This means, their effect on the response is obviously well specified by considering them as a linear component in the index function $v(\bullet)$.

For X4, X5, and X7 non-significant coefficients indicate that either these variables have no influence on the response or that their specification is insufficient. We will now investigate the latter conjecture. As a further graphical tool we use scatterplots for these explanatory variables. In contrast to linear

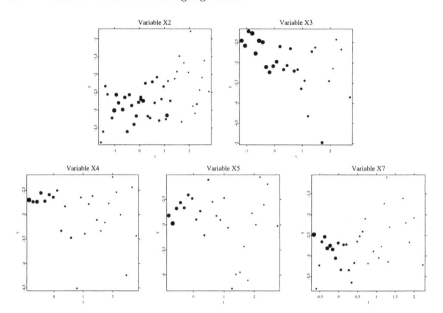

Fig. 3. Marginal dependencies, variables X2 to X5, X7. Thicker bullets correspond to more observations in a class

regression, it is not useful to directly plot Xj vs. Y. However, if we assume model (1) as the underlying, the logits

$$\log\left(\frac{P(Y=1|X)}{P(Y=0|X)}\right)$$

should relate in a linear way to the explanatory variables X. We therefore divide the range of each ot the variables X2 to X5 and X7 into intervals (classes) of similar length and estimate the logits in these intervals using the observed frequencies of $Y = 0$ and $Y = 1$. The class centers are then plotted against the estimated logits. The resulting "scatterplots" are presented in Figure 3. It is obvious that the scatterplots for X2 and X3 follow a linear trend whereas for the other three variables a linear relationship is not obvious. The variables X4, X5 and X7 are hence the most interesting components for considering a nonlinear (nonparametric) modification of the index function (2).

4 Semiparametric Credit Scoring

The logit model (1) is a special case of the the generalized linear model (GLM, see [14]) which is defined as

$$E(Y|X) = G(\beta^\top X)$$

with $G(\bullet)$ denoting a "link" function. Since in our problem Y is binary, it holds
$$E(Y|X) = P(Y = 1|X).$$
Thus, the logit model is a GLM with the logistic distribution function $F(\bullet)$ as link function. This property makes it easy to consider several extensions of the GLM which then hold automatically for the logit model.

The semiparametric modification that we consider here generalizes the linear argument (2) to a partial linear argument. Consider a vector of explanatory variables that splits up into a vector X and a second vector T. The generalized partial linear model (GPLM)
$$E(Y|X,T) = G\{\beta^\top X + m(T)\}$$
allows us to describe the influence of the component T nonparametrically. As before, we assume $G(\bullet)$ to be a known function (here the logistic link F) and β to be an unknown parameter vector. In addition we have to estimate $m(\bullet)$, an unknown smooth function. The parametric component β and the nonparametric function $m(\bullet)$ can be estimated in several ways, for a comparison of estimation algorithms and their numerical properties see [16]. Details on the implementation of these estimators can be found in [15].

We consider the GPLM for several of the metric variables separately as well as for combinations of them. As mentioned earlier, we only consider variables X2 to X5 and X7 to be used within a nonparametric function because of the quasi–discrete structure of X6, X8 and X9. The semiparametric modification of the logit model takes the following form, as indicated here for the example of including X5 in a nonlinear way:

$$P(Y = 1|X) = F\left(\sum_{j=2, j\neq 5}^{24} \beta_j^\top X_j + m_5(X_5)\right).$$

A possible intercept is contained in the function $m_5(\bullet)$.

Table 4 contains the parametric coefficients for the parametric and semiparametric estimates for variables X2 to X9. Coefficients for X10 to X24 are estimated in each of the specified models, but are not listed here. The column headed by "logit" repeats the parametric logit estimates from Table 3.

It turns out, that all linear coefficients vary little over the different estimates. This holds as well for their significance. Variables X4, X5 and X7 are constantly insignificant over all estimates. The semiparametric logit model is estimated by semiparametric maximum-likelihood, a combination of maximizing a classical (parametric) likelihood for estimating β and a smoothed (local) likelihood for estimating the function $m(\bullet)$. The fitted curves for the nonparametric components according to Table 4 can be found in Figure 4 (separate nonparametric functions in X2 to X5 and X7) and Figure 5 (bivariate function in X4 and X5).

Table 4. Parametric coefficients in parametric and semiparametric logit, variables X2 to X9. Bold values are significant at 5%

	logit	nonparametric in						
		X2	X3	X4	X5	X7	X4,X5	X2, X4,X5
constant	**−2.605**	−	−	−	−	−	−	−
X2	**0.247**	−	0.243	0.241	0.243	0.233	0.228	−
X3	**−0.417**	**−0.414**	−	**−0.414**	**−0.416**	**−0.417**	**−0.408**	**−0.399**
X4	−0.062	−0.052	−0.063	−	−0.065	−0.054	−	−
X5	−0.038	−0.051	−0.045	−0.034	−	−0.042	−	−
X6	**0.188**	**0.223**	**0.193**	**0.190**	**0.177**	**0.187**	0.176	**0.188**
X7	−0.138	−0.138	−0.142	−0.131	−0.146	−	−0.135	−0.128
X8	**−0.790**	**−0.777**	**−0.800**	**−0.786**	**−0.796**	**−0.793**	**−0.792**	**−0.796**
X9	**−1.215**	**−1.228**	**−1.213**	**−1.222**	**−1.216**	**−1.227**	**−1.214**	**−1.215**

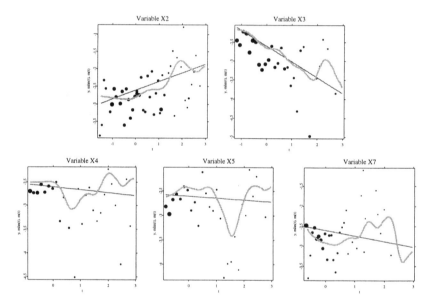

Fig. 4. Estimated curves for variables X2 to X5 and X7. Parametric logit fits (thin dashed lines) and GPLM logit fits (thick solid curves)

For the assessment of whether the semiparametric fit outperforms the parametric logit or not, we present the reported statistical characteristics in Table 5. The deviance is minus twice the estimated log–likelihood of the fitted model in our case. For the logit model, the degrees of freedom just denote

$$df = n - k$$

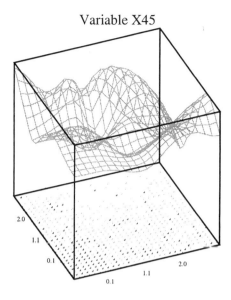

Fig. 5. Bivariate nonparametric surface for variables X4, X5

where n is the sample size and k the number of estimated parameters. In the semiparametric case, a corresponding number of degrees of freedom can be approximated using the trace of the corresponding hat matrix. The deviance and the (approximate) degrees of freedom of the parametric and the semiparametric model can then be used to construct a likelihood ratio test to compare both models [3, 16]. The obtained significance levels from these tests are denoted by α. Finally, we report pseudo R^2 values in the style of McFaddens pseudo R^2 values for the logit case [7, Sec. 21.4.2] representing an analog to the linear regression coefficient of determination.

It is obvious to see that in particular models containing variable X5 in the nonparametric part considerably decrease the deviance and increase the coefficient of determination R^2. Accordingly, the significance level for the test of parametric versus nonparametric modeling decreases.

5 Evaluation of the Scores

For a credit rating system it is important that relevant explanatory variables are detected and enter the model in an optimal way. The semiparametric technique introduced above may help to find transformations of explanatory variables that improve the prediction of defaults.

Table 5. Statistical characteristics in parametric and semiparametric logit fits. Bold values are significant at 10%. Estimation data set

	logit	X2	X3	X4	nonparametric in X5	X7	X4,X5	X2, X4,X5
deviance	2399.26	2393.03	2395.19	2391.29	**2387.17**	**2388.63**	**2372.63**	**2372.43**
df	6118.00	6113.72	6113.57	6113.34	6113.41	6113.61	6103.82	6100.23
α	–	0.210	0.458	0.133	**0.026**	**0.041**	**0.023**	**0.077**
AIC	2523.3	2525.6	2528.0	2524.6	2520.4	2521.4	2525.0	2533.0
Pseudo-R^2	14.7%	14.9%	14.8%	15.0%	15.1%	15.1%	15.6%	15.6%

How can different models (different scores) be compared? The easiest approach is to use misclassification rates. Suppose we have estimated the score $S = S(X)$ for a potential borrower. For example, S may denote

$$S = \sum_{j=2}^{24} \beta_j^\top X_j + \beta_0$$

in the parametric logit model and

$$S = \sum_{j=2, j \neq 5}^{24} \beta_j^\top X_j + m_5(X_5)$$

in the semiparametric logit model when fitting X5 nonparametrically. Typically one predicts

$$\widehat{Y} = \begin{cases} 1 & F(S) > \tau, \\ 0 & \text{otherwise,} \end{cases}$$

where the threshold τ is taken as

$$\tau = 0.5.$$

Considering a range of τ-values allows us to obtain a more detailed picture of the classification of different score values. Table 6 reports misclassified observations from the validation sample (of size 1998) at three different threshold values τ.

The Lorenz curve (cumulative accuracy profile, CAP) visualizes the accuracy of the score with respect to its predictive power for a default. Figure 6 shows the principle of the Lorenz curve. For both axes, sorted score values (from bad=high to good=low) are considered. The horizontal scale shows the percentages of observations above a certain value s, whereas the vertical axes shows percentages of faulty observations above this value s. Mathematically, the Lorenz curve is a plot of

$$P(S > s) \quad \text{versus} \quad P(S > s | Y = 1).$$

Table 6. Misclassifications for $\widehat{Y} = 1$ (default) if $F(S) \leq t$ and $\widehat{Y} = 0$ (non-default) if $F(S) > t$. Validation data set.

threshold τ	logit	nonparametric in						
		X2	X3	X4	X5	X7	X4,X5	X2, X4,X5
0.25	129	133	129	136	130	128	132	130
non-default	41	44	40	49	40	40	46	40
default	88	89	89	87	90	80	86	90
0.5	111	110	111	111	110	108	111	110
non-default	5	5	5	5	5	2	5	4
default	106	105	106	106	105	106	106	106
0.75	107	107	107	107	107	107	107	107
non-default	0	0	0	0	0	0	0	0
default	107	107	107	107	107	107	107	107

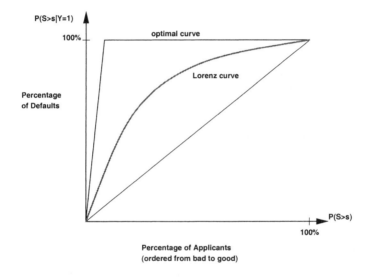

Fig. 6. Principle of the Lorenz curve

Typically, i.e., if the PD is a monotone increasing function of the score, the curve is concave and located above the diagonal. The diagonal can be interpreted as a "worst score": If S and Y have no relation at all, then $P(S > s) = P(S > s|Y = 1)$. The "best score" does a perfect separation of defaults and non-defaults. This leads to the optimal curve shown in Figure 6. A quantitative measure for the performance of a score is based on the area between the Lorenz curve and the diagonal. The Gini coefficient G denotes twice this area. To compare different scores, their accuracy ratios

$$AR = \frac{G}{G_{opt}},$$

i.e., the Gini coefficient G relative to the Gini coefficient of the optimal Lorenz curve can be used. Variants of the Lorenz curve are the receiver operating characteristic (ROC) curve [9] and the performance curve [6, Ch. 4]. See also [20] for a relation between Lorenz curve and ROC and [13] for a general overview on criteria for measuring the accuracy of credit scores.

Lorenz curves can also be used for assessing the impact of single variables. Table 7 shows the AR values for all metric explanatory variables on the estimation data set. We find that those variables which are highly significant in the logit fit (cf. Table 3) also achieve high accuracy ratios. (Note that we used appropriate +/− signs here for each variable, such that the maximal possible AR is reported.)

Table 7. Accuracy ratios of variables X2 to X9. Estimation data set.

	nonparametric in							
	+X2	−X3	−X4	+X5	+X6	−X7	−X8	−X9
AR	0.076	0.168	0.043	0.023	0.024	0.052	0.165	0.107

Let us have a closer look at the Lorenz curves for the three variables X4, X5, X7 which had an obvious nonlinear effect in the score. Figure 7 shows the Lorenz curves and in comparison density estimates separately for defaults and non-defaults. In particular for X5 and X7 we see that the impact of these two variables on Y is non-monotonous: The Lorenz curve crosses the diagonal and the densities cross several times. This means that the nonlinear relationship in the index function $v(\bullet)$ is as well reflected in the Lorenz curve (and vice versa).

Consider now the Lorenz curves and AR values for the fitted logit and semiparametric model. Figure 8 shows the result for the logit fit achieving an AR value of 0.543. Note that most of the performance of the score is contributed by the categorical variables. The continuous variables altogether explain only a small part of the default.

Table 8. Accuracy ratios in parametric and semiparametric logit. Bold values improve the logit fit. Validation data set.

	logit	nonparametric in						
		X2	X3	X4	X5	X7	X4,X5	X2, X4,X5
AR	0.543	0.538	0.543	0.527	**0.556**	0.538	**0.548**	**0.552**

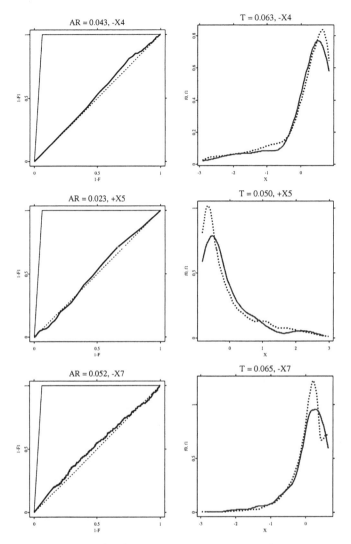

Fig. 7. Lorenz curves (left) and density estimates (right, conditionally on default/non-default) for X4, X5, X7

Table 8 compares the AR performance of the parametric logit fit and the semiparametric logit fit obtained by separately including X2 to X5 nonparametrically. Indeed, the semiparametric model for the influence of X5 improves the performance with respect to the parametric model. The semiparametric models for the influence of X2 to X4 do not improve the performance with respect to the parametric model, though.

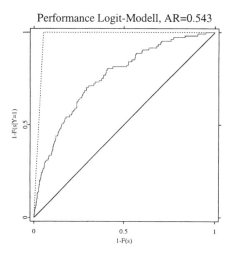

Fig. 8. Lorenz curve for logit model (solid) and optimal curve (dashed)

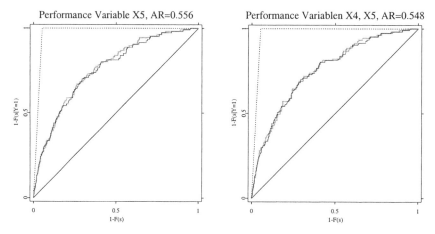

Fig. 9. Performance curves with variables X5 (left) and with variables X4, X5 (right) jointly included nonparametrically. Validation data set.

Figure 9 compares the performance of the parametric logit fit and the semiparametric logit fit obtained by jointly including X4, X5 nonparametrically. This performance curve improves versus nonparametrically fitting only X4, but shows less power versus fitting only X5. Hence, the improvement of using both variables jointly may be explained by the influence of X5 only.

References

1. Arminger, G., Enache, D. & Bonne, T. (1997), 'Analyzing credit risk data: A comparison of logistic discrimination, classification tree analysis, and feedforward networks', *Computational Statistics, Special Issue: 10 Years AG GLM* **12**, 293–310.
2. Banking Committee on Banking Supervision (2001), *The New Basel Capital Accord*, Bank for International Settlements, http://www.bis.org.
3. Buja, A., Hastie, T. & Tibshirani, R. (1989), 'Linear smoothers and additive models (with discussion)', *Annals of Statistics* **17**, 453–555.
4. Fahrmeir, L. & Hamerle, A. (1984), *Multivariate Statistische Verfahren*, De Gruyter, Berlin.
5. Fahrmeir, L. & Tutz, G. (1994), *Multivariate Statistical Modelling Based on Generalized Linear Models*, Springer.
6. Gourieroux, C. & Jasiak, J. (2001), *Econometric Analysis of Individual Risks*, Course Script, http://dept.econ.yorku.ca/~jasiakj.
7. Greene, W. H. (1993), *Econometric Analysis*, 2 edn, Prentice Hall, Englewood Cliffs.
8. Hand, D. J. (2001), 'Modelling consumer credit risk', *IMA Journal of Management mathematics* **12**, 139–155.
9. Hand, D. J. & Henley, W. E. (1997), 'Statistical classification methods in consumer credit scoring: a review', *Journal of the Royal Statistical Society, Series A* **160**, 523–541.
10. Härdle, W. (1991), *Smoothing Techniques. With Implementations in S*, Springer, New York.
11. Härdle, W., Müller, M., Sperlich, S. & Werwatz, A. (2003), *An Introduction to Non- and Semiparametric Models*, Springer, forthcoming.
12. Henley, W. E. & Hand, D. J. (1996), 'A k-nearest-neighbor classifier for assessing consumer credit risk', *Statistician* **45**, 77–95.
13. Keenan, S. & Sobehart, J. (1999), Performance measures for credit risk models, Technical report, Moody's Investor Service, Global Credit Research.
14. McCullagh, P. & Nelder, J. A. (1989), *Generalized Linear Models*, Vol. 37 of *Monographs on Statistics and Applied Probability*, 2 edn, Chapman and Hall, London.
15. Müller, M. (2000), Generalized partial linear models, *in* W. Härdle, Z. Hlávka & S. Klinke, eds, 'XploRe Application Guide', Springer and http://www.xplore-stat.de.
16. Müller, M. (2001), 'Estimation and testing in generalized partial linear models – a comparative study', *Statistics & Computing* **11**, 299–309.
17. Müller, M. & Rönz, B. (2000), Credit scoring using semiparametric methods, *in* J. Franke, W. Härdle & G. Stahl, eds, 'Measuring Risk in Complex Stochastic Systems', Springer.
18. Scott, D. W. (1992), *Multivariate Density Estimation: Theory, Practice, and Visualization*, John Wiley & Sons, New York, Chichester.
19. Silverman, B. W. (1986), *Density Estimation for Statistics and Data Analysis*, Vol. 26 of *Monographs on Statistics and Applied Probability*, Chapman and Hall, London.
20. Sobehart, J. & Keenan, S. (2001), 'Measuring default accurately', *Risk, Credit Risk Special Report* **14**(3), 31–33.

Stable Non-Gaussian Credit Risk Model; The Cognity Approach

Borjana Racheva-Jotova[1], Stoyan Stoyanov[2], and Svetlozar T. Rachev[3]

[1] Faculty of Economics and Business Administration, University of Sofia, Bulgaria
[2] Faculty of Mathematics and Informatics, University of Sofia, Bulgaria
[3] School of Economics, University of Karlsruhe, Germany and University of California, Santa Barbara, USA

Summary. We present a new approach for integrated market and credit risk management for highly volatile financial markets. We will illustrate our approach on Cognity software for evaluation of credit risk. Cognity CreditRisk System comprises two models for credit risk evaluation for complex portfolios of instruments with inherent credit risk – Asset Value Approach (AV Model) and Stochastic Default Rate Model (SDR Model), both based on Stable Distributions. We shall summarize the main features of the current version of Cognity: (i) Risk Drivers Scenarios generation (ii) Estimation of dependence structure between risk drivers and modeling marginal distributions; (iii) Credit risk estimation under AV and SDR models.

1 Introduction

Bravo's *Cognity* is an industry *first* on-line portfolio risk management system based on demonstrably realistic asset return distributions. It replaces the traditional "normal law" that in practice denies the occurrence of important abrupt losses or profits, by the "stable Paretian law" that explicitly recognizes the occurrence of such extreme events.

The system is available in both online and enterprise editions. An online demonstration software is provided at
www.cognity.net/Cognity/presentation/mainDemo and at Bravo's GUEST account http://www.cognity.net/Cognity/jsp/login.jsp?/Cognity/frames.jsp

Cognity is designed for the quantitative hedgefund/risk manager in order to measure, manage, control and optimize market-based VaR across different financial instruments. It can be installed in-house on servers or can be used via the BRAVO *Cognity* portal. *Cognity* measures and manages risks across the entire enterprise with very low integration, hardware and/or administration costs.

Cognity's foundation on "stable Paretian" instead of traditional "normal" simulation techniques allows users to assess and manage market risk much more precisely, accurately measuring the risk of extreme market events. The system includes the traditional models and computations to enable comparisons.

The system includes:

- Mark-to-market
- RiskMetrics, historical and Monte Carlo VaR, Stable Monte Carlo VaR, Stand-alone and Marginal VaR analyses, Shortfall (Conditional VaR) estimation, Relative VaR calculations
- What-If Trade analyses
- Advanced NPV, VaR, Marginal and Stand-alone VaR, stress testing and scenario analysis of market rates/prices, volatilities, correlations, yield curve movements, etc.
- Drill-down risk analysis
- Multiple time horizons and confidence levels
- Covariance matrix estimation
- Scenario Generation
- Flexible reporting, Export/Import features
- Portfolio Optimization (available for the second release of the system)

In this paper we will briefly describe the main features of the Cognity Credit Risk System.

2 Cognity Credit Risk System

COGNITY CreditRisk System is an integrated risk framework based on the value-at-risk concept. The methodology is applicable to all financial instruments worldwide with inherent credit risk and provides a full picture of portfolio credit risks incorporating extreme loss dependencies which can signal over concentration and indicate actions to benefit from diversification in a mark-to-market framework. The system includes all necessary components for active risk management which help determine investment decisions, actions to reduce portfolio risk, consistent risk-based credit limits and rational economic capital.

COGNITY CreditRisk System comprises two models for credit risk evaluation for complex portfolios of instruments with inherent credit risk – Asset Value Approach (Model #1) and Stochastic Default Rate Model (Model #2).

The choice of the model depends on several factors:

- *The nature of the analyzed portfolio* - Asset Value model (AVM) will work best for portfolios with a great number of market driven instruments and mainly corporate obligors. On the opposite, the accuracy of the Stochastic Default Rate Model (DRM) is not influenced by the types of obligors (corporate, retails, etc.).

- *Goals of the analysis* - assessing portfolio risk driven by changes in debt value caused by changes in obligor credit quality (including default) is achievable using the first model. The second model considers only default as a credit event.
- *Availability of the data* – the two models have significant differences in input data requirements.
- *IT Capacity* – The second model requires more computations.

2.1 Cognity CreditRisk-AVM

There are four key steps in the Monte Carlo Approach to Credit Risk Modeling in the Asset-Value Model:

Step 1. *Modelling the Dependence structure* between market risk factors and the credit risk drivers.

Step 2. *Scenario Generation* - each scenario corresponds to a possible "state of the world" at the end of our risk horizon. For our purposes, the "state of the world" is just the credit rating of each of the obligors in our portfolio and the corresponding values of the market risk factors affecting the portfolio.

Step 3. *Portfolio valuation* - for each scenario, we evaluate the portfolio to reflect the new credit ratings and the values of the market risk factors. This step offers us a large number of possible future portfolio values.

Step 4. *Summarize results* - having the scenarios generated in the previous steps, we get an estimate for the distribution of the portfolio value. One may then choose to report any number of descriptive statistics for this distribution.

We provide a schema which summarizes the calculation process and then we shall explain in detail the key steps outlined above.

The general methodology described below is valid for every Monte Carlo Approach to Credit Risk Modelling in the Asset Value Model. We describe the improvements we have introduced in the models.

Step 1. Model dependence structure between market risk factors and the credit risk drivers

The general assumption of the model is that the driver of credit events is the asset value of a company. This means one should be able to model the dependence structure between asset values of the counterparties.

As in CreditMetrics, at this point we assume that the dependence structure between asset values of two firms can be approximated by the dependence structure between the stock prices of those firms.

This fact offers a very natural solution to the problem: if we are successful in modeling dependence structure between the stock prices and all relevant market risk factors (interest rates, exchange rates, etc.), then we accomplish simultaneously two goals:

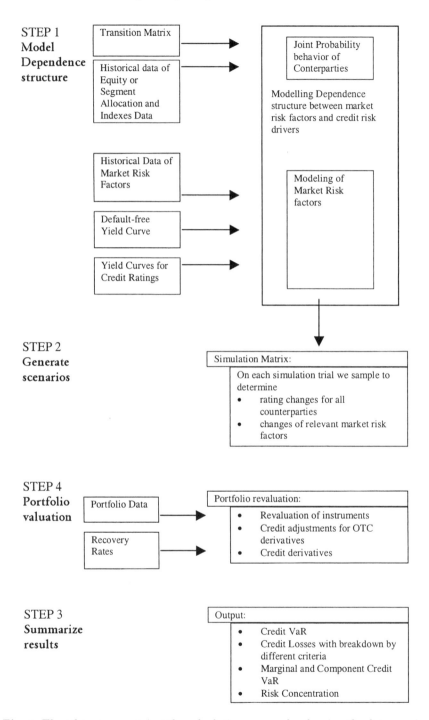

Fig. 1. The schema summarizes the calculation process by showing the data required for each step and highlighting the key components of the model

- We construct dependency between credit risk events of our obligors
- We model dependency between market risk factors and the credit risk drivers.

If one uses as a measure of dependence the correlation between risk factors (as in CreditMetrics), the above task is trivial - all one needs is to estimate the correlation matrix for the stock prices and the relevant market risk factors. This approach has certain disadvantages which will be illustrated with the help of the following example.

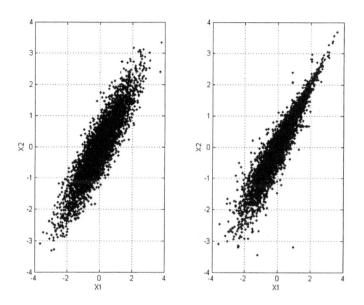

Fig. 2. 5000 simulated data from two bivariate distributions with normal marginals and identical correlation of 0.9 but different dependence structures

Figure 1 shows 5 000 bivariate random draws from two different probability models (X1, X2). The correlation between the two random variables X1 and X2 is identical in both models as well as their marginal distributions – X1 and X2 are normally distributed. Yet it is clear that the dependence structure of the two models is qualitatively different. If we interpret the random variables as financial loss, then adopting the first model could lead to underestimation of the probability of having extreme losses. On the contrary, according to the second model extreme losses have stronger tendency to occur together.

The example above motivates the idea to model the dependence structure with a method more general than the correlation approach.

The correlation is a widespread concept in modern finance and insurance and stands for a measure of dependence between random variates. However this term is very often incorrectly used to mean *any* notion of dependence. Actually correlation is one particular measure of dependence among many. Of course in the world of multivariate normal distribution and, more generally in the world of spherical and elliptical distributions, it is the accepted measure. Yet empirical research shows that real data seldom seems to have been generated from a distribution belonging to this class.

There are at least three major drawbacks of the correlation method:
Let us consider the case of two real-valued random variables X and Y.

1. The variances of X and Y must be finite or the correlation is not defined
This assumption causes problems when working with heavy-tailed data. For instance the variances of the components of a bivariate $t(n)$ distributed random vector for $n \leq 2$ are infinite, hence the correlation between them is not defined.
2. Independence of two random variables implies correlation equal to zero, the opposite, generally speaking, is not correct – zero correlation does not imply independence.
A simple example is the following: Let $X \sim N(0,1)$ and $Y = X^2$. Since the third moment of the standard normal distribution is zero, the correlation between X and Y is zero despite the fact that Y is a function of X which means that they are dependent.
Only in the case of multivariate normal distribution uncorrelattedness and independence are interchangeable notions. This statement is not valid if only the marginal distributions are normal and the joint distribution is non-normal. The example on Figure 1 illustrates this fact.
3. The correlation is not invariant under non-linear strictly increasing transformations $T : \Re \to \Re$ which is a serious disadvantage. In general $corr(T(X), T(Y)) \neq corr(X, Y)$

A more prevalent approach is to model dependency using copulas.

Let us consider a real-valued random vector $X = (X_1, \ldots, X_n)^t$ The dependence structure of the random vector is completely determined by the joint distribution function:

$$F(x_1, \ldots, x_n) = P[X_1 \leq x_1, \ldots, X_n \leq x_n].$$

It is possible to transform the distribution function and as a result to have a new function which completely describes the dependence between the components of the random vector and is not dependent on the marginal distributions. This function is called *copula*.

Suppose we transform the random vector $X = (X_1, \ldots, X_n)^t$ componentwise to have standard-uniform marginal distributions, $U(0;1)$. For simplicity we assume that $X_1, \ldots, X_n$ have continuous marginal distributions $F_1, \ldots, F_n$. The transform $T : R^n -> R^n$, is the following $(x_1, \ldots, x_n)^t -> (F_1(x_1), \ldots, F_n(x_n))^t$. The joint distribution function C of $(F_1(X_1), \ldots, F_n(X_n))^t$ is then called the copula of the random vector $(X_1,\ldots,X_n)^t$ or the multivariate distribution F. It follows that

$$F(x_1,\ldots,x_n) = P[F_1(X_1) \leq F_1(x_1),\ldots,F_n(X_n) \leq F_n(x_n)] = C(F_1(x_1),\ldots,F_n(x_n))$$

The use of copulas offers the following advantages:

- The nature of dependency that can be modelled is more general. In comparison, only linear dependence can be explained by the correlation.
- Dependence of extreme events might be modelled
- Copulas are indifferent to continuously increasing transformations (not only linear as it is true for correlations):

If $(X_1,\ldots,X_n)^t$ has copula C and $T_1,\ldots,T_n$ are increasing continuous functions, then $(T_1(X_1),\ldots,T_n(X_n))^t$ also has copula C.

This is extremely important in Asset-Value models for Credit Risk, because this property postulates that the asset values of two firms shall have exactly the same copula as the stock prices of these two companies. The latter is true if we consider the stock price of a company as a call option on its assets and if the option pricing function giving the stock price is continuously increasing with respect to the asset values.

Cognity Credit Risk Module supplies both models for describing dependence structure:

- The simplified approach using correlations like a measure for dependency
- The copula approach.

As a conclusion to this part of the discussion it is worth saying that in case there is no information about stock prices for a given obligor we employ the idea of segmentation described in CreditMetrics. The essence of this approach is that the user determines the percentage of the allocation of obligor volatility among the volatilities of certain market indexes and explains the dependence between obligors by the dependence of the market indexes that drive obligors' volatility.

Step 2. Scenario generation

In this section, we discuss how to generate scenarios of future credit ratings for the obligors in our portfolio and simultaneously for the changes of the market risk factors. Each set of future credit ratings and values of the market risk factors corresponds to a possible "state of the world" at the end of our risk horizon.

We shall rely heavily on the asset value model. The steps to scenario generation are as follows:

1. Establish asset-return thresholds for the obligors in the portfolio.
2. Generate scenarios of asset returns and the market risk factors using appropriate distribution – this is an assumption to be imposed.

3. Map the asset return scenarios to credit rating scenarios.

If we are using the multivariate normal distribution as a probability model for the log-returns of asset values and market risk factors, generating scenarios is a simple matter of generating correlated, normally distributed variables. There is a well known algorithm:

1. Compute the covariance matrix Σ from historical data
2. Use Cholesky factorization (or another method) to find the matrix A in the following decomposition of the covariance matrix: $\Sigma = AA^T$
3. To generate one scenario from $N(0, \Sigma)$, compute AY where $Y = (Y_1, Y_2, \ldots, Y_n)^T$ is a n-dimensional vector the components of which are independent normally distributed with zero mean and unit variance and the matrix A is from the previous step.

However, it is a widely accepted critique of the normal distribution that it fails to explain certain properties of financial variables - fat tails and excess kurtosis.

The family of Stable distributions that we are proposing contains as a special case the Gaussian (Normal) distribution. However, Non-Gaussian Stable models do not possess the limitations of the normal one and all share a similar feature that differentiates them from the Gaussian one – heavy probability tails. In addition they are completely described by four parameters which control, in addition to the variability and mean, the degrees of heavy tails and skewness. Thus they can model greater variety of empirical distributions including skewed ones.

A very important advantage is that stable distributions form a family that contains the normal distribution as a special case. Thus, all of the beneficial properties of the normal distribution which make it so popular within financial theory are also valid for the stable distributions as well, namely:

- The sum of independent identically distributed (iid) stable random variables is again stable. This property allows us to build portfolios.
- Stable distributions are the only distributional family that has its own domain of attraction - that is a large sum of iid random variables will have distribution that converges to a stable one. This is a unique feature, which means that if a given stock price/rate is reflected by many small shocks, then the limiting distribution of the stock price can only be stable (that is Gaussian or non-Gaussian stable).

The graphs below illustrate the density fit of a Gaussian (normal), and stable (non-Gaussian) distributions to the empirical (sample) distribution of 1 week and 1 year EURIBOR rate.

As it is demonstrated in the figures above, stable distributions provide much more realistic models for financial variables which can capture the kur-

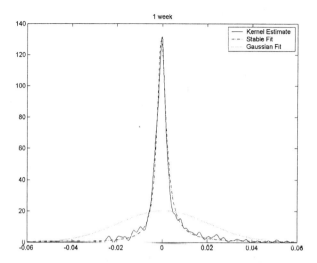

Fig. 3. Density fit of a Gaussian (normal), and stable (non-Gaussian) distributions to the empirical (sample) distribution of 1 week

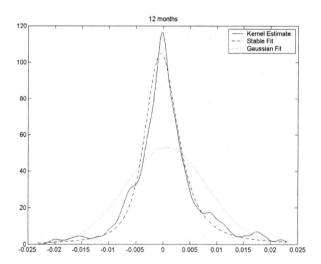

Fig. 4. Density fit of a Gaussian (normal), and stable (non-Gaussian) distributions to the empirical (sample) distribution of 1 year EURIBOR rate

tosis and the heavy-tailed nature of financial data.[4] We assume that the probability model for asset returns is the family of Stable laws.

---

[4] More information on Stable Distributions can be found in Rachev, Mittnik (2000), Stable Paretian Models in Finance.

When using stable distributions the following sub-steps are required (for full examination refer to BRAVO Risk Management Group (2000) – Stable Models in Finance):

1. Estimate the parameters of the stable distribution for each factor (asset values or market risk factor) using historical data. There are several approaches; maximum likelihood estimation provides the most accurate results.
2. This first sub-step models the marginal distributions of our risk drivers.
3. Employ the dependence structure model in Section 1 and construct a multivariate distribution the marginals of which are stable with estimated parameters from step 2.
4. Generate scenarios sampled from the multivariate probability model developed in step 3.

Once we have scenarios for the asset values, we only need to assign credit ratings for each scenario. This is done by comparing the asset value in each scenario to the rating thresholds.

Step 3. Portfolio valuation

For non-default scenarios, the portfolio valuation step consists in applying a valuation model for each particular position within the portfolio over each scenario. The yield curve corresponding to the credit rating of the obligor for this particular scenario should be used.

For default scenarios, the situation is slightly different. There are two approaches dealing with the recovery rate required for default scenarios:

- Assume constant recovery rates – then the value of a position in case of a default scenario is simply the nominal amount times recovery rate
- Allow the recovery rate to be a random variable.

As discussed in many empirical analysis recovery rates are not deterministic quantities but rather display a large amount of variation. This variation of value in the case of default is a significant contributor to risk.

Recovery rates are modelled using Beta distribution with a specified mean and standard deviation

In this case, for each default scenario for a given obligor, we should generate a random recovery rate for each particular transaction with defaulted obligor. The value of a given position in case a particular default scenario is realized will be different.

Step 4. Summary of the results

Having the portfolio value scenarios generated in the previous steps, we obtain an estimate for the distribution of the portfolio values. We may then choose to report any number of descriptive statistics for this distribution. The calculation of statistics is one and the same for both models of *Cognity*. For example, mean and standard deviation of future portfolio value can be obtained from the simulated portfolio values using sample statistics. Because of the skewed nature of the portfolio distribution, the mean and standard deviation may not be good measures of risk. Since the distribution of values is not normal, we cannot infer percentile levels from the standard deviation. Given the simulated portfolio values, we can compute better measures, for example empirical quantiles, or mean shortfall.

To this point, we have considered only statistics, which describe the portfolio distribution. We would also like to consider individual assets and to ascertain how much risk each asset contributes to the portfolio. To this end, we will describe marginal statistics.

In general, the marginal statistic for a particular asset is the difference between that statistic for the entire portfolio and that statistic for the portfolio not including the asset in question. Thus, if one wishes to compute the marginal tenth percentile of the i-th asset in our portfolio, one can take

$$q_{10}\left[(Q_1, ..., Q_i, ..., Q_n) \begin{pmatrix} I_{1,1} & ... & I_{1,j} & ... & I_{1,m} \\ ... & ... ... & ... ... \\ I_{i,1} & ... & I_{i,j} & ... & I_{i,m} \\ ... & ... ... & ... ... \\ I_{n,1} & ... & I_{n,j} & ... & I_{n,m} \end{pmatrix}\right]$$

$$-q_{10}\left[(Q_1, ..., 0, ...Q_n) \begin{pmatrix} I_{1,1} & ... & I_{1,j} & ... & I_{1,m} \\ ... & ... ... & ... ... \\ I_{i,1} & ... & I_{i,j} & ... & I_{i,m} \\ ... & ... ... & ... ... \\ I_{n,1} & ... & I_{n,j} & ... & I_{n,m} \end{pmatrix}\right]$$

where Q_i is the amount of i-th position, $I_{i,j}$ is the j-the simulated value for the i-th instrument and q_{10} represents the tenth percentile of the values in question.

This marginal figure may be interpreted as the amount by which we could decrease the risk of our portfolio by removing the i-th position.

2.2 Cognity CreditRisk-DRM

There are five key steps in the Monte Carlo Approach to Credit Risk Modeling based on Stochastic Modeling of Default Rate:

Step 1. Build econometric models for default rates and for the explanatory variables. An econometric model is evaluated for the default probability of a segment based on explanatory variables (macro-factors, indexes, etc.) using historical data for default frequencies in a given segment and historical time series for the explanatory variables.

Step 2. Generate scenarios. Each scenario corresponds to a possible "state of the world" at the end of our risk horizon. Here, the "state of the world" is a set of values for the market variables and for the explanatory variable defined in (1)

Step 3. Estimate default probabilities under each scenario for each of the segments using the scenario values of the explanatory variables and the model estimated in (1). *Simulate sub-scenarios for the status of each obligor* (default/non-default) based on the estimated default probability.

Step 4. Portfolio valuation. For each scenario, revalue the portfolio to reflect the new credit status of the obligor and the values of the market risk factors. This step generates a large number of possible future portfolio values.

Step 5. Summarize results. Having the scenarios generated in the previous steps, we possess an estimate for the distribution of portfolio values. We may then choose to report any descriptive statistics for this distribution.

We provide a schema describing the calculation process and then we continue with a detailed description of the key steps outlined above.

Step 1. Build the econometric models

This first step is in fact the most challenging and critical task of the model. Two crucial models should be defined and estimated:

- *An econometric model for default probability* of a segment based on explanatory variables like macro-factors, indexes, etc.
- *Time series model for explanatory variables.*

Default probability models should be evaluated for each user-defined segment. The segment definitions can be flexible enough based on the following criteria

- credit rating
- industry
- region
- size of the company, provided the time series of default rates are available for each of the segments.

The explanatory variables that might be suitable to represent the systematic risk of the default rates in the chosen country-industry-segments depend on the nature of the portfolio and might be

- industry indices
- macro variables (GDP, unemployment)
- long-term interest rates, exchange rates, etc.

The first task is to define a model for default probability of a segment based on explanatory variables (macro-factors, indexes, etc.) using historical data for default frequencies in a given segment and historical time series for

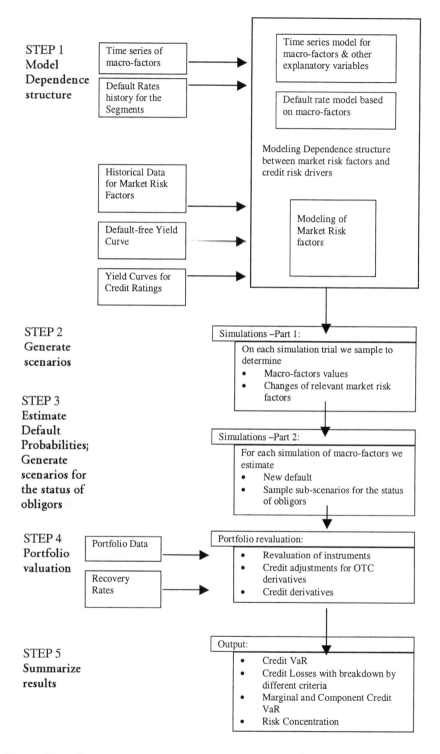

Fig. 5. The schema summarizes the process showing the data required for each step and highlighting the key components of the model

the explanatory variables. In other words we choose a function f and estimate its parameters such that

$$DF_{s,t} = f(X_{1,t}, \ldots X_{N,t}) + u_t \qquad (1)$$

where $DF_{s,t}$ is the default frequency in the segment s for the time period t, $X_{i,t}$ is the value of the i-th explanatory variable at time t, $i = 1 \ldots N$

The second model is a time-series model for explanatory variables. The usual way to model dependent variables (as also suggested in CreditPortfolio-View^{TM}) is to employ some kind of ARMA (p,q) model. That is the same as assuming that

$$X_t = a_0 + \sum_{i=1}^{p} a_i X_{t-i} + \sum_{j=1}^{q} b_j e_{t-i} + e_t, \text{ where } e_t \text{ is } N(0, \sigma^2), \qquad (2)$$

It is important to note that the proper modelling of default rate will depend very much on the proper modelling of dependent variables: From (1) and (2) we can write

$$DF_{s,t} = f\left(\sum_{i=1}^{p} a_{1,i} X_{1,t-i} + \sum_{j=1}^{q} b_{1,j} e_{1,t-i} + e_{1,t}, \ldots, \sum_{i=1}^{p} a_{n,i} X_{n,t-i}\right.$$

$$\left. + \sum_{j=1}^{q} b_{n,j} e_{n,t-i} + e_{n,t}\right) + u_t = g\left(X_{1,t-1}, \ldots, X_{1,t-p_1}, \ldots, X_{n,t-1}, \ldots, X_{n,t-p_n},\right.$$

$$\left. e_{1,t}, \ldots e_{1,t-q_1}, \ldots, e_{n,t}, \ldots e_{n,t-q_n}\right) + u_t \qquad (3)$$

The following charts show the US unemployment levels for the period 1961-1995, the differenced series of the unemployment levels, the standardized differences and finally – the empirical density of the residuals after applying ARMA model together with the fitted normal distribution.

We can observe that the real distribution of residuals deviates form the assumption of the model – residuals are not normal. They are

- skewed
- with fatter tails
- with higher peak around the centre of the distribution
- and there is a volatility clustering (see chart 3)

Thus the improper use of normal residuals in (3) will end up with "incorrect" scenarios (simulations) for the possible default rates.

For the modelling of macro-factors, we propose the following more general model – stable Vector AR(1)_ARCH – type model.

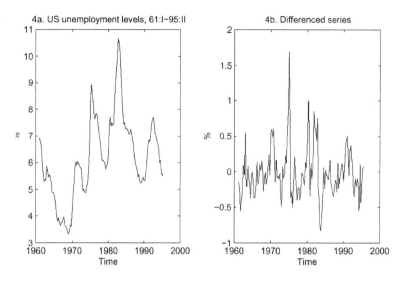

Fig. 6. US unemployment levels for the period 1961-1995 and the differenced series of the unemployment levels

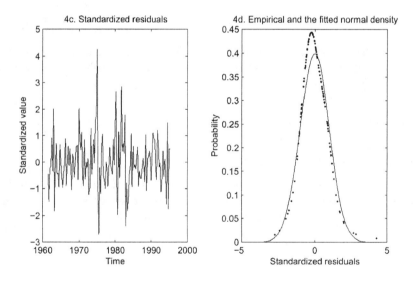

Fig. 7. Standardized differences and the empirical density of the residuals after applying ARMA model together with the fitted normal distribution

Under this model

$X_t = A_1 X_{t-1} + E_t$, where $X_t = (X_1, ..., X_n)'$ is the vector of explanatory variables, A_1 is n by n-matrix, $E_t = (e_{1,t}, .., e_{n,t})'$ is the vector of residuals which are modelled by multivariate stable ARCH model.

The following two charts show the difference between modeling residuals using the traditional normal distribution and using the proposed family of stable non-Gaussian distributions on different macro-series.

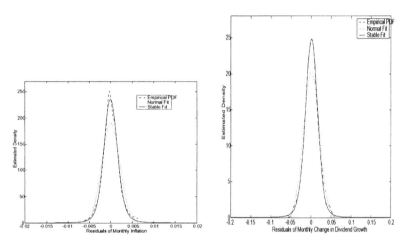

Fig. 8. Difference between modeling residuals using the traditional normal distribution and using the proposed family of stable non-Gaussian distributions on different macro-series

Employing stable residuals results in:
- fatter tails of the residuals
- higher variability of default rates

The ARCH component of the model takes care of volatility clustering. Moreover, since the model is a vector autoregressive model we will eventually succeed in modelling *joint* behavior of macro-factors.

Note 1: Relevant market variables (like interest rates) are also included in the model.

Step 2. Generation of the scenarios

Each scenario corresponds to a possible "state of the world" at the end of our risk horizon. Here, the "state of the world" is a set of values for the market variables and for the explanatory variable defined in Step 1.

The scenarios are simulated according to the vector-autoregressive model discussed in Step 1. The important feature of the model is that the macro-factors and market risk factors are modelled by a *joint* probability distribution

(See note 1 above). The latter means that the realizations of the market risk factors depend on the realizations of the macro-factors.

Step 3. Estimation of the default probabilities

In the third step, the default probabilities are estimated under each scenario for each of the segments using the simulated values of the explanatory variables and the model estimated in Step 1. Then we simulate sub-scenarios for the status of each obligor (default/non-default) based on the estimated default probability.

For each scenario set j in Step 2 we estimate the probability of default for each segment based on the model estimated in Step 1

$$P_{s,j} = f_s(X_{1,j}, \ldots X_{N,j}) + u_j \qquad (4)$$

where $P_{s,j}$ is the default probability for the segment s under the j-th simulation.

Now, for each scenario j, we generate independent sub-scenarios for each counterparty state (default or no-default) based on its probability of default $P_{s,j}$.

$$XY_1, \ldots\ldots\ldots\ldots, XY_j, \ldots\ldots\ldots\ldots, XY_{SimCount}$$

Counterparty 1 $\quad I_{1,1}^j \ldots\ldots I_{1,K}^j$

$\ldots\ldots\ldots\ldots\ldots$

Counterparty L $\quad I_{L,1}^j \ldots\ldots I_{L,K}^j$

where $I_{L,k}^j$ is the state of the counterparty L under the j-th scenarios set for the k-th sub-scenario and

$$I_{L,k}^j = \begin{cases} 1 = \text{default} \\ 0 = \text{no default, } s \end{cases}$$

$E(I_{L,k}^j) = P_{s,j} = f(X_{1,j}, \ldots X_{N,j})$, where s is the segment of the L-th counterparty.

At this point, we have obtained the full set of scenarios describing the possible "states of the world" in terms of market risk factor values and obligors' statutes (Default or Non-default).

Step 4. Portfolio valuation

For each scenario, we revalue the portfolio to reflect the new credit status of the obligor and the values of the market risk factors. The result is a large number of possible future portfolio values.

This step is similar to the corresponding step in the AVM described in the previous section. The only difference is that in the current model there is a simplification – it considers only default or non-default status for an obligor.

Step 5. Summary of the results

At this point, we have created a number of possible future portfolio values. The final task is then to synthesize this information into meaningful risk estimates. We can use any of the descriptive statistics from the previous section.

3 Conclusion

As a conclusion we remark that our approach is capable of measuring credit risks in all kinds of exposure including market driven instruments like swaps, forwards, etc. This is only because we propose an *integrated* market and credit risk framework.

Only an integrated model could describe, for example, both the correlations of swap exposures across a portfolio (capturing, for instance, that swaps based on the same interest rate would tend to go in- or out-of-the-money together), and the correlations between credit and market moves (for instance, that swap counterparties might be more likely to default in one interest regime than in another). That makes *Cognity* Credit Risk a unique product for risk management of large variety of exposures including off-balance-sheet items.

The purpose of each risk management system is to provide effective techniques that simultaneously comprise protection against market crashes, issue early warning signals, and serve as an objective advisor to investment decision-making. Credit Risk Management itself demands various complex problems to be resolved. We believe that with *Cognity CreditRisk Software* we offer financial industry the first new generation product that will establish new standards in the qualitative and quantitative risk management.

References

1. S.T. Rachev, J-R.Kim, and S.Mittnik, Stable Paretian Models in Econometrics: Part 1, Mathematical Scientists, 1999, 24, 24 - 55.
2. S. T. Rachev, J-R.Kim, and S. Mittnik, Stable Paretian Models in Econometrics: Part 2, Mathematical Scientists, 1999, 24, 113 - 127.
3. S. T. Rachev, S. Han, Portofolio management with stable distributions, Mathematical Methods of Operations Research, 2000, 51, 341 - 352.
4. G. Götzenberger, S. T. Rachev, and E. Schwartz, Performance Measurements: The Stable Paretian Approach Applied Mathematics Reviews, Vol. 1, World Scientific Publishing, 2000, 329 - 406.

5. S. T. Rachev, Y. Tokat, Asset and Liability Management: Recent Advances, Handbook of Analytic-Computational Methods in Applied Mathematics, 2000, 859 - 908.
6. S. T. Rachev, G. Samorodnitsky, Long strange segments in a long range dependent moving average, Stochastic Processes and Their Applications, 2001, 93, 119-148.
7. I. Khindanova, S. T. Rachev, and E. Schwartz, Stable Modeling of Value at Risk, Mathematical and Computer Modelling, 2001,34,1223-1258
8. P. Mansfield, S. T. Rachev, and G. Samorodnitsky, Long Strange Segments of a Stochastic Process, Annals of Applied Probability, 2001, 11, 878-921.
9. S.Mittnik, S. T. Rachev, and E.Schwartz, Value-at-risk and Asset Allocation with Stable Return Distributions, The German Statistical Review (Allgemeines Statistisches Archiv - ASTA), 86, 1-15, 2002, special issue "Statistical and Econometric Risk Analysis of Finance Markets"
10. S. Mittnik, M.S. Paolella, and S. T. Rachev, Stationarity of Stable Power-GARCH Processes Journal of Econometrics, 106, 97-107, 2002.
11. Y. Tokat, S. T. Rachev, and E. Schwartz, The Stable non-Gaussian Asset Allocation: A Comparison with the Classical Gaussian Approach to appear in Journal of Economic Dynamics and Control, 2002.
12. S. Ortobelli, S. T. Rachev, E. Schwartz, and I. Huber, Portfolio Choice Theory with non-Gaussian Distributed Returns to appear in Handbook of Heavy Tailed Distributions in Finance, North Holland Handbooks of Finance (Series Editor W. T. Ziemba), 2001.
13. S. T. Rachev, W. Römisch, Quantitative stability in stochastic programming: The method of probability metrics to appear in Mathematics of Operations Research, 2002.
14. S. T. Rachev, E. Schwartz, and I. Khindanova, Stable Modeling of Market and Credit Value at Risk to appear in the Handbook of Heavy Tailed Distributions in Finance, North Holland Handbooks of Finance (Series Editor W. T. Ziemba), 2002.
15. S. T. Rachev, S.Mittnik, Stable Paretian Models in Finance, Series in Financial Economics and Quantitative Analysis, Wiley & Sons, 2000.

An Application of the CreditRisk$^+$ Model

Thomas Rempel-Oberem, Rainer Klingeler, and Peter Martin

ifb AG, Neumarkt, Germany

Summary. The present paper discusses a novel approach to the determination of default events, which has been developed for two different credit institutions. It is based on the CreditRisk$^+$ model, however, in a simulation environment. In order to meet the particular requirements and standards of the respective credit institutions the original model had to be modified and extended. It is suited for evaluating the risk structure of two portfolios with middle-class obligors and premium creditworthiness obligors, respectively.

1 Introduction

Two different basic approaches are currently used for the description of credit risk and its management: analytical models and simulation approaches, each of them representing a series of specific advantages and disadvantages. In every day practice the most striking difference between the two approaches is the calculation speed when evaluating large portfolios. On the one hand, analytical models offer the possibility of gaining insight into the risk structure of very large and complex portfolios within a pretty short period of time. Thus, the impact of many possible steering directives can be simulated, and sound proposals for future actions are submitted. However, some assumptions have to be made prior to the valuation, e.g. about the character of the loss distribution. On the other hand, simulation approaches sacrifice performance for the sake of significance, as so-called "long-tail" confidence levels require large computational efforts. Thus, the development of new and computationally more efficient frameworks is subject to leading edge research.

The results from two projects that aim at the introduction of a CreditRisk$^+$-based portfolio model in two different credit institutions will be presented in this paper. When developing portfolio models on behalf of a client instead in the context of basic research it is most important to develop very individual concepts. The intention is to reveal specific solutions suited best to meet the particular requirements of the respective credit institution.

As a first step, the risk contribution of an individual obligor to the entire portfolio has to be monitored in order to derive efficient steering directives. The main risk-driving factors have to be identified. The portfolio model is supposed to launch effective steering mechanisms for the total risk. The second step is to realize the directives by trading credit risk in order to reorganize and hedge the portfolio. The latter topic will be briefly commented on in the final section. In particular, the impact of selected steering actions on the overall risk situation has to be simulated.

Another building block in the risk-management framework is the pricing of default. The model presented here does not take account of option theoretic models, but rather employs intensity models. The arrival of default is described in terms of a rate equation which might involve quite a number of parameters. The present discussion is limited to an inhomogeneous model. A stochastic model is calibrated using market data, e.g. bond prices and empirical credit quality transition histories. The discussion addresses the probability of events rather than their impact.

The paper is organized as follows. In Section 2 the basic concepts of CreditRisk$^+$ will be briefly introduced, however, in the framework of a simulation approach. The default of an obligor is considered to be an unpredictable event. A more detailed view of the simulation will be given in Section 3. Section 4 reveals a validation of the input parameters, and careful attention is paid to the default probability and its volatility. In Section 5 the model will be extended to an intensity-based model, and variations of the default risk are explicitly taken into account. In Section 6 the simulation of credit quality migration will be briefly commented on.

2 CreditRisk$^+$ in a Simulation Approach

The present paper demonstrates how the basic concepts of the CreditRisk$^+$ model are realized in a simulation approach. This approach is in contrast to the original approach of Credit Suisse First Boston which is an analytical one. The original model adopts Poisson-distributed default events that are extended to loss events. Moreover, the volatility of the mean value of the default rate is taken into account via convolution using a gamma distribution.

On the one hand, the simulation approach suffers from inferior performance with respect to the analytical approach. Performance is sacrificed for the sake of an explicit consideration of correlations between the respective risk factors. On the other hand, the implementation of modifications and extensions to the model is straightforward when using the simulation approach. In particular the latter advantage has a couple of charming aspects. The basics of this approach are depicted in Figure 1. Careful attention has to be paid to the fact that the default event itself is modeled independently for each obligor.

The state of each obligor in the portfolio is monitored at the beginning of the planning period and by the end of this period, namely at the risk

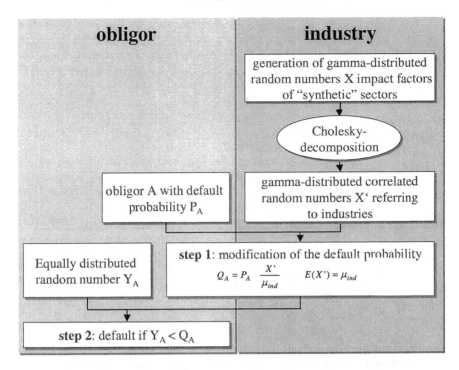

Fig. 1. Scheme of the simulation approach. To the left, obligor related parameters are presented. To the right, industry related parameters are shown. The default is modeled independently for each obligor (see step 2).

horizon. This is usually one year. According to the default probability of the obligors two states are possible by the end of the year: default and survival. This implies that the default time is discrete, and default can only occur in one-year steps.

The question whether the obligor defaults or not in one particular simulation run is answered in two steps. The first step is the modification of the default probability of each obligor. Starting from the default probability at the beginning of the planning period, the individual default probability of one particular obligor is modified according to a gamma-distributed random number. This gives the obligors default probability at the risk horizon. When modeling default rates the use of normal-distributed random numbers is second only to gamma-distributed random numbers since the latter distribution does not have any density at values below zero. When using the normal distribution, non-zero density at negative default rates might occur. In order to avoid this non-plausible side effect, negative values of the default probability have to be mapped to zero or to positive values. However, mapping to zero creates cusps in the density, which in an asset value approach causes unlikely transitions of low-rated obligors to premium creditworthiness.

In a second step, an equally distributed random number determines whether the obligor defaults or not, simply by comparison with the obligors individual default probability at the risk horizon. This mechanism is also shown in Figure 2. It is worth emphasizing that the default events occur independently from each other.

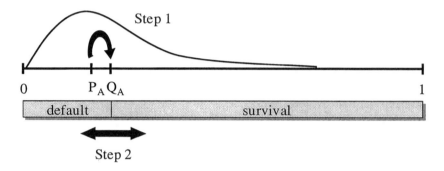

Fig. 2. Modeling the default rate and the default event. Step 1: A gamma-distributed random number modifies the default probability P_A of obligor A (distribution assumption according to CreditRisk$^+$). Step 2: An equally distributed random number determines whether the obligor A defaults or not.

The correlations between individual obligors result from their affiliation to a specific sector. The variation of the default probability of obligors belonging to the same sector is perfectly positively correlated with a correlation coefficient equal to one, whereas the correlation for obligors belonging to different sectors might be different. This is depicted in Figure 3.

The classification of obligors according to industries rather than to independent sectors is considered to be capable of delivering superior results. In commercial practice the issuing of credits is heavily based upon industry-related rating. It is very common to allow for affiliation to more than one industry, for example according to manufacturing and sales. In order to obtain precise numbers needed for the model historical time series of the relevant industries are taken into consideration. In particular, historical default events are evaluated and industrial default probabilities are deduced. Moreover, volatilities of default probabilities and correlations between different industries can be obtained. These numbers define the variance-covariance matrix.

As mentioned above, the simulation approach offers the possibility of implementing correlations between obligors explicitly. Via multiplication with the Cholesky matrix, uncorrelated and independent random numbers can be

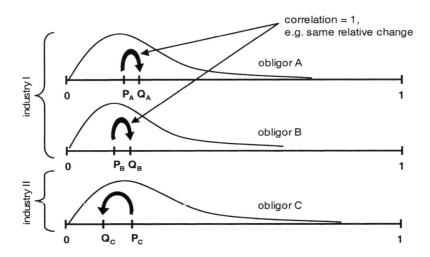

Fig. 3. The modification of the default rates of obligors belonging to the same industry (A and B) is correlated perfectly. However, default of individual obligors occurs independently (see Figure 2 step 2).

transformed into correlated ones that meet the variance and correlations postulated by the variance-covariance matrix. While the uncorrelated random numbers can be interpreted in terms of impact factors of synthetic and independent sectors according to the CreditRisk+ model, the correlated random numbers obtained via Cholesky decomposition refer to industries.

The default probability P_A of each obligor A is determined by the rating. For each obligor it is assumed that a linear dependence between the individual volatility of the default probability σ_A and the volatility of the default probability σ_{ind} of the industry the obligor is part of is given. μ_{ind} is the mean default rate of the respective industry. Moreover, the correlation of two obligors refers to the correlation of the respective industries.

$$\sigma_A = P_A \cdot \frac{\sigma_{ind}}{\mu_{ind}}$$

3 The Simulation Step

The number of simulations that have to be run has to be considered carefully. On the one hand, depending on the level of confidence required, the number of simulations must not fall below a certain number in order not to fail statistical significance. For example, when observing the evolution of the portfolio at a confidence level of 0.99, the run of 1000 simulations allows for the neglect of the ten worst cases, e.g. ten values represent the lower limit of the confidence interval and determine the probabilistic prediction. The 11th worst scenario is the first one not to be neglected, and it is the one that determines the Value At Risk of the portfolio. One might argue about the number of scenarios that might fall below the confidence level. However, around ten the limiting case is at hand. Correspondingly, higher levels of confidence require more simulations in order not to fall short of the statistical significance requirements. On the other hand, simulations are very time consuming.

The first step in each simulation step is to create a gamma-distributed random number $\in\ ]0,1[$ for each industry. Using the Cholesky matrix, this vector of uncorrelated numbers is transformed into a vector of correlated numbers $\mathbf{Y}$. These numbers, together with the obligors individual default probability P_A according to the rating, are crucial for the default rate P at the risk horizon during the current simulation step. For each obligor a straight proportional dependence between the individual default rate at the risk horizon and the individual default probability P_A according to the rating is assumed. The proportionality constant is the random number of the obligors industry, scaled to the total default rate in this industry P_{ind}.

$$P = P_A \cdot \frac{Y_{ind}}{P_{ind}}$$

For each of the N obligors in the portfolio, the default event itself is determined by generating a set of N equally distributed random numbers, one for each obligor, and comparing it to the obligors individual default rate at the risk horizon. The loss due to default is accumulated for each simulation step, and the loss distribution at the respective loss is increased by 1 / number of simulations.

However, the application of this scheme to different example portfolios raises the suspicion that a most effective portfolio control must incorporate effects beyond default event driven credit risk. For example, a more sophisticated approach would be to consider the individual default probability as a function of time, e.g. to take into account upgrades and downgrades. While identifying downgrades with loss and upgrades with profit a most differentiated view of the portfolio is possible. Thus, not only the risk of default itself has to be taken into account, but also the risk arising from changes in the creditworthiness of an obligor. This reflects the fact that junk bonds inhere less value than premium creditworthiness bonds, even if they do not default. This topic will be commented on in Section 5.

4 Validation of the Input Parameters

Portfolio models are usually based on the individual default probability of obligors, e.g. according to external ratings. A very attractive and elegant way of deducing credit quality is to evaluate historical data of bond prices [Li; Duffie and Singleton (1997)]. A very basic approach to this topic is discussed here.

The first step to deduce default probabilities from historical bond price data is to identify today's bond price V(t₀) with the sum of all future cash flows CF(t$_i$) (i = 1, ...,n) coming from the bond that have been discounted using risk charged discount factors. The risk charged discount factor is a product of the risk free discount factor $e^{-r(t) \cdot t}$ and the probability S(t) that the obligor does not default before the cash flow is done. The survival rate S(t) is solution to the rate equation under the assumption that for all times the ratio between the number of surviving obligors and the number of defaulting obligors is a constant which is called the hazard rate. It reflects the concept of default probability within the next infinitesimal time step dt. In particular, the hazard rate does not depend on time:

$$\frac{dS(t)}{dt} = -\lambda \cdot S(t)$$
$$S(t) = \exp(-\lambda \cdot t)$$

The case of a constant hazard rate has been considered by Jarrow and Turnbull, who followed classical arguments from the Poisson process. The current price of the bond is obtained:

$$V(t) = \sum_{i=1}^{n} CF(t_i) \cdot \exp\left[-\left[r(t_i) + \lambda\right] \cdot t_i\right]$$

This equation is used to deduce the default intensity λ and the default rate $1 - S(t)$. The risk-charged discount factor is adjusted such that $V(t)$ reflects the defaultable exposure. Then a defaultable bond can be regarded as risk-free. The concept of recovery has been introduced by Duffie and Singleton as a proportion of the predefault value of the defaultable bond. The application of the scheme yields reasonable results, as explicitly shown in the following example.

Daily values of a certain bond in the course of one year have been examined. According to the external rating a3 the obligors probability of default is 0.108%. The current evaluation using the constant default rate model (recovery rate equal to zero) gives a probability of default of 0.072% along with a 0.059% standard deviation.

5 Intensity Model

In the following section an intensity model is motivated using the default intensity as explained in the preceding section (see e.g. Madan; Duffie and

Gârleanu; Jarrow, Lando, and Yu). In particular, the default intensity is modeled using a stochastic process. For example, the comparison of two empirical two-year transition matrices, one as direct observation, the other as product of one-year matrices, will yield a deviation between them. Thus, when calculating transition probabilities, inhomogenity has to be taken into account. The concept of hazard rate of default has been extended to time-dependent hazard rates and stochastic processes by Madan and Unal by adapting it to Brownian filtration.

$$\frac{dS(t)}{dt} = -\lambda(t) \cdot S(t)$$

$$S(t) = \exp\left[-\int_0^t \lambda(s) \cdot ds\right]$$

The question is how to model and how to calibrate the default intensity. A rather common way of doing this is to adopt the so-called jump-diffusion process (Duffie and Singleton (1999)). It is the synthesis of three components that jointly determine the fate of the default intensity during each next time step dt:

$$d\lambda = \kappa(\theta - \lambda(t))dt + \sigma dW(t) + \Delta J(t)$$

The first term is a mean reverting term similar to those used in interest rate modeling (James and Webber (2000) and Rebonato (1996)). Its significance is that the farther away the default intensity is from a certain constant mean value θ, the faster it is bent backwards to the mean value again. The strength of the restoring force is described by κ, the reversion rate. The second term represents the standard diffusion term also known as Brownian motion. The volatility of the motion is described by σ. These two terms are influenced by the obligors industry and affect each obligor from the same industrial branch in the same manner, whereas the third term reflects the individual fate of the obligor.

This term represents the jump process. The point of time at which an obligor is struck by some event in a way that his individual default probability changes to high values instantaneously is modeled by a Poisson distribution, and the height of the jump is exponentially distributed.

The current state of the simulation model is depicted in Figure 4. The modification of the default rate depends on the jump process and the rating as well as on the industrial random number.

6 Credit Quality Migration

As already mentioned in Section three, the consideration of the individual default probability as a function of time might be a crucial extension to the

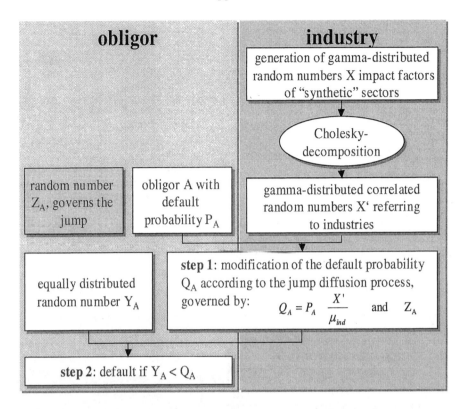

Fig. 4. Scheme of the simulation approach. To the left, obligor related parameters are presented. To the right, industry related parameters are shown. The modification of the default rate depends on the jump process and the rating as well as on the industrial random number.

default-based model. Thus, a third simulation step is appended to the default model discussed in the previous sections. Even if an obligor does not default, a modified rating class due to a modified default probability is assigned to the obligor. This modification can be deduced by comparison between the default probability that emerges from the jump diffusion process and the transition matrix. Thus, the obligor might join a different rating class at the risk horizon. As a result of the model, a migration matrix for transitions between rating classes can be deduced. This simulated migration matrix can be compared to empirical ones.

Now the calibration stage has to be entered. The model parameters κ, θ, σ and J of the jump-diffusion process have to be varied simultaneously in order to match the empirical migration matrix as best as possible. The question of what "as best as possible" means is not addressed here. Nevertheless, as a result from the calibration process, one or several sets of parameters are available and can be used to obtain the distribution of default probabilities as

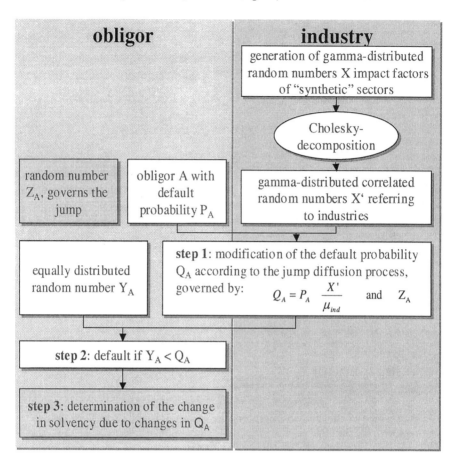

Fig. 5. Scheme of the simulation approach. To the left, obligor related parameters are presented. To the right, industry related parameters are shown. The modification of the default rate depends on the jump process and the rating as well as on the industrial random number. Moreover, in case of survival the credit quality at the risk horizon is taken into account.

a projection of the jump-diffusion process to the risk horizon. For example, the parameters can be chosen such that each jump is devastating no matter what rating the respective obligor has. On the other hand, the parameters might be chosen in a way that premium bonds recover, whereas junk bonds default. The model explicitly allows for multiple jumps and correlated processes.

7 Summary and Outlook

CreditRisk+ represents both a straightforward and convincing approach to model the credit risk of a portfolio. The implementation of the concept within a simulation framework offers many possibilities for modifications. Thus, in the context of specific credit institutions adequate solutions show up. The quality of the model and of possible modifications will become obvious when verified on the basis of empirical data.

References

1. Jessica James and Nick Webber, Financial Engineering: Interest Rate Modelling, John Wiley & Sons, Ltd., Chichester, England (2000)
2. Riccardo Rebonato, Financial Engineering: Interest-Rate Option Models, John Wiley & Sons, Ltd., Chichester, England (1996)
3. Darrell Duffie and Kenneth Singleton, Modeling Term Structure of Defaultable Bonds, Review of Financial Studies 12, 687-720 (1999)
4. D. Madan and H. Unal, Pricing the Risks of Default, Review of Derivatives Research 2/3, 121-160 (1998)
5. D. Madan and H. Unal, A Two-Factor Hazard Rate Model for Pricing Risky Debt and the Term Structure oof Credit Spreads, Journal of Financial and Quantitative Analysis (2000)
6. R. Jarrow and S. Turnbull, Pricing Derivatives on Financial Securities Subject to credit Risk, Journal of Finance, 53-85 (1995)
7. Darrell Duffie and Kenneth Singleton, Simulating Correlated Defaults, Working Paper, Graduate School of Business, Stanford University (1998)
8. David Li, Constructing a Credit Curve, Credit Risk – Models and Management, D. Shimko (Ed.), Risk Books (1999)
9. Dilip B. Madan, Pricing the Risks of Default, Working Paper, Robert H. Smith School of Business, University of Maryland (1996)
10. Darrell Duffie and Nicolae Gârleanu, Risk and Valuation of Collateralized Debt Obligations, Working Paper, Graduate School of Business, Stanford University (1999)
11. Robert A. Jarrow, David Lando, and Fan Yu, Default Risk and Diversification: Theory and Applications (2001)
12. CreditRisk$^+$, Credit Suisse First Boston – Financial Products, London, Technical Document

Internal Ratings for Corporate Clients *

Ingo Schäl

zeb/rolfes.schierenbeck.associates, Münster, Germany

1 Purpose and Principles of Building Rating Systems

1.1 Rating Systems as Crucial Controlling Instruments for Credit Risk Management

In Germany credit transactions with corporate clients suffer from significant credit defaults. The number of insolvencies in Germany are steadily increasing and have more than doubled between 1993 and 2001 (see Figure 1). In 2002 financial institutions are supposed to be confronted with approximately 40.000 insolvencies in the corporate sector.[2]

Against the background of increasing insolvencies measurement, pricing and management of credit risk become more and more important. Sophisticated models support the process of credit risk controlling: rating systems focus on a single obligor and measure the probability of an obligor going into default. The expected default frequency is the basis for pricing credit risk. The pricing of credit risk focuses on a single credit. The expected default frequency, the expected exposure at default and the expected recovery rate are the key parameters for the calculation of risk differentiated premiums. The resulting risk premium (expected loss) is part of a preliminary costing and a key parameter in a credit portfolio model. A portfolio model focuses on a portfolio of

---

* The author would like to thank his colleagues at zeb/rolfes.schierenbeck.associates, in particular Torsten Pyttlik and Christian Helwig, for their support and helpful comments for the issue of this paper.
[2] See Rinker, A.; "Auswirkungen von Basel II auf Kreditinstitute und Mittelstand"; Statement; press-conference of zeb/rolfes.schierenbeck.associates; Frankfurt a.M. February 19th 2002; Rinker, A./Jansen, S.; "Basel II als Chance begreifen"; press release of zeb/rolfes.schierenbeck.associates; Münster/Frankfurt a.M. February 19th 2002; Kirmße, S.; "Die Bepreisung und Steuerung von Ausfallrisiken im Firmenkundengeschäft der Kreditinstitute - Ein optionspreistheoretischer Ansatz"; Frankfurt a. M. 1996; p. 1-7

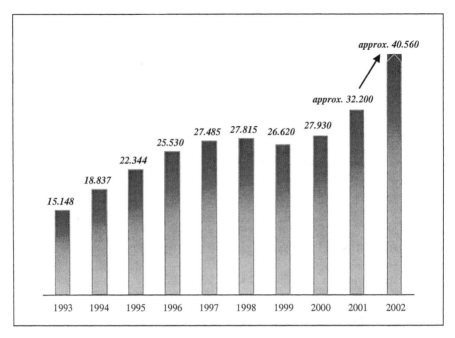

Fig. 1. Insolvencies of corporates in Germany (Statistisches Bundesamt; press release; Wiesbaden 5th April 2002; Statistisches Bundesamt; "Statistisches Jahrbuch" of several volumes; Stuttgart; Verband der Vereine Creditreform e.V.; "Insolvenzflut bei Unternehmen und Verbrauchern hält an"; press release; Neuss June 19th 2002)

credits and takes correlations between defaults of obligors into account. Given the distribution of obligors over rating classes and the expected loss resulting from the risk-premium-calculation and the unexpected loss (credit-value-at-risk) resulting from a portfolio model, a financial institution can perform an active credit portfolio management.[3]

Figure 2 illustrates the three levels of credit risk controlling.

This paper focuses on rating systems for corporate clients. As described in Figure 2 rating systems are the basis for credit risk management. Rating systems have advanced a lot in recent years and are frequently used by financial institutions to measure credit risk.

Since external ratings only exist for a small range of obligors, mainly large and quoted companies, financial institutions develop internal rating systems for obligors with no available external ratings. The new Basel Capital Accord stimulates the development of internal rating systems since financial institu-

---

[3] See Jansen, S.; "Bankinterne Ratingansätze im Firmenkundengeschäft", in: Schierenbeck, H./Rolfes, B. (editor) "Ausfallrisiken – Quantifizierung, Bepreisung und Steuerung"; Frankfurt am Main 2001; p. 97; Schäl, I.; "Kreditderivate"; Ulm 1999; p. 103-108; Schierenbeck, H.; "Ertragsorientiertes Bankmanagement (1) "; 7th edition; Wiesbaden 2001; p. 293 ff

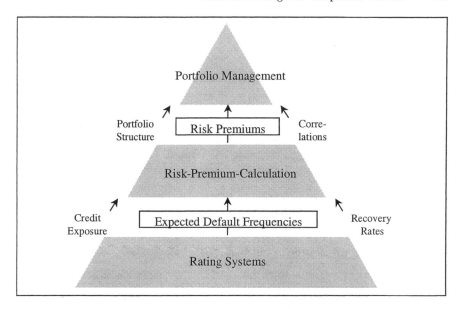

Fig. 2. Levels of credit risk controlling (According to Jansen, S.; "Bankinterne Ratingansätze im Firmenkundengeschäft", in: Schierenbeck, H./Rolfes, B. (editor); "Ausfallrisiken – Quantifizierung, Bepreisung und Steuerung"; Frankfurt am Main 2001; p. 97)

tions are allowed to use their internal ratings for the calculation of regulatory capital within the internal ratings based approach.[4]

In this paper the author focuses on his own practical experience with building rating systems. In Chapter 1.2 the main principles for building rating systems are introduced. After defining an "obligor in default" (Chapter 1.3) elements of rating systems for corporate clients are discussed in further detail (Chapter 2). Statistical methods for the evaluation of financial ratios will be the main focus of this chapter. Chapter 3 covers the quality of rating systems and calibration. In Chapter 3.1 the author describes how the quality of rating systems can be measured. Since the determination of expected default frequencies (EDFs) is crucial for pricing credit risk, the calibration on EDFs

---

[4] See Bank for International Settlements; "The New Basel Capital Accord"; Consultative Document issued by the Basel Committee on Banking Supervision; Basel January 2001; p. 32 ff; Jansen, S.; "Auswirkungen von Basel II auf Kreditinstitute und Mittelstand"; Statement; press-conference of zeb/rolfes.schierenbeck.associates; Frankfurt a.M. February 19th 2002; Rolfes, B./Emse, C.; "Rating-basierte Ansätze zur Bemessung der Eigenkapitalunterlegung von Kreditrisiken"; Forschungsbericht des European Center for Financial Services; August 2000; p. 4-16

is mentioned in Chapter 3.2. The paper concludes with basic requirements for a backtesting procedure (3.3).

1.2 The 10 Principles of Building Rating Systems

When building and using sophisticated rating systems certain rules have to be observed. The following principles emerged to be generally accepted and go along with economic and regulatory requirements. Although other catalogues of principles, which might be more or less related, can be found in the literature[5], the author believes that the following ten principles are of particular relevance.

Principle 1: A rating system must refer to well-defined customer segments and has to take all specific criteria into account.

Principle 2: Rating systems must be objective and consistent.

Principle 3: Rating systems must have a sufficient separating power.

Principle 4: The development of rating systems has to rely on a historic data base, which has to meet high standards in both quantity and quality.

Principle 5: The rating result has to be connected with a probability of default.

Principle 6: The rating scale of each customer segment should be mapped on a masterscale which is valid for the entire financial institution.

Principle 7: The calibrated expected default frequencies have to be reviewed via a periodic backtesting procedure.

Principle 8: Rating systems have to be integrated into credit risk controlling instruments.[6]

Principle 9: The rating process has to be consistent with regulatory requirements and has to be integrated into credit processes in a reasonable way.[7]

Principle 10: Rating systems have to be comprehensive and easy to apply.

These ten principles shall serve as a guideline throughout this paper. They will be explained in further detail within the appropriate Chapters.

---

[5] See Krahnen, J. P./Weber, M.; "Generally accepted rating principles: a primer"; Frankfurt a.M./Mannheim November 22th 1999; p. 8-14

[6] See Chapter 1.1

[7] See Bundesaufsichtsamt für das Kreditwesen; "Mindestanforderungen an das Kreditgeschäft der Kreditinstitute"; Rundschreiben an alle Kreditinstitute in der Bundesrepublik Deutschland; Draft; Bonn February 20th 2002

1.3 Definition of "Obligor in Default"

Building a rating system means to specify the probability of going into default for each obligor within a well-defined customer segment (*principle 1, principle 5*[8]). In agreement with the Basel Committee[9] the probability of default usually refers to a one-year-horizon.

Before being able to estimate a probability of default, obligors have to be clustered into two mutually exclusive groups, namely those being in default and those not being in default. Thus a clear definition for an obligor in default has to be given. The Basel Committee on Banking Supervision submits the following definition: An obligor is in default if at least one of the following four propositions is true.

1. "it is determined that the obligor is unlikely to pay its debt obligations (principal, interest, or fees) in full";[10]
2. "a credit loss event associated with any obligation of the obligor, such as a charge-off, specific provision, or distressed restructuring involving the forgiveness or postponement of principal, interest, or fees";[11]
3. "the obligor is past due more than 90 days on any credit obligation";[12]
4. "the obligor has filed for bankruptcy or similar protection from creditors".[13]

If *none* of the four propositions is true, the obligor is *not* in default.

While propositions 2, 3 and 4 are measurable and well-known criteria for an obligor in default, proposition 1 is subject to individual interpretation. Thus proposition 1 can hardly be used to be implemented as a criterion of default by a rating system. The author recommends to refer to propositions 2, 3, and 4 when building a rating system. In addition a financial institution should ensure that specific provisions are made when it believes that proposition 1 has become true. Thus proposition 1 becomes part of proposition 2 and a rating system refers to measurable rating definitions, namely proposition 2, 3, and 4.

---

[8] See Chapter 1.2
[9] See Bank for International Settlements; "The New Basel Capital Accord"; Consultative Document issued by the Basel Committee on Banking Supervision; Basel January 2001, paragraph 263
[10] Bank for International Settlements; "The New Basel Capital Accord"; Consultative Document issued by the Basel Committee on Banking Supervision; Basel January 2001, paragraph 272
[11] ibidem
[12] ibidem
[13] ibidem

2 Elements of Rating Systems for Corporate Clients

The elements of a rating system have to take all relevant criteria of a customer segment into account (*principle 1*). In this paper the focus is on middle-sized corporates with a turnover smaller than EUR 6 million.

The rating of a corporate obligor requires an in-depth analysis of a wide range of relevant areas of assessment. This analysis can be divided into the analysis of the past and present development of a corporate obligor and the analysis of the future development (see Figure 3).

The analysis of the past and present development is mainly driven by assessing financial statements about the corporate's past development. Areas of assessment are financial statements, personal assets (where appropriate), the corporate's information policy and its account management. The analysis of the future development examines whether management quality and corporate planning promise to generate substantial cash inflows to cover its debt obligations. Relevant areas of assessment are industrial sector/area of business, corporate planning and quality of management.

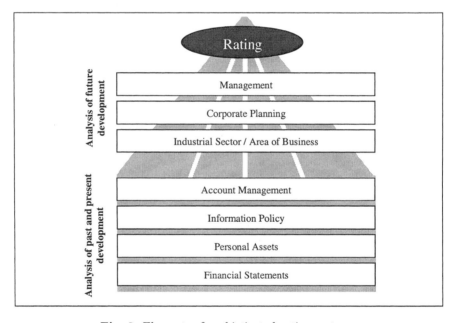

Fig. 3. Elements of sophisticated rating systems

2.1 Financial Statements

The assessment of financial statements is the heart of every rating system. It gives a deep insight into the economic situation of a company. Usually more than 50% of the final rating will be determined by the analysis of financial statements.

Advanced statistics are used to find a scoring function with a high separating power and a relevant set of variables (financial ratios), usually five up to fifteen. An extensive catalogue of more than 200 financial ratios, produced from balance sheet data, can be analysed and reduced to the relevant set.[14]

The statistical analysis aims to eliminate financial ratios without statistic or economic relevance and to combine the remaining ratios to a scoring function. The separating power of a scoring function can be measured by power curves (see Chapter 3.1).

A scoring function with a high separating power can be achieved by the following three steps (see Figure 4): the analysis of single ratios, the analysis of groups of ratios, and the analysis of different scoring functions.

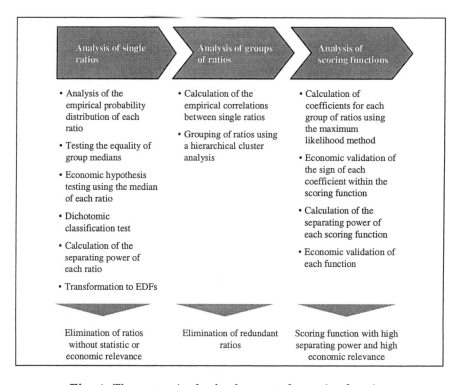

Fig. 4. Three steps in the development of a scoring function

---

[14] See Hüls, D.; "Früherkennung insolvenzgefährdeter Unternehmen", Düsseldorf, 1995, p. 75 ff for a definition of approximately 250 financial ratios

Analysis of Single Ratios

Statistical methods can be used in order to build a scoring function. The methods heavily depend on the distributions of the financial ratios that have to be analyzed. Frequently used statistical methods like multivariate discriminant analysis models[15] postulate normally distributed ratios. To verify this assumption the K-S-Test by Kolmogoroff-Smirnov and the "test on significant divergence from the skewness and the kurtosis of a normal distribution" can be used.

The K-S-Test checks the null hypothesis H_o: "The financial ratio is normally distributed". Here the cumulative frequency of the empirical distribution is compared to the frequency function of a normal distribution. The test on skewness and kurtosis refers to the null hypotheses H_o: "The skewness of the ratio's distribution is zero" and H_o: "The kurtosis of the ratio's distribution is zero". A normally distributed ratio has got a skewness and a kurtosis which equal zero.[16]

Generally speaking the K-S-Test is more conservative meaning that the hypothesis of a normal distribution is rejected more frequently.[17] However an extensive analysis of several ratios has shown that financial ratios are rarely normally distributed.[18] This has to be kept in mind when entering into advanced statistical methods (see Chapter 2.1 Analysis of scoring functions).

The following five steps are necessary in order to identify a set of ratios with high statistic and economic relevance:

1. Testing the equality of group medians
2. Examination of the working hypothesis
3. Dichotomic classification test
4. Calculation of the separating power
5. Transformation to "expected default frequencies" (EDFs)

Testing the equality of group medians

A basic requirement for incorporating a financial ratio into the final scoring function is its ability to distinguish between an obligor in default and an obligor not being in default. Analyzing a historic data base, it can be said retrospectively which obligor was in default according to the definition of Chapter 1.3.

---

[15] See Baetge, J.; "Früherkennung von Kreditrisiken"; in: Rolfes, B./ Schierenbeck, H./Schüller, S. (editors); "Risikomanagement in Kreditinstituten"; 2nd edition; Frankfurt am Main 1997; p. 194-206

[16] See Hüls, D.; "Früherkennung insolvenzgefährdeter Unternehmen"; Düsseldorf 1995; p. 111-119

[17] See Bleymüller, J./Gehlert, G./Gülicher, H.; "Statistik für Wirtschaftswissenschaftler"; 9th edition; Munich 1994

[18] The author's experience is supported by Hüls, D.; "Früherkennung insolvenzgefährdeter Unternehmen"; Düsseldorf 1995; p. 119-121

In order to identify relevant ratios the medians of a ratio for the groups of default and non-default obligors are compared. This can be accomplished by graphing the group medians for several time periods. The more the spread between the two curves widens before the point of default, the better a financial ratio separates between default and non-default obligors. Also, for economic reasons the two curves must not intersect (*principle 2*). The time periods in the following figure are defined as follows[19]:

> t-1 one year before default
> t-2 two years before default
> t-3 three years before default

In the example shown in Figure 5 the liquidity ratio would fail the test while the equity ratio passes it.

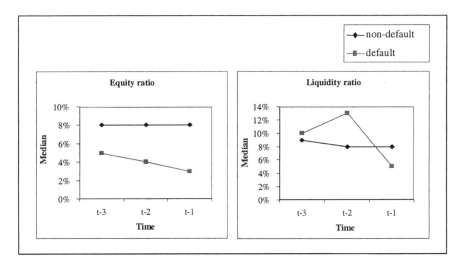

Fig. 5. Comparison of group medians

To verify whether there is a significant widening in the spread of the ratios of the group of default and non-default obligors the following null hypothesis H_o is tested: "The median of the ratio distribution is equal in both groups". This test is based on a chi-squared distribution with one degree of freedom.[20]

Financial ratios which are not able to distinguish between default and non-default obligors should be excluded from further analysis.

---

[19] Definition for defaulted obligors. It should be clear that there is no evident t_0 for a non-default obligor. Thus the time periods for non-default obligors are defined by a reference point of time (e.g. 31.12.2001).

[20] See Sachs, L.; "Angewandte Statistik", 9th edition, Berlin 1999, p. 488

216 Ingo Schäl

Examination of the working hypothesis

From economic considerations each credit analyst has an economic idea whether a financial ratio is higher (or smaller) for a solvent or an insolvent company. Thus for each ratio an economic working hypothesis can be postulated in the following way: "The financial ratio is on average higher (smaller) for a solvent company than for an insolvent company"[21]. For instance, insolvent companies have on average a smaller equity ratio than solvent companies.

These working hypotheses are verified by comparing the medians of the financial ratios for default and non-default obligors (*principle 2*). For example, the median of the equity ratio for non-default obligors should be higher than for default obligors within the sample. Otherwise the ratio will not be subject to any further considerations.

Dichotomic classification test

The result of this test is an indicator for the separating power of a financial ratio. For each ratio the number of default and non-default cases above and below the median of the whole sample will be counted. Regarding an economic working hypothesis (see above) non default obliors are expected to have a ratio above (below) the median and default obligors are expected to have a ratio below (above) the median. The α-error (type I error) and the β-error (type II error)[22] can be derived easily (see Figure 6).

When developing a rating system the used data sample usually consists of much more non-default than default obligors. Therefore the median of the whole sample is close to the group median of the non-default obligors for all financial ratios. This leads to a β-error of approximately 50% for all ratios. In contrast, the α-error will vary widely. For this reason the α-error is a good indication for the separating power of a financial ratio. In Figure 6 the equity ratio would be preferred to the debt ratio, since the α-error is much smaller.

Calculation of the separating power

In addition, the separating power of each ratio will be calculated directly using power curves (see Chapter 3.1). For further selections ratios with a high separating power will be preferred.

Transformation to "expected default frequencies" (EDFs)

Financial ratios have different ranges of values. For example, the values of a debt ratio range from 0% to 100%, whereas the cash-flow profitability ratio (cash-flow/total capital) also has values below 0% and above 100%. In

---

[21] This does not apply to ratios which cannot be assumed to be a monotone function of the expected default frequency.
[22] See Freund, J. E.; "Mathematical Statistics"; 5th edition; Englewood Cliffs 1992; p. 427-428

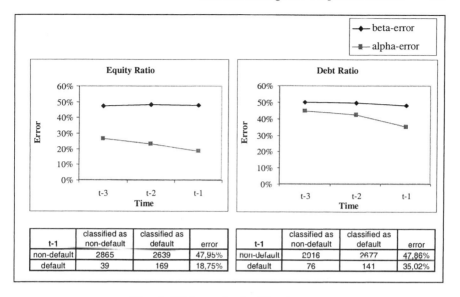

Fig. 6. Dichotomic classification test

order to make these ranges comparable to each other the ratios are transformed to EDF. For this transformation the range of values of a ratio will be discretised in disjoint intervals. Then, for each interval the empirical default rate of the data sample (EDF) and an average for the ratio values will be calculated. These pairs of values are plotted in a graph and fitted by a continuous, monotone curve (e.g. a modified logit function with four parameters). This transformation extends the original range of values in which the change from default to non-default takes place to the range of corresponding EDFs and supports a better distinction between default and non-default obligors. Ratios having an EDF-fit with a steep sigmoid form should be preferred for further analyses, because this indicates a sufficient separating power.

At the same time the transformation compresses the border areas of the range of values which reduces the influence of extreme values. In contrast to non-EDF-transformed ratios, the domination of the scoring function by extreme values and the calculation of wrong results can be avoided using this method.

In the following figure the return on equity ratio would be chosen for further analysis and the liquidity ratio would not be subject to any further consideration.

The result of the analysis of single ratios is a reduced catalogue of financial ratios which can be used to develop a scoring function.

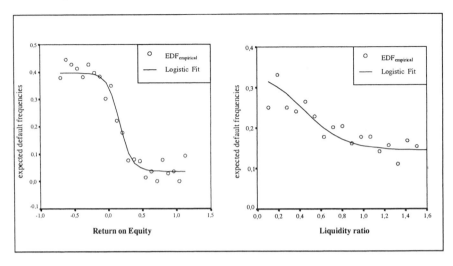

Fig. 7. EDF transformation of ratios and logistic fit with 4 parameters

Analysis of Groups of Ratios

Choosing a set of ratios for a scoring function, strong correlations between ratios should be avoided. Otherwise the optimisation of the coefficients of the ratios within the scoring function could lead to instable and suboptimal results.

To identify correlated ratios, all ratios are grouped together using a hierarchical cluster analysis. Groups of ratios, so-called clusters, are built so that they have high correlations within a cluster but low correlations with ratios from other clusters. The clusters can easily be pictured using dendrograms (see Figure 8). A small scaled cluster distance signifies a high correlation, a large scaled cluster distance signifies a low correlation. For example, in Figure 8 ratios 1, 2 and 3 are highly correlated, while ratios 13 and 14 are lowly correlated. Proceeding with extracting ratios for the resulting scoring function from each cluster, the financial ratio with the highest separating power will be chosen.

Thus the number of ratios can be further reduced. For the remaining financial ratios the resulting correlations have to be surveyed. This can be done reproducing a Spearman correlation matrix. The Spearman correlation uses ranks of ratio values only and is therefore independent of the exact values. The Spearman correlation also produces adequate results for ratios which are not normally distributed.[23] Thus a Spearman correlation is usually preferred to a Pearson correlation.

---

[23] See Freund, J. E.; "Mathematical Statistics"; 5th edition; Englewood Cliffs 1992; p. 601-602; Lothar Sachs; "Angewandte Statistik"; 9th edition; Heidelberg 1999; p. 476-477, 511-515, 601-603

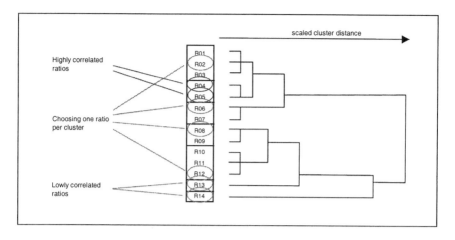

Fig. 8. Grouping ratios with hierarchical cluster analysis

Analysis of Scoring Functions

Having reduced the number of relevant ratios, several different subsets of the remaining (transformed) ratios can be used to build different scoring functions.

For the analysis of financial statements a model is needed to estimate a probability of default. The author suggests to use the approach of a logistic regression, which frequently leads to a high separating power. Whereas a linear regression models dependent variables with unbounded values, a logistic regression can be applied to dependent variables, e.g. a probability of default, with values in the interval (0;1).

We have seen that we cannot assume normally distributed ratios, which is a basic assumption for a linear discriminant analysis.[24] In contrast, the approach of a logistic regression is not that much dependent on this assumption.

The use of n ratios delivers a scoring function of the following type[25]:

$$Logistic\_function\,(r_{i=1,\ldots,n})$$
$$= \frac{1}{1 + EXP\left[-\left(\sum_{i=1,\ldots,n} coeff_i \cdot EDF_i(r_i) + const.\right)\right]},$$

with EDF_i being the EDF-transformation of ratio r_i.

This value $Logistic\_function(r_{i=1,\ldots,n})$ models the probability of a specific obligor (with the ratios $r_{i=1,\ldots,n}$) belonging to the group of default obligors.

---

[24] See Backhaus/Erichson/Plinke/Weiber, "Multivariate Analysemethoden", 9th edition, Heidelberg 2000, page 167
[25] See Backhaus/Erichson/Plinke/Weiber, "Multivariate Analysemethoden", 9th edition, Heidelberg 2000, page 109

This approach transforms a set of n ratios r_i into a value within the interval (0;1).

The coefficients $coeff_i$ of the ratios r_i are optimised using the maximum likelihood method.[26]

For different sets of ratios different scoring functions might be produced and the pros and cons of each function have to be taken into account. Each scoring function will be examined regarding to:

1. the sign of each coefficient
2. the separating power and statistical validation
3. the economic validation.

Sign of each coefficient

Since the result of the logistic regression function is an estimated probability of belonging to the group of default obligors, all coefficients' signs must be positive. Otherwise $Logistic\_function(r_{i=1,...,n})$ would be decreasing when a transformed ratio (EDF_i) increases (all other EDF_i being fixed), thus leading to unreasonable results. Consequently an economically reasonable scoring function using ratios transformed to EDF must satisfy:
$coeff_i > 0$ for all i (*principle 2*).

Separating power and statistical validation

If there are several scoring functions with plausible coefficients, the scoring function with the highest separating power is preferable (*principle 3*). To be sure that the scoring function has not been "overfitted" to the data sample used to develop the scoring function, the separating power has to be examined on another data sample, which is disjoint to the development data sample (*principle 4*). If there is a significant decrease in its separating power on this validation sample, the scoring function should not be considered any further.

Economic validation

All scoring functions have to be examined with respect to their economic content. Since acceptance by the potential users within a financial institution is crucial to a successful rating system (*principle 9, principle 10*), this is an important issue and should not be underestimated. A highly sophisticated and statistically optimised rating system will not work if credit analysts are not willing to use it in an appropriate way. Relevant and generally accepted economic information about debt, liquidity or financial condition should be covered to some extent by the ratios included in the scoring function.

Having gone through the three steps of building a scoring function it is up to the management of a financial institution to decide in favour of a specific function. This will be a scoring function with high statistical relevance (separating power, *principle 3*) and high economic relevance (*principle 1*).

---

[26] See Backhaus/Erichson/Plinke/Weiber, "Multivariate Analysemethoden", 9th edition, Heidelberg 2000, page 112-114

2.2 Personal Assets

The structural differences between limited liability companies[27] on the one hand and business partnerships[28] on the other have to be considered when assessing financial statements[29]. In case of insolvency partners of a business partnership might be liable to unlimited extent (depending on their legal status) while partners of a limited liability company are not.

Depending on the fiscal system of a political economy, financial ratios (like equity ratios) of limited liability companies and business partnerships might differ considerably. Thus, financial ratios have to be assessed in consideration of the corporate's legal form. This can be done by building different scoring functions for different legal forms, or by including an assessment of the personal assets of partners liable to unlimited extent.

The latter should be preferred since it reflects the whole economic situation of a corporate better. To assess the personal assets of a corporate's partner, personal assets should be opposed to private liabilities and revenues should be opposed to expenses.

Once a decision is made in favour of incorporating personal assets for specified legal forms, there must not be a suffrage for a credit analyst, whether to consider personal assets or not. It should be clear that substantial personal assets can improve an obligor's rating whereas negative implications for the rating can be caused by substantial private debts.

The consideration of personal assets for business partnerships delivers a significant separating power (additional marginal Accuracy Ratio (AR)[30] up to 8%), especially for smaller companies with a total turnover smaller EUR 4 Million. The bigger a company the less important are personal assets for the final rating.

2.3 Information Policy

Internal rating systems will have to come along with a more cooperative relationship between a financial institution and its obligors. Financial institutions will require much more information from its obligors than they used to. An obligor is required to produce financial statements, which have to be produced also for periods of less than a year, its business strategy, a corporate planning and information about its current business activities.[31] Without that knowledge a financial institution has got a limited information basis about an obligor and will be more conservative assessing its degree of creditworthiness.

---

[27] by German law: "Kapitalgesellschaften"
[28] by German law: "Personengesellschaften" and "Einzelunternehmer"
[29] See Hamer, E.; "Die unterschätzte Gefahr für den Mittelstand"; in: Frankfurter Allgemeine Zeitung, October 23th 2001
[30] See Chapter 3.1
[31] See Lindemann, B.; "Auf dem Weg zu einem mittelstandsfreundlichen Rating"; in: Frankfurter Allgemeine Zeitung; May 7th 2001

On the other hand obligors will be awarded to maintain an open relationship with their financial institution, the award expressed in terms of a better rating and thus a better credit condition.

2.4 Account Management

The analysis of the obligor's account management usually has a high separating power, since it gives current information about the obligor.[32] While financial statements suggest the obligor's ability to manage its financial affairs in the past, the analysis of the account management suggests this ability in the present.

The daily changing account balances have to be surveyed concerning debit and credit balances. Continuous overdrawings of an account have to be assessed in a negative way. An overdrawing over 90 days necessarily leads to a classification into a default class whatever the assessment of the other areas (see Chapter 1.3).

2.5 Industrial Sector/Area of Business

After the discussion of financial statements, personal assets, information policy and account management the analysis of the obligor's past and present development is completed. Now the obligor's future development is anticipated. This is done by assessing the industrial sector, corporate planning, and the quality of management

A corporate's business activity is embedded into a national economy and into regional markets with their own specific peculiarities. Thus the industrial sector and the area of business as well as the obligor's dependency on both of them have to be analyzed.[33]

Economic trends, economic policy measures and technological changes are relevant criteria for the analysis of the industrial sectors. According to the author's experience economic trends and policy measures are less important for the valuation of the credit-worthiness of a medium-sized corporate than for a big or quoted corporate.[34] For medium-sized corporates their competitive position and market share in the *regional* market is much more relevant.

---

[32] See Kirmße, S./ Jansen, S.; "BVR-II-Rating: Das verbundeinheitliche Ratingsystem für das mittelständische Firmenkundengeschäft – Bericht aus dem BVR-Projekt "Management-Informationssystem zum Risikocontrolling""; in: Bank-Information und Genossenschaftsforum; 28 (2001); 2; p. 70

[33] See Kirmße, S.; "Die Bepreisung und Steuerung von Ausfallrisiken im Firmenkundengeschäft der Kreditinstitute - Ein optionspreistheoretischer Ansatz"; Frankfurt a. M. 1996; p. 4-6

[34] See also Weber, M./Krahnen J. P./Voßmann, F.; "Risikomessung im Kreditgeschäft: Eine empirische Analyse bankinterner Ratingverfahren"; in: Zeitschrift für betriebswirtschaftliche Forschung; special edition 41; 1998; p. 131

Great importance also has to be attached to the dependency on suppliers and consumers.[35]

2.6 Corporate Planning

Corporate planning builds the basis for future business activities. It should be discussed between a financial institution and its obligor periodically. If an obligor does not regularly generate corporate plannings, an institution should be more conservative assessing its degree of creditworthiness. When corporate plannings are delivered to a financial institution, their quality and consistency have to be assessed. The budgeted balance sheet and the profit and loss account can be analyzed similarly to the assessment of financial statements.

2.7 Management

The assessment of a corporate's management covers the expertise of the business managers, their strategic planning, the efficiency of organizational structures, the follow-up regulation, the control mechanisms and the controlling system.[36]

For medium-sized corporates the assessment of managing directors is particularly important. Here the future business activities depend on the business sense, the business knowledge, and the personal characteristics of a small number of managers.

3 Quality of Rating Systems and Calibration

3.1 Estimation of the Separating Power

A statistical measure for the separating power of a rating system is the Accuracy Ratio (AR)[37] resulting from a Cumulative Accuracy Profile (CAP). A rating system with a high separating power is a system enabling a financial institution to forecast defaults with precision. Chapter 1.3 elaborated on the definition of an obligor in default. Looking at a data sample retrospectively it

---

[35] See Jansen, S.; "Bankinterne Ratingansätze im Firmenkundengeschäft", in: Schierenbeck, H./Rolfes, B. (editor); "Ausfallrisiken – Quantifizierung, Bepreisung und Steuerung"; Frankfurt am Main 2001; p. 102-103

[36] See Jansen, S.; "Bankinterne Ratingansätze im Firmenkundengeschäft", in: Schierenbeck, H./Rolfes, B. (editor); "Ausfallrisiken – Quantifizierung, Bepreisung und Steuerung"; Frankfurt am Main 2001; p. 102-103

[37] In the German literature the AR is ususally called "PowerStat". See Jansen, S.; "Bankinterne Ratingansätze im Firmenkundengeschäft", in: Schierenbeck, H./Rolfes, B. (editor); "Ausfallrisiken – Quantifizierung, Bepreisung und Steuerung"; Frankfurt am Main 2001; p. 107-110

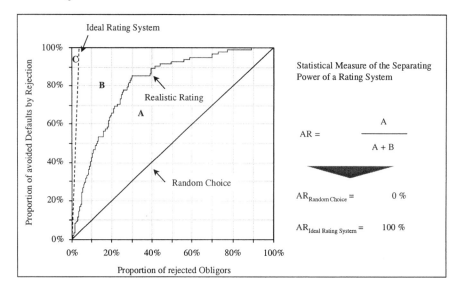

Fig. 9. CAP-plot used to evaluate the separating power of a rating system

is clear to which obligor the definition applies. The AR is an ex-post measure of the separating power.

To plot the so-called CAP curve, which is a power curve, the obligors of the data sample are ranked from riskiest to safest according to the rating of the system which has to be tested. The CAP plots the proportion of (hypothetically) avoided defaults depending on the proportion of rejected obligors according to their ranking (see Figure 9).[38]

In Figure 9 the CAP-plot for a realistic rating is plotted. It should be clear that a random rejection of obligors – ignoring their rating – results in a CAP-plot with the form of a straight diagonal line. The CAP-plot for an ideal rating system results from a hypothetic rating system, which can exactly forecast the case of default. From these power curves three areas A, B and C can be obtained (see Figure 9). The Accuracy Ratio is defined as the quotient of the area between the power curve of the rating system to be tested and the diagonal (area A) and the area between the power curve of an ideal rating system and the diagonal (areas A plus B).[39] The square measure of C equals

---

[38] See Keenan, S.C./Sobehart, J.R.; "Performance Measures for Credit Risk Models"; New York 1999; p. 5-8

[39] See Jansen, S.; "Bankinterne Ratingansätze im Firmenkundengeschäft", in: Schierenbeck, H./Rolfes, B. (editor); "Ausfallrisiken – Quantifizierung, Bepreisung und Steuerung"; Frankfurt am Main 2001; p. 107-110; Kirmße, S./ Jansen, S.; "BVR-II-Rating: Das verbundeinheitliche Ratingsystem für das mittelständische Firmenkundengeschäft – Bericht aus dem BVR-Projekt "Management-Informationssystem zum Risikocontrolling""; in: Bank-Information und Genossenschaftsforum; 28 (2001); 2; p. 69

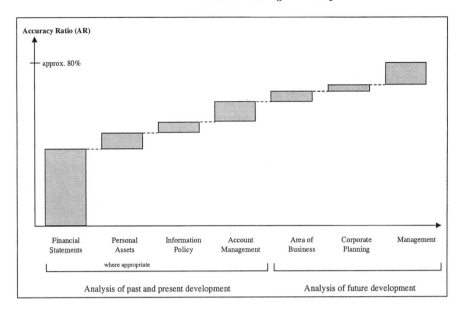

Fig. 10. Exemplary results of a rating system for corporate clients

a half of the percentage of defaults in the sample. The square measure of A can be obtained via numerical integration.[40] Then the square measure for B can easily be obtained and the AR can be calculated.

"Differences in the proportion of defaulters/non-defaulters in the data sets used to test each model affects the relative performance of each model."[41] The AR measures the absolute performance of a model.

A rating system with a significant separating power results in an AR of 65% up to 100%. Figure 10 shows a typical result of a corporate rating system with high separating power. The assessment of financial statements usually results in a separating power of 50-60%. Further fields of assessment like account management or management can increase this separating power significantly.

3.2 Calibration of Expected Default Frequencies (EDFs)

Masterscale

The masterscale is an important controlling instrument. Via the masterscale a rating system is linked to a credit pricing system and to a system of portfolio management.

---

[40] See Keenan, S.C./Sobehart, J.R.; "Performance Measures for Credit Risk Models"; New York 1999; p. 7-8

[41] Keenan, S.C./Sobehart, J.R.; "Performance Measures for Credit Risk Models"; New York 1999; p. 7

The definition of a masterscale contains the notation of each rating class and the assignment of a probability of default (see column three in Figure 11) (*principle 5, principle 6*). According to the Basel Committee this probability refers to a one year horizon[42]. Since the risk premium depends on the probability of default, several rating classes should be considered for a better distinction of different risks. The Basel Committee suggests a masterscale including at least 7 classes for obligors not being in default and one default class.[43] It is common practice to establish up to 30 classes (including default and non-default classes).

Although a financial institution might assign segment-specific rating scales to individual customer segments (column two in Figure 11) these segments have to be transformed to the masterscale which is valid for the entire financial institution and for every customer segment (*principle 7*). This makes different credit risks comparable: As far as the probability of default is concerned a quoted company can be compared with a private customer.[44]

Calibration of the Masterscale

A rated obligor has to be ranged in a corresponding class within the masterscale. Therefor an assignment of the scores resulting from the rating system to the classes in the masterscale is necessary. According to the Basel Committee these have to be classes with a probability of default not less than 0,03% for corporate obligors[45]. This step is called calibration and is usually done using expected default frequencies.

For estimating the expected default frequencies for each rating class a financial institution should use its own historical data. If the available data base is not sufficient, external data or data from a data pool might be used.[46]

---

[42] See Bank for International Settlements; "The New Basel Capital Accord"; Consultative Document issued by the Basel Committee on Banking Supervision; Basel January 2001, paragraph 180

[43] See Pluto, K.; "Internes Rating – Kredite an Unternehmen und Privatpersonen"; Speech on the conference of Deutsche Bundesbank "Basel II – die internationalen Eigenkapitalregeln in der aktuellen Diskussion"; Frankfurt a. M. March 13th 2002

[44] See Kirmße, S./ Jansen, S.; "BVR-II-Rating: Das verbundeinheitliche Ratingsystem für das mittelständische Firmenkundengeschäft – Bericht aus dem BVR-Projekt "Management-Informationssystem zum Risikocontrolling""; p. 71; Kramer, J. W.; "Ratingsysteme in Banken – Grundlage einer ertrags- und risikoorientierten Steuerung"; in: Bank-Information und Genossenschaftsforum; 28 (2001); 9; p. 9

[45] See Bank for International Settlements; "The New Basel Capital Accord"; Consultative Document issued by the Basel Committee on Banking Supervision; Basel January 2001, paragraph 180

[46] See Jansen, S.; "Ertrags- und volatilitätsgestützte Kreditwürdigkeitsprüfung im mittelständischen Firmenkundengeschäft der Banken, Entwicklung eines optionspreistheoretischen Verfahrens der Bonitätsanalyse und empirische Überprüfung

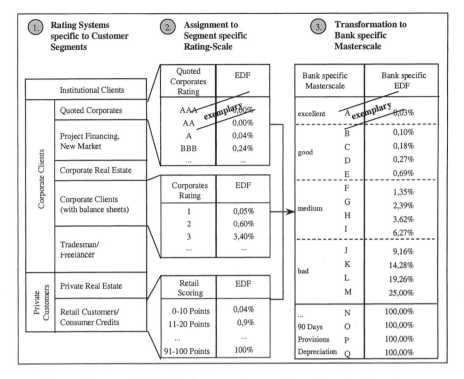

Fig. 11. An exemplary masterscale

According to the Basel Committee the historical observation period should be at least 5 years.[47]

In each rating class the number of defaults is observed and compared with the total number of obligors in that class. An obligor is defined as being in default when the definition of Chapter 1.3 applies.

"A bank is allowed to use an average of individual default-probability estimates for borrowers in a given grade using statistical default prediction models subject to the adherence to the formulated minimum requirements."[48]

The calculated probabilities of default have to be verified at least on a yearly basis.

der Anwendungsmöglichkeiten am Beispiel des Kreditportefeuilles einer Sparkasse"; Frankfurt am Main 2001; p. 67-75

[47] See Bank for International Settlements; "The New Basel Capital Accord"; Consultative Document issued by the Basel Committee on Banking Supervision; Basel January 2001; paragraphs 56-283

[48] Bank for International Settlements; "The New Basel Capital Accord"; Consultative Document issued by the Basel Committee on Banking Supervision; Basel January 2001; paragraph 280

3.3 Backtesting

Financial institutions have to establish a backtesting procedure to validate "the accuracy and consistency of rating systems" and the estimation of expected default frequencies (EDFs) (*principle 7*).[49] A rating system has to run through a backtesting procedure at least once a year.[50] For the backtesting of a rating system basically two things have to be done: The backtesting of separating power and the backtesting of expected default frequencies.

Backtesting of the Separating Power

The separating power of a rating system has to be surveyed throughout time. If the separating power is extremely decreasing, the reasons for the decrease have to be found. In that case the separating power of single criteria has to be analyzed again. If single criteria figure out to have a decreasing separating power, the model needs to be adjusted. This can be done by adjusting the corresponding coefficients. If basic inconsistencies in the model are discovered, alternatives for the structure of the scoring functions and for the number and kind of criteria have to be found (*principle 2*).

Backtesting of Expected Default Frequencies (EDFs)

For a backtesting of expected default frequencies a table of defaults per rating class has to be examined:

In this example thirteen (non-default) rating classes are considered. At the beginning of a year a financial institution determines the distribution of its obligors over the thirteen rating classes. Given the expected default frequencies from the masterscale, the expected defaults for the forthcoming year can be calculated.

At the end of the year the realized defaults in each rating class are counted. A financial institution has to define rules when differences between expected and realized defaults have to be considered as significant. Statistical tests have to be found to test the significance of these differences. According to the present discussion among statisticians and practitioners no standards seem to emerge.

A backtesting procedure will not only focus on a one year horizon but will also examine rating results over several years.

---

[49] Bank for International Settlements; "The New Basel Capital Accord"; Consultative Document issued by the Basel Committee on Banking Supervision; Basel January 2001; paragraph 302

[50] See Bank for International Settlements; "The New Basel Capital Accord"; Consultative Document issued by the Basel Committee on Banking Supervision; Basel January 2001; paragraph 305

Rating Class	Number of Obligors	EDF	Expected Defaults	Realized Defaults	Difference
A	9	0,03%	0	0	0
B	83	0,10%	0	0	0
C	349	0,18%	1	0	1
D	521	0,27%	1	3	-2
E	583	0,69%	4	4	0
F	473	1,35%	6	5	1
G	236	2,39%	6	10	-4
H	169	3,62%	6	5	1
I	85	6,27%	5	4	1
J	47	9,16%	4	4	0
K	24	14,28%	3	5	-2
L	13	19,26%	3	2	1
M	4	25,00%	1	3	-2
Total	2596	1,57%	40	45	-5

Fig. 12. Backtesting of expected default frequencies

References

1. Backhaus, K./Erichson, B./Plinke, W./Weiber, R.; *Multivariate Analysemethoden*; 9th edition, Heidelberg 2000
2. Baetge, J.; *Früherkennung von Kreditrisiken*; in: Rolfes, B./ Schierenbeck, H./Schüller, S. (editors); *Risikomanagement in Kreditinstituten*; 2nd edition; Frankfurt am Main 1997; p. 191-221
3. Bank for International Settlements; *The New Basel Capital Accord*; Consultative Document issued by the Basel Committee on Banking Supervision; Basel January 2001
4. Bleymüller, J./Gehlert, G./Gülicher, H; *Statistik für Wirtschaftswissenschaftler*; 9th edition; Munich 1994
5. Bundesaufsichtsamt für das Kreditwesen; *Mindestanforderungen an das Kreditgeschäft der Kreditinstitute*; Draft; Rundschreiben an alle Kreditinstitute in der Bundesrepublik Deutschland; Bonn February 20th 2002
6. Freund, J. E.; *Mathematical Statistics*; 5th edition; Englewood Cliffs 1992
7. Hamer, E.; *Die unterschätzte Gefahr für den Mittelstand*; in: Frankfurter Allgemeine Zeitung; October 23th 2001
8. Hüls, D.; *Früherkennung insolvenzgefährdeter Unternehmen*; Düsseldorf 1995
9. Jansen, S.; *Auswirkungen von Basel II auf Kreditinstitute und Mittelstand*; Statement; press-conference of zeb/rolfes.schierenbeck.associates; Frankfurt a.M. February 19th 2002
10. Jansen, S.; *Bankinterne Ratingansätze im Firmenkundengeschäft*, in: Schierenbeck, H./Rolfes, B. (editor); *Ausfallrisiken – Quantifizierung, Bepreisung und Steuerung*; Frankfurt am Main 2001; p. 95-125
11. Jansen, S.; *Ertrags- und volatilitätsgestützte Kreditwürdigkeitsprüfung im mittelständischen Firmenkundengeschäft der Banken, Entwicklung eines optionspreistheoretischen Verfahrens der Bonitätsanalyse und empirische Überprüfung der Anwendungsmöglichkeiten am Beispiel des Kreditportefeuilles einer Sparkasse*; Frankfurt am Main 2001
12. Keenan, S.C./Sobehart, J.R.; *Performance Measures for Credit Risk Models*; New York 1999

13. Krahnen, J. P./Weber, M.; *Generally accepted rating principles: a primer*; Frankfurt a.M./Mannheim November 22th 1999
14. Kramer, J. W.; *Ratingsysteme in Banken – Grundlage einer ertrags- und risikoorientierten Steuerung*; in: Bank-Information und Genossenschaftsforum; 28 (2001); 9; p 4-11
15. Kirmße, S.; *Die Bepreisung und Steuerung von Ausfallrisiken im Firmenkundengeschäft der Kreditinstitute - Ein optionspreistheoretischer Ansatz*; Frankfurt a. M. 1996
16. Kirmße, S./ Jansen, S.; *BVR-II-Rating: Das verbundeinheitliche Ratingsystem für das mittelständische Firmenkundengeschäft – Bericht aus dem BVR-Projekt Management-Informationssystem zum Risikocontrolling*; in: Bank-Information und Genossenschaftsforum; 28 (2001); 2; p. 67-71
17. Lindemann, B.; *Auf dem Weg zu einem mittelstandsfreundlichen Rating*; in: Frankfurter Allgemeine Zeitung; May 7th 2001
18. Pluto, K.; *Internes Rating – Kredite an Unternehmen und Privatpersonen*; Speech on the conference of Deutsche Bundesbank *Basel II – die internationalen Eigenkapitalregeln in der aktuellen Diskussion*; Frankfurt a. M. March 13th 2002
19. Rinker, A.; *Auswirkungen von Basel II auf Kreditinstitute und Mittelstand*; Statement; press-conference of zeb/rolfes.schierenbeck.associates; Frankfurt a.M. February 19th 2002
20. Rinker, A./Jansen, S.; *Basel II als Chance begreifen*; press release of zeb/rolfes.schierenbeck.associates; Münster/Frankfurt a.M. February 19th 2002
21. Rolfes, B./Emse, C.; *Rating-basierte Ansätze zur Bemessung der Eigenkapitalunterlegung von Kreditrisiken*; Forschungsbericht des European Center for Financial Services; August 2000
22. Sachs, L.; *Angewandte Statistik – Anwendung statistischer Methoden*; 9th edition; Heidelberg 1999
23. Schäl, I.; *Kreditderivate*; Ulm 1999
24. Schierenbeck, H.; *Ertragsorientiertes Bankmanagement (1)*; 7th edition; Wiesbaden 2001
25. Statistisches Bundesamt (Hrsg.); Statistisches Jahrbuch 2001 für die Bundesrepublik Deutschland; Stuttgart 2001
26. Statistisches Bundesamt; press release; Wiesbaden 5th April 2002
27. Verband der Vereine Creditreform e.V.; *Insolvenzflut bei Unternehmen und Verbrauchern hält an*; press release; Neuss June 19th 2002
28. Weber, M./Krahnen J. P./Voßmann, F.; *Risikomessung im Kreditgeschäft: Eine empirische Analyse bankinterner Ratingverfahren*; in: Zeitschrift für betriebswirtschaftliche Forschung; special edition 41; 1998; p. 117-142

Finding Constrained Downside Risk-Return Efficient Credit Portfolio Structures Using Hybrid Multi-Objective Evolutionary Computation

Frank Schlottmann and Detlef Seese

Institut AIFB, Universität Karlsruhe, Germany

Summary. In contemporary credit portfolio management, the portfolio risk-return analysis of financial instruments using certain downside credit risk measures requires the computation of a set of Pareto-efficient portfolio structures in a non-linear, non-convex setting. For real-world problems, additional constraints, e. g. supervisory capital limits, have to be respected. Particularly for formerly non-traded instruments, e. g. corporate loans, a discrete set of decision alternatives has to be considered for each instrument. The main result of this paper is a new, fast and flexible framework for solving the above issues using a hybrid heuristic method that combines multi-objective evolutionary and problem-specific local search methods in a unique way. We incorporate computational complexity in our considerations and consider proper genetic modelling of portfolio credit risk related problems. Moreover, we analyse empirical results showing that our method is superior in convergence speed to a non-hybrid evolutionary approach and finds risk-return efficient portfolios within reasonable time.

1 Introduction

The intensive development of quantitative portfolio credit risk models since the late 1990s and the increasing trade in financial instruments for transferring credit risk like credit default swaps, asset backed transactions etc. are major reasons for a growing importance of credit portfolio risk-return analysis and optimisation. Beyond that, there will possibly be more demand for credit portfolio optimisation as soon as the supervisory capital requirements for banks will be changed due to proposals of the Basel Committee, e. g. by setting new capital weights on some credit risk exposure types, providing supervisory capital relief for risk mitigation and establishing additional regulatory capital requirements for operational risk (cf. Basel Committee for Banking Supervision [3] and subsequent publications from the Bank of International Settlements).

In this paper, we will focus on an algorithmic framework for the calculation of discrete risk-return efficient sets for credit portfolios with respect to constraints, e. g. imposed by changes of supervisory capital regulations or internal reallocation of risk capital. This kind of portfolio management is of great importance especially for, but not limited to, many German and European banks since the typical largest exposures to credit risk for small and medium size universal banks are loans given to companies not having direct access to the capital market.

In contrast to the methods for the computation of the efficient frontier for a given set of alternative stock market investments based on the portfolio's variance and related measures, usually a non-linear, non-convex downside risk measure like the Credit-Value-at-Risk is preferred for portfolio credit risk-return analysis, therefore requiring a different method of computation. Moreover, this computational problem often cannot be modelled using real-valued variables, since typically neither the decision alternatives allow an arbitrary amount of each credit risk exposure to be traded nor it is possible to obtain a short position providing a hedge for each arbitrarily chosen exposure from a given portfolio. In addition to that, e. g. the capital requirements for credit risk exposures imposed by the banking supervision authorities are an important constraint to be considered in the computation of efficient credit portfolio structures, and these capital requirements are going to change significantly in the future according to recent proposals of the Basel Committee.

For our considerations, the concept of Pareto-optimality is essential, i. e. efficient structures are Pareto-optimal concerning the two distinct (and in many cases contrary) objective functions specifying the aggregated risk and the aggregated return of each potential credit portfolio structure for a given set of alternatives. Therefore, we are interested in multiple, feasible non-dominated solutions to the constrained portfolio credit risk-return optimisation problem that are comparable to the efficient frontier in stock portfolio investment analysis, but in a discrete search space having many local optima and particularly using multiple target functions not required to be linear, quadratic or convex. In this context, a feasible non-dominated solution is a portfolio structure that does not violate the constraints, and for which we cannot find any other feasible solution being better in all two target function values.

We introduce a novel approach to such problems that combines recent constrained multi-objective evolutionary computation methodology and problem specific knowledge to create a hybrid algorithm providing rapid discovery of a set of efficient credit portfolio structures with respect to constraints for a given instance of the above problem.

The paper is organised as follows: In the first section, we describe our portfolio credit risk optimisation problem and analyse its structure from a computational perspective. After a short look to traditional methods for multi-objective optimisation, we give a short introduction to multi-objective evolutionary algorithms. Then we derive a proper genetic modelling for portfolio credit risk problems before giving an overview and discussing some elements of our hybrid evolutionary algorithm framework for the computation of constrained risk-return efficient credit portfolio structures. The next section introduces some specific credit portfolio model features which are exploited by our implementation of the proposed framework. We provide details of our implementation before describing the parameters and the results of an empirical study where we compare our hybrid implementation with a non-hybrid approach on several test problems.

2 Basic Definitions and Results

In this section, we will present the basic terminology of the constrained discrete credit portfolio optimisation problem which is to be solved by our hybrid method. The following definitions and results are necessary to understand the structure and the computational complexity of the problem. For a detailed coverage of the graph and complexity methodology used below see Seese & Schlottmann [28], [29] and [30].

Definition 2.1 *Given are $m > 1$ investment alternatives and a time horizon T. Each investment alternative (obligor) $i \in \{1, \ldots, m\}$ incorporates the risk of default and it is characterised by the following data which is considered to be constant within the time period $(0, T)$:*

- *net exposure e_i (loss given default of investment i)*
- *expected rate of return r_i based on e_i (net of cost)*
- *expected cumulative default probability p_i within time horizon $(0, T)$*
- *capital requirement percentage w_i based on e_i.*

The dependence structure between joint defaults of investment alternatives $i, j \in \{1, \ldots, m\}$ can be modelled by an undirected graph $G = (V, E)$ and a function $h : E \to \Re$, where $V = \{1, \ldots, m\}$ is the set of investment alternatives, $E = V \times V$ is the complete edge set of potential default dependencies between investment alternatives (i, j) and $h : E \to \Re$ is a function expressing the strength of the dependency between each pair of two investment alternatives (i, j). The function h can be a correlation based function or more general, a copula based function, see e. g. Frey & McNeil [11] for details about mathematical modelling of default dependencies.

In the following text, we will abbreviate the respective set of scalar variables e_i, r_i, p_i, w_i of all obligors by vectors $e := (e_1, \ldots, e_m)^T$, $r := (r_1, \ldots, r_m)^T$, $p := (p_1, \ldots, p_m)^T$, $w := (w_1, \ldots, w_m)^T$.

The investor that has to decide about holding a subset of the investment alternatives in her portfolio can e. g. be a bank that wants to optimise its loan portfolio containing m different obligors. According to the next definition, we assume a fixed risk capital budget for the investments that can e. g. be given by the bank's maximum supervisory capital which is required to be provided by the bank due to the supervisory regulations (cf. e. g. Basel Committee for Banking Supervision [3]).

Definition 2.2 *A capital budget of the investor is given by $K > 0$.*

We need the following definition to describe possible portfolio structures that can be constructed for the given investment alternatives.

Definition 2.3 *A portfolio structure is given by a vector*

$$x = (x_1, x_2, \ldots, x_m)^T, x_i \in \{0, e_i\}.$$

Since every x_i can only take the values 0 or e_i, the investor has to decide whether to hold the whole net exposure e_i in her portfolio. In many real world portfolio optimisation problems the decision is e. g. either keeping the obligor i in the credit portfolio or selling the entire net exposure of obligor i to a risk buyer. This is particularly true for formerly non-traded instruments like corporate loans in a bank's credit portfolio. Even if there are more than two decision alternatives for each potential investment i, the decision variables will still consist of a finite, discrete number of choices.

Facing these decision alternatives, an investor has to consider two conflicting objective functions: the aggregated return and the aggregated risk from her portfolio. Usually, there is a trade-off between both objectives since any rational investor will ask for a premium (additional return) to take risk.[1]

Definition 2.4 *The aggregated expected return from a portfolio structure x is calculated by*

$$ret(x, p, r) = \sum_{i=1}^{m} r_i x_i - \sum_{i=1}^{m} p_i x_i = \sum_{i=1}^{m} (r_i - p_i) x_i. \quad (1)$$

This is a common net risk adjusted return calculation since the aggregated expected losses are subtracted from the portfolio's aggregated net return.

Definition 2.5 *The aggregated downside risk $risk(x, p; h)$ from the portfolio structure x for the investor is calculated such that the risk measure $risk(x, p; h)$ satisfies the following condition:*

$$\forall i \in \{1, \ldots, m\} : p_i \equiv 0 \Rightarrow \forall x, h : risk(x, p; h) \equiv 0 \quad (2)$$

---

[1] Since there is currently no perfect and transparent capital market for trading illiquid financial instruments like loans we do not distinguish explicitly between systematic and idiosyncratic (obligor specific) risk here. In a capital market equilibrium, there is only a premium for taking systematic risk according to the CAPM (see e. g. Sharpe [31]).

The condition specified in Definition 2.5 expresses the natural property of the risk measure that there is no credit risk for the investor if the default probabilities of all investment alternatives are equal to zero. Of course, $risk(...)$ is required to satisfy additional properties to be a downside risk measure. Artzner et al. [2] provide definitions and a discussion of different risk measures and their properties. For our empirical study described later, we will use the Credit-Value-at-Risk downside risk measure that satisfies condition (2). It is a very common measure of credit risk in contemporary credit risk management.

Definition 2.6 *For a given portfolio structure x the Credit-Value-at-Risk (CVaR) at the arbitrary, but fixed confidence level $\alpha \in (0,1)$ is obtained by calculating*

$$risk(x,p;h) := q_{pf}^{\alpha}(x,p;h) - \mu_{pf}(x,p) \tag{3}$$

where $q_{pf}^{\alpha}(x,p;h)$ is the α-percentile of the cumulative distribution function of aggregated losses calculated from the portfolio for the given parameters x,p and the dependency structure specified by h. Moreover, $\mu_{pf}(x,p)$ is the expected loss calculated by $\mu_{pf}(x,p) = \sum_{i=1}^{m} x_i p_i$.

For our theoretical considerations in this section, we do not need a specification of the calculation procedure for the cumulative distribution function of aggregated losses or q_{pf}^{α}. We will return to these details in the section describing the portfolio credit risk model used by our implementation of the proposed hybrid algorithm.

Definition 2.7 *The required capital of a given portfolio structure x is*

$$cap(x,w) := \sum_{i=1}^{m} x_i w_i.$$

Definition 2.8 *A portfolio structure x is feasible if and only if*

$$cap(x,w) \leq K.$$

The following definition is essential for the concept of Pareto-optimality.

Definition 2.9 *Given are two distinct feasible portfolio structures x and y. x dominates y if and only if one of the following cases is true:*

$$ret(x,p,r) > ret(y,p,r) \wedge risk(x,p;h) \leq risk(y,p;h) \tag{4}$$

$$ret(x,p,r) \geq ret(y,p,r) \wedge risk(x,p;h) < risk(y,p;h) \tag{5}$$

If x dominates y, we will denote this relationship by $x >_d y$.

This means that a feasible portfolio structure x is better than a feasible portfolio structure y if and only if x is better in at least one of the two criteria and not worse than y in the other criterion. It is obvious that a rational investor will prefer x over y if $x >_d y$.

Definition 2.10 *Given is the set S of all possible portfolio structures for the specified data from Definition 2.1 and the subset $S' \subseteq S$ of all feasible structures in S. A solution $x \in S'$ is a feasible global non-dominated portfolio structure if and only if it satisfies the following condition:*

$$\forall y \in S' : \neg (y >_d x). \tag{6}$$

To choose between the best investment alternatives using her preferences or utility function, a rational investor is interested in finding the set of non-dominated portfolio structures that has maximum cardinality. This set is the Pareto-optimal set which is comparable to the efficient frontier of Markowitz [19], but in a discrete decision space.

Problem 1 *The problem of finding the set of feasible Pareto-efficient portfolio structures having maximum cardinality for the set of investment alternatives S can be formulated as: Calculate the set*

$$PE^* := \arg \max_{PE \subseteq S'} \{|PE|\} \tag{7}$$

where

$$PE := \{x \in S' : \forall y \in S' : \neg (y >_d x)\}. \tag{8}$$

Now we want to focus on the computational complexity of Problem 1. Therefore, we consider the usual computing model of Turing machines, so we temporarily assume all variables to be rational numbers.

Lemma 2. *Assuming all scalar variables to be rational numbers, the corresponding decision problem to Problem 1 is NP-hard.*

Proof. See appendix.

Corollary 2.11 *Unless $P = NP$, there is no exact algorithm that calculates PE^* within polynomial computing time (measured by the size of the input m).*

The next section will describe approaches for the approximation of PE^* for portfolio data given according to Definition 2.1.

3 A Framework for Hybrid Multi-Objective Evolutionary Computation of Pareto-Efficient Credit Portfolio Structures

3.1 Traditional Methods for Solving Multi-Objective Optimisation Problems

A well-known traditional method of solving multi-objective optimisation problems like Problem 1 is weighted sum scalarisation. In the case of our optimisation problem this e. g. implies that the two objective functions for the credit portfolio's aggregated return and aggregated risk would have to be transformed into a single objective function g by adding the two original objective functions using appropriate weights c_1, $c_2 \in \Re$ (see e. g. Ehrgott [8], pp. 55-76 for details):

$$g(x,p,r;h) = c_1 ret(x,p,r) + c_2 risk(x,p;h) \qquad (9)$$

These and other traditional methods (see e. g. Ehrgott [8], pp. 77ff. for an overview) of solving multi-objective problems and handling constraints work well on linear or convex problems in a continuous setting, i. e. problems consisting of linear or at least convex, continuous objective functions, and constraints also satisfying these properties. However, the portfolio optimisation problems based on downside risk measures usually inhibit non-linear, non-convex objective functions, e. g. if the concept of Value-at-Risk or Credit-Value-at-Risk is used – see e. g. Pflug [22] for the mathematical properties of such downside risk measures. It is an obvious fact that the above mentioned methods relying on convexity in the objective function space produce suboptimal results at least for some instances of such problems (cf. [8], p. 77). Moreover, we have to deal with a discrete optimisation problem consisting of a fixed number of distinct choices. So we apply an alternative approach to our discrete optimisation problem. In the following paragraphs we will give a short introduction to evolutionary approaches to such problems.

3.2 Evolutionary Approaches to Multi-Objective Optimisation

Since the first reported implementation and test of a multi-objective evolutionary approach, the Vector Evaluated Genetic Algorithm (VEGA) by Schaffer [24] in 1984, this special branch of Evolutionary Algorithms (EAs) has attracted many researchers dealing with non-linear and non-convex multi-objective optimisation problems. After the introduction of VEGA, many different EAs have been proposed for multi-objective optimisation problems, see e. g. Deb [7] for an overview.

In general, a Multi-Objective Evolutionary Algorithm (MOEA) is a randomised heuristic search algorithm reflecting the Darwinian 'survival of the fittest principle' that can be observed in many natural evolution processes, cf.

e. g. Holland [16]. At each discrete time step t, a MOEA works on a set of solutions $P(t)$ called population or generation. A single solution $x \in P(t)$ is an individual. To apply a MOEA to a certain problem the decision variables have to be transformed into genes, i. e. the representation of possible solutions by contents of the decision variables has to be transformed into a string of characters from an alphabet Σ. The original representation of a solution is called phenotype, the genetic counterpart is called genotype.

For evaluation of each genotype in a population the MOEA requires a quality measure (fitness function) for evaluation of every possible solution (not necessarily feasible if the problem is constrained) that is usually based on the quality of the corresponding phenotype. The individuals from the population $P(t)$ are selected for survival into the next generation after application of variation operators (see below) according to their fitness values. The fitness function and the selection scheme of most MOEAs differ substantially from the fitness functions and selection procedures of single-objective EAs by incorporating special mechanisms for preserving diversity of solutions in the search space (since one is interested in finding a representative Pareto-efficient set containing different solutions) and for selection of solutions that cannot be directly compared using a total order in a multi-dimensional fitness space.

The selected individuals from the current population $P(t)$ are modified using genetic variation operators (see e. g. Fogel & Michalewicz [10] for an overview). A standard variation operator for discrete decision variables is the one point crossover, i. e. the gene strings of two selected individuals are cut at a randomly chosen position and the resulting tail parts are exchanged with each other to produce two new offspring. This operation is performed with crossover probability p_{cross} on individuals selected for reproduction. The main goal of crossover is to move the population through the space of possible solutions.

In analogy to natural mutation, the second standard variation operator in most MOEAs changes the genes of selected individuals randomly with probability p_{mut} (mutation rate) per gene to allow the invention of new, previously undiscovered solutions in the population. Its second task is the prevention of the MOEA stalling in local optima as there is always a positive probability to leave a local optimum if the mutation rate is greater than zero.

After this short introduction of MOEAs we will now describe the structure and the details of our Hybrid Multi-Objective Evolutionary Algorithm (HMOEA) which incorporates the above general features of MOEAs. Moreover, since our focus is on the development of a flexible framework for discrete credit portfolio optimisation problems, which should not be restricted to a certain downside risk measure and which should support non-linear, non-convex constraints as well we propose a hybrid method that particularly uses ideas that can be found in several MOEA schemes as well as an additional problem-specific local search operator and a problem-specific preprocessing stage.

3.3 Genetic Modelling of Portfolio Credit Risk Related Problems

The first question when applying an EA to a problem is to choose a proper genetic representation of the decision variables. For portfolio credit risk optimisation problems like Problem 1, we assume that the decision variables x_i will be arranged in a vector to obtain gene strings representing potential solutions. The resulting genotypes consist of real-valued genes which are connected to strings and take either value 0 or e_i depending on the absence or presence of investment alternative i in the current solution. So we obtain strings of length m that represent some of the 2^m combinations of possible (but neither necessarily feasible nor necessarily optimal) portfolio structures.

Since the one point crossover is the first standard variation operator in many EAs (and we also use it in our approach), we should keep some important issues of portfolio credit risk modelling in mind to choose a well adapted genetic representation for the phenotypes.

The one point crossover cuts two gene strings at a random position and crosses the tails of the strings to produce two offspring with crossover probability p_{cross}. The probability of two genes i, j (these variables represent the index of the genes associated to investment alternative i and j, respectively) from one individual being cut by the crossover increases proportional to the distance $|i - j|$ between the two genes in the gene string as the cut position is determined by a draw from a uniform distribution over $m - 1$ cut possibilities:

$$p(\text{crossover cuts gene } i \text{ and } j) = \frac{1}{m-1} |i - j| \tag{10}$$

For better results of the crossover operator we must ensure that there is a high probability of good partial solutions being recombined with other solutions and not being destroyed by the crossover's cut operation. More formally, we search for a permutation $\pi(i)$ of the portfolio data represented by our genes ensuring a high probability of success for crossover. Therefore, we have to remember that the degree of dependence between two different investment alternatives i, j plays the central role in aggregated portfolio credit risk calculations according to Definition 2.5.

In our Problem 1, the dependence structure has no influence on the other objective function given in Definition 2.4, the aggregated return, so it is sufficient to concentrate solely on the risk objective function when considering the possible influence between different gene positions.

According to Definition 2.1, the dependence structure between investment alternatives is determined by the function $h(i, j)$. Without any specific assumptions on the structure of the dependencies, we cannot provide a general algorithm for finding an optimal permutation. However, we can determine the maximum strength of the dependency of an investment alternative from all others and build a permutation based on this measure. This greedy algorithm ensures that the genes of the more dependent investment alternatives are located closely to each other, and it has a very low computational complexity compared to combinatorial problems arising from the question of finding the best of $m!$ possible permutations.

Definition 3.1 *The maximum strength of the dependency of investment alternative i from all other investment alternatives $j \neq i$ is given by*

$$s(i) := \max_{j \neq i} \{h(i,j)\}.$$

We can build our requested permutation $\pi(i)$ by calculating and sorting the strength values from Definition 3.1. Since sorting the strength values requires only $O(m \log m)$ computational steps, this calculation can be performed very efficiently.

Of course, the above considerations are also applicable to portfolio credit risk problems having continuous decision variables, and they can be adapted to other choices of crossover operators. After discussing the genetic modelling of portfolio credit risk problems, we will now give an overview of our hybrid approach for solving Problem 1 that also incorporates the above genetic modelling principles.

3.4 Overview of our Hybrid Multi-Objective Algorithm

Since many of the general MOEA concepts in the literature were designed and tested for optimisation problems having continuous decision variables and do not respect structural properties of our Problem 1, we have designed a problem-specific algorithm that provides a framework for finding constrained Pareto-efficient credit portfolio structures using non-linear, non-convex downside risk measures and discrete decision variables. Table 1 shows an overview of our Hybrid Multi-Objective Evolutionary Algorithm (HMOEA).

The first operation in our HMOEA scheme is the permutation of the given investment alternatives according to our considerations from section 3.3 to obtain an adequate genetic representation of our decision variables. After that, the initial population $P(0)$ will be generated by random initialisation of every individual to obtain a diverse population in the search space of potential solutions.

We propose the use of an elite population $Q(t)$ in our algorithm that contains the feasible, non-dominated solutions found so far at each population step t. Rudolph & Agapie [23] have shown that the existence of such an elite population ensures convergence of an EA to the global Pareto-efficient set (in their work the elite population it is called 'archive population'). Furthermore, the algorithm can be terminated at any time by the user without losing the best feasible solutions found so far. This enables real-time optimisation tasks. Remember that we consider constrained optimisation problems so current members of the population $P(t)$ might not be feasible at an arbitrary interruption time.

Table 1. HMOEA scheme

HMOEA
Input: e, p, r, h, w, K
1: Define gene position of each investment alternative i according to permutation $\pi(i)$ based on dependency structure
2: $t := 0$
3: Generate initial population $P(t)$
4: Initialise elite population $Q(t) := \emptyset$
5: Evaluate $P(t)$
6: **Repeat**
7:   Select individuals from $P(t)$
8:   Recombine selected individuals (variation operator 1)
9:   Mutate recombined individuals (variation operator 2)
10: Apply local search to mutated individuals (var. operator 3)
11: Create offspring population $P'(t)$
12: Evaluate joint population $J(t) := P(t) \cup P'(t)$
13: Update elite population $Q(t)$ from $J(t)$
14: Generate $P(t+1)$ from $J(t)$
15: $t := t + 1$
16: **Until** $Q(t) = Q(\max\{0, t - t_{diff}\}) \vee t > t_{max}$
Output: $Q(t)$

The evaluation of $P(t)$ in line 5 and $J(t)$ in line 12 is based on the non-domination concept proposed by Goldberg [13] and explicitly formulated for constrained problems e. g. by Deb [7]. In our context, it leads to the following type of domination check (cf. [7], p. 288) which extends Definition 2.9 by the cases 13 and 14.

Definition 3.2 *Given are two distinct portfolio structures x and y. x constraint-dominates y if and only if one of the following cases is true:*

$$cap(x,w) \leq K \wedge cap(y,w) \leq K \wedge$$
$$ret(x,p,r) > ret(y,p,r) \wedge risk(x,p;h) \leq risk(y,p;h) \quad (11)$$
$$cap(x,w) \leq K \wedge cap(y,w) \leq K \wedge$$
$$ret(x,p,r) \geq ret(y,p,r) \wedge risk(x,p;h) < risk(y,p;h) \quad (12)$$

$$cap(x,w) \leq K \wedge cap(y,w) > K \quad (13)$$

$$cap(x,w) > K \wedge cap(y,w) > K \wedge cap(x,w) < cap(y,w). \quad (14)$$

If x constraint-dominates y, we will denote this relationship by $x >_c y$.

The first two cases in Definition 3.2 refer to the cases from Definition 2.9 where only feasible solutions were considered. Case 13 expresses a preference for feasible over infeasible solutions and case 14 prefers the solution that has lower constraint violation.

The non-dominated sorting procedure in our HMOEA uses the dominance criterion from Definition 3.2 to classify the solutions in a given population, e. g. $P(t)$, into different levels of constraint-domination. The best solutions which are not constraint-dominated by any other solution in the population, obtain fitness value 1 (best rank). After that, only the remaining solutions are checked for constraint-domination, and the non-constraint-dominated solutions among these obtain fitness value 2 (second best rank). This process is repeated until each solution has obtained an associated fitness rank.

In line 7 from Table 1, the selection operator is performed using a binary tournament based on Definition 3.2. Two individuals x and y are randomly drawn from the current population $P(t)$, using uniform probability of $p_{sel} := \frac{1}{|P(t)|}$ for each individual. These individuals are checked for constraint-domination according to Definition 3.2 and if, without loss of generality, $x >_c y$ then x wins the tournament and is considered for reproduction. If none of the two solutions dominates the other, they cannot be compared using the constraint-domination criterion, and the winning solution is finally determined using a draw from a uniform distribution over both possibilities.

The first two variation operators are the standard one point crossover and the standard mutation operator as described in section 3.2. Our third variation operator in line 10 of Table 1 represents a problem-specific local search procedure that is applied with probability p_{local} to each selected solution x after crossover and mutation. This local search procedure can exploit the structure of a given solution x to perform an additional local optimisation of x towards the global, feasible Pareto-efficient set, e. g. by using a so-called hill climbing algorithm that changes x according to local information about our objective functions in the region around x. We consider this to be a significant improvement compared a standard, non-hybrid MOEA since the randomised search process of the MOEA can be guided a bit more towards the global, feasible Pareto-efficient set and therefore, such a local search operator can improve the convergence speed of the overall algorithm towards the desired solutions. This is particularly important for real-world applications, where speed matters when large portfolios are to be considered. In addition to these arguments, some portfolio credit risk models provide additional local structure information for a current solution x beyond the objective function values that can be exploited very efficiently from a computational complexity's perspective. An example underlining this fact will be provided in section 4 of this article.

By applying the variation operators to the selected individuals we obtain an offspring population $P'(t)$. The members of the joint population $J(t)$ containing all parent solutions from $P(t)$ and all offspring solutions from $P'(t)$ are evaluated using the non-dominated sorting procedure described above. After that, the elite population $Q(t)$ is updated.

Before finishing the population step t and setting $t \to t+1$ the members of the new parent population $P(t+1)$ have to be selected from $J(t)$ since

$|J(t)| > |P(t+1)|$ by definition of $J(t) := P(t) \cup P'(t)$. Since elitist EAs, which preserve the best solutions from both parents and offspring, usually show better convergence properties, we also use this mechanism in our algorithm. Besides elitism, we also need a diversity preserving concept to achieve a good distribution of the solutions in the whole objective space. We incorporate the concept of crowding-sort proposed in Deb [7], p. 236. This diversity-preserving mechanism is favourable over other proposals, e. g. niche counting based on Euclidean ε-regions in the decision variable space or the objective function space since the crowding-sort does not require an additional parameter $\varepsilon > 0$ which is difficult to estimate particularly for our discrete, non-linear and non-convex Problem 1, and in our case the crowding sort has a smaller computational complexity of $O(|J(t)|\log|J(t)|)$ compared to the quadratic complexity which is required by other mechanisms (cf. Deb [7], p. 237).

The algorithm terminates if $Q(t)$ has not been improved for a certain number t_{diff} of population steps or if a maximum number of t_{max} population steps has been performed.

We will now provide some references to related research in the area of portfolio optimisation problems, particularly using downside risk measures.

3.5 Other Approaches to Similar Problems

In the existing literature, the main focus of portfolio selection and optimisation has been in the area of stock portfolio investments where Markowitz [19] created the standard framework for calculating efficient frontiers of investment alternatives. Based on his mean-variance approach, many different calculation procedures have been suggested, see e. g. Elton & Gruber [9] for an overview.

For downside risk measures like the Value-at-Risk which is similar to the Credit-Value-at-Risk from Definition 2.6, different approaches have been proposed even in a single objective function setting where the expected return from a portfolio is to be maximised with respect to a fixed level of downside risk. This is due to the mathematical properties of such percentile-based downside risk measures, cf. our remarks and references in Section 2. For example, Gilli & Kellezi [12] used the Threshold Accepting heuristic to approximate the efficient set of a stock portfolio in a Value-at-Risk based setting. A comparison of different heuristic approaches to constrained stock portfolio optimisation problems was e. g. performed by Chang et al. [5].

Concerning credit portfolios, Andersson & Uryasev [1] proposed the use of simplex algorithms under a tail conditional expectation risk measure (Conditional Value-at-Risk) in a simulation model framework. Lehrbass [18] proposed the use of Kuhn-Tucker optimality constraints for a credit portfolio optimisation problem having real-valued variables.

In the next section we will describe more details of our implementation of the hybrid evolutionary framework and present results of an empirical study where the performance of the implementation is investigated.

The first work proposing the use of Evolutionary Algorithms for solving credit portfolio optimisation problems related to Problem 1 was Schlottmann & Seese [26] in 2001. In that work, a hybrid EA was introduced to solve a constrained optimisation problem that was build upon a single objective function combining the aggregated return and the aggregated risk of a credit portfolio.

4 An Implementation and Empirical Test of the HMOEA Framework Using CreditRisk+

4.1 Portfolio Credit Risk Models and Overview of CreditRisk+

When implementing an algorithm for risk-return optimisation of credit portfolios, the first question is how to model the dependencies between joint changes of the credit quality of different investment alternatives, so we have to choose a model that provides us a function h according to Definition 2.1 and beyond that, an algorithm for computing $risk(...)$ according to Definition 2.5.

In the literature, there are different alternatives for modelling the dependencies between obligors and for calculating portfolio credit risk measures. Among these alternatives, CreditMetrics (see Gupton et al. [15]), CreditRisk+ (see CreditSuisse Financial Products [6]) Wilson's model (see Wilson [32], [33] and the KMV option based approach (see Kealhofer [17]) are intensively discussed in many academic and application-oriented publications. Since we will set our focus on the default risk of loan portfolios in our empirical study described later, we will concentrate on CreditRisk+. However, our hybrid framework is compatible with any other portfolio credit risk model providing a loss (in case of a default mode model) or a profit-loss distribution (in case of a mark-to-market model).

In the following paragraphs, we will give a brief description of the CreditRisk+ General Sector Analysis model for a one year horizon here that concentrates on the main issues concerning our algorithm (see CreditSuisse Financial Products [6], pp. 32-57 for a more detailed derivation of the model). It is an actuarial approach that uses an intensity based modelling of defaults, i. e. the default of each obligor in the portfolio is considered to be a stopping time of a hazard rate process expressed by a Poisson-like process. In case of a default event, the amount of credit exposure (net exposure) lent to the defaulting obligor will be entirely lost.

Given is the data of m obligors in the portfolio (cf. Definition 2.1). Particularly, each obligor has a net exposure e_i, an associated annual mean default rate p_i (typically, p_i is small: $0 < p_i < 0.1$) and an annual default rate volatility $\sigma_i \geq 0$. Furthermore, there is a total of n independent sectors as common risk factors, where the first sector ($k = 1$) is obligor specific, i. e. in this sector there is no implicit default correlation between obligors ($k = 1, .., n$ below unless otherwise noted). The obligors are allocated to the sectors according to sector weights $\Theta_{ik} \in [0,1], \forall i : \sum_{k=1}^{n} \Theta_{ik} = 1$.

The probability generating function (abbreviated PGF) for the losses from the entire portfolio is defined by

$$G(z) := \sum_{i=0}^{\infty} prob(\text{aggregated losses} = i \cdot L) \cdot z^i \quad (15)$$

where L is a constant defining net exposure bands of constant width and $prob(\ldots)$ represents the probability of losing i times the value of L from the whole portfolio.

Since the sectors are independent this can be decomposed to

$$G(z) = \prod_{k=1}^{n} G_k(z) \quad (16)$$

where $G_k(z)$ is the PGF for the losses from the portfolio in sector k.

To obtain the approximated cumulative loss distribution function for the portfolio a recurrence relation, the recursion algorithm by Panjer [20], can be applied to evaluate the coefficients of the PGF (for a more detailed background see Panjer & Willmot [21]). After that, risk figures, e. g. the 99th percentile, can be calculated.

An interesting feature of the model concerning the portfolio optimisation task are the marginal risk contributions of obligor i to the standard deviation of portfolio credit risk:

$$RC_i^{\sigma} := e_i \frac{\partial \sigma_{pf}}{\partial e_i} = \frac{e_i p_i}{\sigma_{pf}} \left(e_i + \sum_{k=1}^{n} \left(\frac{\sigma_k}{\mu_k} \right)^2 e_i p_i \Theta_{ik} \right) \quad (17)$$

where σ_{pf} is the portfolio standard deviation derived from the PGF of the portfolio losses, μ_k, σ_k are sector specific parameters calculated directly from the input parameters $e_i, p_i, \sigma_i, \Theta_{ik}$ using formula (18) below (note that $\sigma_1 = 0$ by definition of sector 1).

$$\forall k : \mu_k := \sum_{i=1}^{m} \Theta_{ik} p_i, \sigma_1 := 0, \forall k > 1 : \sigma_k := \sum_{i=1}^{m} \Theta_{ik} \sigma_i \quad (18)$$

Alternatively, by setting $\sigma_k := \omega_k \mu_k$ for $k > 1$ using parameters (variation coefficients) $\omega_k, k = 2, \ldots, n$ only μ_k has to be calculated according to (18) and in this case, no obligor-specific default rate volatilities σ_i are required to calculate the sector specific parameters.

To calculate an approximation for the risk contribution e. g. to the 99th percentile, a scaling factor is defined in the following manner:

$$\xi_{pf} := \frac{q_{pf}^{0.99} - \mu_{pf}}{\sigma_{pf}} \qquad (19)$$

where $\mu_{pf}, \sigma_{pf}, q_{pf}^{0.99}$ are the expectation, standard deviation and 99th percentile of the portfolio loss distribution, respectively.

The figures calculated by applying formula (17) can be used as a basis for the approximate risk contribution to the 99th percentile by scaling the risk contribution obtained from (17) according to ξ_{pf} and adding it to the obligor specific expected loss:

$$RC_i^{0.99} := e_i p_i + \xi_{pf} RC_i^{\sigma} \qquad (20)$$

We use these figures to ensure a computationally efficient calculation within our local search variation operator in the HMOEA implementation described in the next subsection. The calculation of (20) for all investment alternatives $i \in \{1, ..., m\}$ requires only $O(mn)$ additional operations after the calculation of the coefficients of the PGF from formula (16) which is mandatory to evaluate the $risk()$ target function for each individual. Note that the number of sectors n is constant in a given problem instance and usually small ($n < 10$), so the computation of (20) requires only linear computing time measured by the number of investment alternatives m.

4.2 Further Implementation Details Referring to the HMOEA Scheme

In the CreditRisk$^+$ model, the volatilities of the obligors' default probabilities in conjunction with the common risk factors of all obligors replace a direct modelling of the default correlation $\rho(i, j)$ for two investment alternatives i, j. Therefore, for the calculations according to this portfolio credit risk model, no explicit default correlations are required. However, in CreditSuisse Financial Products [6], p. 56ff. the following implicit default correlation formula is provided:

$$\rho(i,j) \approx \sqrt{p_i p_j} \sum_{k=1}^{n} \Theta_{ik} \Theta_{jk} \left(\frac{\sigma_k}{\mu_k}\right)^2 \qquad (21)$$

By setting $h(i,j) := \rho(i,j)$ according to (21), we obtain the complete dependence structure required for our Definition 2.1 and the subsequent results from our theoretical considerations in Section 2 of this article. Moreover, this explicit definition of the dependence structure can be exploited to provide an adequate genetic modelling of the decision variables for the given portfolio data using the maximum strength values from Definition 3.1.

To create a local search operator required by an implementation of the HMOEA scheme, we use the following local search target function that uses the quotient between aggregated net return and aggregated risk to evaluate a given portfolio structure x:

$$f(x,p,r;h) := \frac{ret(x,p,r)}{risk(x,p;h)} \qquad (22)$$

Considering Definitions 2.4 and 2.6 as well as the CreditRisk$^+$ calculation method for the 99th percentile $q_{pf}^{0.99}(x,p,\sigma,\Theta)$ of the cumulative distribution of aggregated losses from the portfolio structure x under the given data p, σ, Θ, r the function f can be written as:

$$f(x,p,\sigma,\Theta,r) := \frac{\sum\limits_{i=1}^{m} x_i(r_i - p_i)}{q_{pf}^{0.99}(x,p,\sigma,\Theta) - \sum\limits_{i=1}^{m} x_i p_i} \qquad (23)$$

If we maximise this function f we will implicitly maximise $ret(x,p,r)$ and minimise $risk(x,p;h)$, and this will drive the portfolio structure x towards the set of global Pareto-efficient portfolio structures (cf. the domination criteria specified in Definition 2.9). In addition to that, we have to consider our constraints to ensure the local search variation operator keeps the portfolio structure x feasible or moves an infeasible portfolio structure x back into the feasible region. An overview of our local search operator scheme based on these considerations is shown in Table 2.

The partial derivative d_j for obligor j required in line 12 of Table 2 can be calculated using the following formula (a proof is provided in the appendix):

$$d_j := \frac{x_j(r_i - p_i)(\xi_{pf}\sigma_{pf}) - \left(\sum\limits_{i=1}^{m} x_i(r_i - p_i)\right)(\xi_{pf} RC_j^{\sigma})}{x_j(\xi_{pf}\sigma_{pf})^2} \qquad (24)$$

If the current solution x from $P(t)$ to be optimised with probability p_{local} is infeasible because the capital restriction is violated (cf. line 2 in Table 2), the algorithm will remove the investment alternative having the minimum gradient component value from the portfolio (lines 14 and 15). This condition drives the hybrid search algorithm towards feasible solutions. In case of a feasible solution that is to be optimised, the direction of search for a better solution is determined by a draw of a uniformly distributed (0,1)-random variable (cf. lines 5 and 6). This stochastic behaviour helps preventing the local search variation operator from stalling into the same local optima. The local search algorithm terminates if the current solution cannot be modified further, if it is already included in the populations $P(t)$ or $Q(t)$ or if no improvement considering the violation of constraints or the target function can be made.

Remembering the fact that the risk contributions, and therefore, the partial derivatives d_j can be calculated in linear time for an individual which has already a valid fitness evaluation this yields a very fast variation operator.

Table 2. Local search operator scheme

Local search operator

Input: $e, p, \sigma, \Theta, r, w, K, P(t)$

1: **For each** $x \in P(t)$ apply the following instructions with probability p_{local}
2:    **If** $cap(x, w) > K$ **Then**
3:      $D := -1$
4:    **End If**
5:    **If** $cap(x, w) \leq K$ **Then**
6:      Choose between $D := 1$ or $D := -1$ with uniform probability 0.5
7:    **End If**
8:    Initialisation $\forall i : \widehat{x}_i := x_i$
9:    **Do**
10:      Copy $\forall i : x_i := \widehat{x}_i$
11:      $ret_{old} := \sum_{i=1}^{m} x_i(r_i - p_i)$ and $risk_{old} := q_{pf}^{0.99}(x, p, \sigma, \Theta) - \sum_{i=1}^{m} x_i p_i$
12:      For each x_j calculate the partial derivatives $d_j := \frac{\partial}{\partial x_j} f(x, p, \sigma, \Theta, r)$
13:      **If** $D = -1$ **Then**
14:        Choose the minimal gradient component $\widehat{i} := \arg\min_{j} \{d_j \,|\, x_j > 0\}$
15:        Remove this exposure: $\widehat{x}_{\widehat{i}} := 0$
16:      **Else**
17:        Choose the maximal gradient component $\widehat{i} := \arg\max_{j} \{d_j \,|\, x_j = 0\}$
18:        Add this exposure to portfolio: $\widehat{x}_{\widehat{i}} := e_{\widehat{i}}$
19:      **End If**
20:      $ret_{new} := \sum_{i=1}^{m} \widehat{x}_i(r_i - p_i)$ and $risk_{new} := q_{pf}^{0.99}(\widehat{x}, p, \sigma, \Theta) - \sum_{i=1}^{m} \widehat{x}_i p_i$
21:    **While** $(\exists i : \widehat{x}_i > 0) \land (\exists j : \widehat{x}_j = 0) \land \widehat{x} \notin P(t) \land \widehat{x} \notin Q(t) \land$
       $((D = -1 \land cap(\widehat{x}, w) > K) \lor (D = 1 \land cap(\widehat{x}, w) \leq K \land$
       $(ret_{new} > ret_{old} \lor risk_{new} < risk_{old})))$
22:    Replace x in $P(t)$ by its optimised version
23: **End For**

Output: $P(t)$

The next section contains the parameters and test cases for an empirical test of the implemented hybrid framework for credit portfolio risk-return analysis and optimisation.

4.3 Specification of Test Cases, Parameters and Performance Criteria

Besides our Hybrid Multi-Objective Evolutionary Algorithm (HMOEA), we have also implemented a simple enumeration algorithm that investigates all possible portfolio structures to determine the feasible global Pareto-efficient set PE^* having maximum cardinality for small instances of our Problem 1,

i. e. the latter algorithm serves as a proof for the globally optimal portfolio structures that should be discovered by the other search algorithms. For all instances considered in this article, we compared the results of the HMOEA to the respective results of a non-hybrid MOEA that incorporates all features of the HMOEA except for the local search operator which is disabled in the non-hybrid algorithm. Particularly, the MOEA also benefits from all problem specific algorithmic features that we have proposed for the HMOEA in the previous sections, e. g. the presence of the elite population and the preprocessing algorithm. All tests of the above implementations were carried out on a standard desktop PC (800 MHz single CPU). For all evolutionary algorithms, we performed 20 independent runs of the same algorithm on the respective test problem using different pseudorandom number generator seeds.

Although more test cases were examined during development of the system (e. g. for estimating the parameters like $|P(t)|$, p_{cross} etc.) we focus on the following sample loan portfolios in this paper. The structure of these portfolios is analogous to real world data.[2]

Our first test data set is named portfolio m20n2. It consists of $m = 20$ investment alternatives which are allocated to $n = 2$ sectors. The capital budget restriction is assumed to be 50% of the maximum capital requirement that will be required if all investment alternatives are held in the portfolio. The detailed structure of portfolio m20n2 is provided in the appendix.

The medium size portfolio m45n2 contains $m = 45$ investment alternatives allocated to $n = 2$ sectors. The capital restriction is $K := 80000$ which is about 71% of the sum of all investment alternatives' capital requirements.

The largest problem instance named portfolio m100n3 contains $m = 100$ investment alternatives allocated to $n = 3$ sectors. A capital restriction is set to about 56% of the sum of all investment alternatives' capital requirements.

In all test cases, we chose a quite common parameter setting of $p_{cross} := 0.95$ and $p_{mut} := \frac{1}{m}$, which is reported to work well in many other EA studies, and this was also supported by test results during our development of the HMOEA and the non-hybrid MOEA.

The choice of p_{local} can be made by the respective user of the HMOEA depending on his or her preferences: If one is interested in finding better solutions in earlier populations, the probability shall be set higher, and in this case more computational effort is spent by the algorithm on the local improvement of the solutions. However, the local search optimisation pressure should not be too high since one is usually also interested in finding a diverse set of solutions. Therefore, a choice of $0 < p_{local} \leq 0.1$ appears to be adequate, and this is supported by our tests.

For the portfolio m20n2 data set, we chose $|P(t)| := 30$ individuals per population, and $p_{local} := 0.005$. The evolutionary process was stopped after $t_{max} := 1000$ population steps. For the non-hybrid MOEA all these parame-

---

[2] All test portfolios can be retrieved via http://www.aifb.uni-karlsruhe.de/CoM/HMOEA/tests.html.

ters were set equally except for $p_{local} := 0$ which means that there is no third variation operator in the non-hybrid MOEA.

In the portfolio m45n2 test, we set $|P(t)| := 40$ individuals per population and the probability for the third variation operator in the HMOEA was set higher to $p_{local} := 0.05$ due to the larger search space. The other parameters were set analogously to the portfolio m20n2 test case. In addition to the investigation of the detailed results for $p_{local} = 0.05$, we will compare the average results of our chosen performance metrics for different settings of p_{local} to show the influence of the third variation operator on the results in the HMOEA.

In the portfolio m100n3 test, we set $|P(t)| := 50$ individuals per population and the probability for the third variation operator in the HMOEA was set to $p_{local} := 0.1$ to reveal the significant differences between the hybrid and the non-hybrid approach. Again, the other parameters were set according to the portfolio m20n2 test case.

Particularly, to achieve a better comparison between the evolutionary algorithms we used the same initial population for both the HMOEA and the non-hybrid MOEA given a specific pseudorandom generator seed. This means we used the same 20 (randomly determined) initial populations for both algorithms on a test data set to obtain a fair basis for the comparison of the results.

For all test cases, we have calculated performance measures of the algorithms based on the set coverage metric from Zitzler [34]. In our context, the set coverage metric is defined as follows:

Definition 4.1 *Given are two sets of portfolio structures PE_1, PE_2 which are approximations for PE^* defined in Problem 1. The pair of set coverage metric values $C_{1,2} := (C_1, C_2)$ is calculated by*

$$C_1 := C(PE_2, PE_1) = \frac{|\{x \in PE_1 | \exists y \in PE_2 : y >_c x\}|}{|PE_1|} \quad (25)$$

$$C_2 := C(PE_1, PE_2) = \frac{|\{y \in PE_2 | \exists x \in PE_1 : x >_c y\}|}{|PE_2|} \quad (26)$$

This metric provides us a criterion for comparing two different sets of solutions produced by different algorithms. We have chosen this metric since it allows the comparison of approximation sets having different cardinalities, and particularly in our larger test cases, we do not need PE^* for the evaluation of the results. An algorithm 2 calculating PE_2 is considered to be better in convergence to PE^* than an algorithm 1 that computes PE_1 if $C_1 > C_2$, i. e. if the fraction of solutions in PE_2 which are dominated by solutions from PE_1 is smaller than the fraction of solutions in PE_1 that are dominated by solutions from PE_2. To be more transparent, we investigate both the nominator and the denominator of (25) and (26) separately. Therefore, two important goals of multi-objective approximation algorithms are evaluated: Finding an approximation set whose elements are very close to corresponding members of PE^* and which also has a high cardinality. So we can compare both the quantity and the quality of two alternative approximations for PE^*.

In addition to the evaluation of these goals, we compare the maximum spread (cf. Zitzler [34]) for each calculated approximation set for PE^* according to the next definition.

Definition 4.2 *Given is an approximation set of portfolio structures PE_1 for PE^* defined in Problem 1. The maximum spread value $\delta(PE_1)$ is obtained by evaluation of*

$$\delta(PE_1) := \sqrt{dist_{ret}^2 + dist_{risk}^2} \qquad (27)$$

where

$$dist_{ret} := \max_{x \in PE_1} (ret(x, p, r)) - \min_{x \in PE_1} (ret(x, p, r))$$

$$dist_{risk} := \max_{x \in PE_1} (risk(x, p; h)) - \min_{x \in PE_1} (risk(x, p; h)).$$

The maximum spread allows a comparison between different approximation sets based on the largest Euclidean distance between two solutions in the two-dimensional objective function space. We have chosen this additional metric because the set coverage metric does not cover the largest spread between the found solutions which is also a goal in multi-objective optimisation. A larger spread is preferable, i. e. an approximation set PE_1 is better than another set PE_2 concerning this criterion if $\delta(PE_1) > \delta(PE_2)$.

Of course, we calculate the set coverage metric and the maximum spread from the members of the elite population $Q(t)$ for a fixed value t after running the respective algorithms. We will present the results in the next subsection.

4.4 Empirical Results

In all test cases, the approximation set calculated by the non-hybrid MOEA is denoted by PE_1, the approximation set from running the HMOEA is denoted by PE_2. First of all, for the portfolio m20n2 test data set, we compare the result PE_2 of one HMOEA run to PE^* which was obtained by a complete enumeration of the search space that required approximately 72 minutes. In contrast to this, each run of the HMOEA (as well as a run of the non-hybrid MOEA) required about 3 minutes for the computation of an approximation set PE_2. Figure 1 shows both results. It is easy to check by visual inspection that PE_2 is a good approximation set for PE^* since all points of PE^* (indicated by circles) are approximated by mostly identical or at least very close points of PE_2 which are marked by a respective 'x' in Figure 1.

Table 3 shows the detailed results of the HMOEA and the non-hybrid MOEA for this small portfolio. The second and third column display the nominator of the respective C metric value whereas the fourth and the fifth column

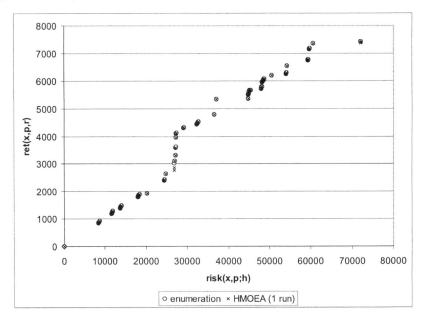

Fig. 1. Comparison of PE^* and PE_2 for portfolio m20n2

contain its denominator value. The results indicate that both algorithms find quite similar solutions as we expect it when considering the very small local search variation operator probability $p_{local} = 0.005$ for the HMOEA in this case. However, the quality of the solutions found by the HMOEA concerning the set coverage metric is on average a bit better than the quality of the solutions by the non-hybrid algorithm which is indicated by the smaller number of dominated solutions in PE_2 (column 3) compared to PE_1 (column 2). Furthermore, the number of runs where less solutions from PE_2 are dominated by solutions from PE_1 (10 runs) is higher than vice versa (6 runs). In addition to this slightly better performance, the HMOEA found more solutions on average and in more runs (9 runs) than the other algorithm (7 runs), therefore the average value of the set coverage metric is about two times better for the HMOEA. Concerning the maximum spread values, both algorithms yielded quite the same results, but the non-hybrid approach is slightly better in this criterion due to the higher number of runs (6 versus 4) where it had higher spreads than the HMOEA.

Summarising the results in the small constrained test case, both algorithms found a good approximation set for PE^* within a few minutes. The HMOEA shows a higher convergence speed but this is at the cost of a slightly reduced maximum spread of the approximation set compared to the non-hybrid algorithm in our test runs. This is mainly due to the fact that the intended higher convergence pressure towards feasible, global non-dominated solutions caused

Table 3. Comparison of PE_1 and PE_2 for portfolio m20n2 (better values are in bold face)

run #	nom (C_1)	nom (C_2)	denom (C_1)	denom (C_2)	C_1	C_2	$\delta(PE_1)$	$\delta(PE_2)$
1	0	2	81	80	**0.0000**	0.0250	51337.58	51337.58
2	2	1	75	79	0.0267	**0.0127**	51337.58	51337.58
3	1	1	77	81	0.0130	**0.0123**	51337.58	51337.58
4	1	3	79	78	**0.0127**	0.0385	51191.22	**51337.58**
5	0	1	76	76	**0.0000**	0.0132	51337.58	51337.58
6	3	3	77	79	0.0390	0.0380	51191.22	**51337.58**
7	6	1	81	77	0.0741	**0.0130**	**51337.58**	51191.22
8	3	0	80	81	0.0375	**0.0000**	51337.58	51337.58
9	5	1	80	76	0.0625	0.0132	**51337.58**	51191.22
10	1	3	76	78	**0.0132**	0.0385	51191.22	**51337.58**
11	1	2	81	77	**0.0123**	0.0260	51337.58	51337.58
12	6	0	75	79	0.0800	**0.0000**	**51337.58**	51191.22
13	1	0	77	77	0.0130	**0.0000**	51191.22	51191.22
14	1	1	80	80	0.0125	0.0125	51337.58	51337.58
15	2	0	72	80	0.0278	**0.0000**	51285.10	**51337.58**
16	4	1	80	76	0.0500	0.0132	**51337.58**	51191.22
17	4	0	79	76	0.0506	**0.0000**	**51337.58**	51191.22
18	0	1	77	77	**0.0000**	0.0130	51337.58	51337.58
19	2	2	79	83	0.0253	**0.0241**	51337.58	51337.58
20	1	0	77	78	0.0130	**0.0000**	51191.22	51191.22
avg.	2.20	**1.15**	77.95	**78.40**	0.0282	**0.0146**	**51298.37**	51286.35

by the local search operator leads to early discovery of isolated Pareto-optimal solutions which might strongly dominate the population in the relatively small search space. However, this is not a general disadvantage of the hybrid algorithm, since we have to remind at this point that there is a trade-off between the two goals of finding globally optimal solutions very fast and discovering a diverse set of solutions, and this conflict is to be faced by any algorithm that solves instances of Problem 1. We have put more weight on the first criterion in conjunction with the discovery of feasible solutions during development of the hybrid algorithm, and the maximum spread of the HMOEA is very close to the globally optimal maximum spread of PE^*, so the slightly lower maximum spread of PE_2 is not critical.

Beyond that, we will show now that in the other test cases, which have larger search spaces that grow exponentially in the number of investment alternatives m, the hybrid approach can exploit its advantages more significantly and usually yields both a better set coverage metric value and a higher maximum spread. To underline this claim, Table 4 shows the results in the portfolio m45n2 test case. A run of the non-hybrid MOEA required about 7

Table 4. Comparison of PE_1 and PE_2 for portfolio m45n2 (better values are in bold face)

run #	nom (C_1)	nom (C_2)	denom (C_1)	denom (C_2)	C_1	C_2	$\delta(PE_1)$	$\delta(PE_2)$
1	118	**67**	428	**446**	0.2757	**0.1502**	65566.46	**70334.59**
2	157	**73**	450	449	0.3489	**0.1626**	**73449.55**	73330.93
3	80	88	450	445	**0.1778**	0.1978	66534.15	**70936.96**
4	160	**67**	450	450	0.3556	**0.1489**	67954.26	**73449.55**
5	115	**62**	417	**448**	0.2758	**0.1384**	65866.56	**73330.93**
6	129	**48**	441	450	0.2925	**0.1067**	68265.32	**70334.59**
7	120	**72**	450	450	0.2667	**0.1600**	66537.58	**70936.96**
8	151	**76**	450	450	0.3356	**0.1689**	68940.85	**70936.96**
9	137	**61**	450	447	0.3044	**0.1365**	67126.12	**73330.93**
10	136	**57**	449	450	0.3029	**0.1267**	65764.79	**72536.10**
11	130	**72**	450	442	0.2889	**0.1629**	66349.65	**70936.96**
12	119	**79**	437	450	0.2723	**0.1756**	66236.72	**70334.59**
13	130	**84**	450	430	0.2889	**0.1953**	68427.92	**70334.59**
14	170	**58**	450	447	0.3778	**0.1298**	68010.23	**73475.61**
15	121	**65**	450	446	0.2689	**0.1457**	69228.15	**72732.18**
16	139	**71**	436	**449**	0.3188	**0.1581**	69674.10	**70936.96**
17	140	**96**	448	450	0.3125	**0.2133**	69223.61	**70334.59**
18	171	**62**	450	450	0.3800	**0.1378**	68316.28	**73112.27**
19	163	**62**	**442**	441	0.3688	**0.1406**	72742.12	**73465.07**
20	110	**85**	442	450	0.2489	**0.1889**	66938.45	**70334.59**
avg.	134.8	**70.25**	444.5	**447**	0.3031	**0.1572**	68057.64	**71772.80**

min. 30 sec. to compute the approximation set PE_1 in this setting, and the HMOEA required approximately 8 minutes to calculate PE_2.

The hybrid approach is better in all averages of the performance metrics for our medium size test case. Except for one of the 20 independent runs, the HMOEA always found remarkably better solutions than the other algorithm (cf. the second and the third column). Moreover, the hybrid algorithm found quite the same number of solutions in all runs like the non-hybrid MOEA, therefore the set coverage metric value is significantly better for the HMOEA due to the better quality of the found solutions. In contrast to the results presented above for the smaller portfolio, the maximum spread values of PE_2 are also much better than the respective values of PE_1 except for one run so the hybrid approach is favourable concerning both performance criteria defined in Section 4.3.

The influence of the local search variation operator on the results in our medium size test case is indicated in Figures 2 and 3 where we have plotted the average performance metric values depending on different settings of p_{local}. We have plotted ordinary least squares (OLS) regression lines in each figure to estimate the linear trend of the performance metric values depending on the choice of of p_{local}.

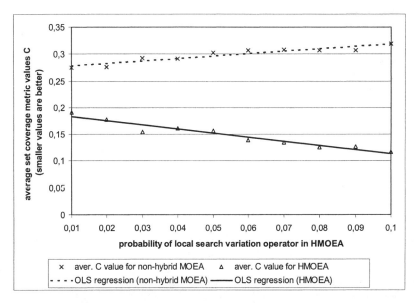

Fig. 2. Average set coverage metric values depending on p_{local} for portfolio m45n2

Since both algorithms rely on probabilistic variation operators, we cannot expect perfect monotony of the performance metrics depending on the variation of p_{local}. However, the linear regression lines in both figures clearly indicate the influence of the local search variation operator. For the set coverage metric, a higher value of p_{local} typically leads to a higher quality of the solutions discovered by the HMOEA compared to the solutions discovered by the non-hybrid MOEA. Remembering the fact that smaller set coverage metric values are preferable, this is indicated by both the negative slope of the regression line for the set coverage metric values of the HMOEA and the positive slope of the other regression line for the non-hybrid algorithm in Figure 3. Of course, different settings of p_{local} do not influence the maximum spread of the non-hybrid MOEA whereas the hybrid algorithm benefits from higher p_{local} values since the slope of the regression line is positive.

Beyond this analysis of the influence of p_{local} on the results, we can see in the portfolio m45n2 test case that the hybrid approach is preferable if the convergence speed, the quality of the found solutions and the maximum spread in the objective function space matters. In addition to our above discussion of the detailed results for $p_{local} = 0.05$, this is underlined by the fact that for

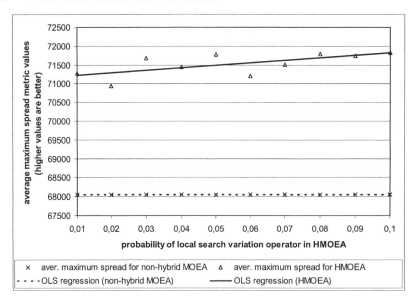

Fig. 3. Average maximum spread values depending on p_{local} for portfolio m45n2

each tested value of p_{local} all average values of the performance metrics shown in Figure 3 were better in the hybrid case.

For the largest portfolio of our test data sets the results of the HMOEA and the non-hybrid MOEA after $t_{max} = 1000$ population steps are displayed in Table 5. A single run of the HMOEA required about 16 minutes for the calculation of PE_2 due to the high value of $p_{local} = 0.1$ which was chosen to reveal the differences between both algorithms, whereas the non-hybrid algorithm terminated within 11 minutes.

In the portfolio m100n3 test case, the HMOEA clearly outperforms the non-hybrid approach in every average of the performance criteria. Beyond that, even in each single run the hybrid approach found more solutions than the non-hybrid algorithm (compare the fifth to the fourth column), and the quality of the found solutions is also better concerning the number of dominated solutions (see third versus second column). Therefore, $C_1 > C_2$ in each test run. Obviously, this leads to a significant difference between the average set coverage metric values where the HMOEA is more than three times better than the non-hybrid approach. Concerning the maximum spread, the HMOEA is better in the average over all test cases, and there is only one case where the hybrid approach is slightly worse than the non-hybrid approach whereas in all other cases, the hybrid approach produces an approximation set PE_2 that has a larger maximum spread value than the other approximation set PE_1.

As a consequence, the results of all tests, and particularly the medium and large test cases, support our claim that the hybridisation of the MOEA im-

Table 5. Comparison of PE_1 and PE_2 for portfolio m100n3 (better values are in bold face)

run #	nom (C_1)	nom (C_2)	denom (C_1)	denom (C_2)	C_1	C_2	$\delta(PE_1)$	$\delta(PE_2)$
1	287	82	477	550	0.6017	**0.1491**	62852.17	**72033.78**
2	267	100	497	536	0.5372	**0.1866**	69200.63	**71744.75**
3	273	104	471	563	0.5796	**0.1847**	62622.74	**70284.81**
4	235	111	432	543	0.5440	**0.2044**	65324.67	**72033.78**
5	249	108	487	554	0.5113	**0.1949**	61872.88	**70284.81**
6	268	87	468	526	0.5726	**0.1654**	61469.33	**70284.81**
7	192	101	430	576	0.4465	**0.1753**	55946.92	**73594.82**
8	235	118	447	570	0.5257	**0.2070**	61485.23	**70284.81**
9	260	94	466	572	0.5579	**0.1643**	64166.60	**72033.78**
10	245	88	458	532	0.5349	**0.1654**	57477.45	**70284.81**
11	236	115	468	549	0.5043	**0.2095**	67783.41	**73594.82**
12	275	97	447	549	0.6152	**0.1767**	63082.70	**73594.82**
13	215	123	478	542	0.4498	**0.2269**	60424.41	**70284.81**
14	339	56	456	579	0.7434	**0.0967**	**66232.23**	65885.00
15	294	109	462	547	0.6364	**0.1993**	65921.33	**66057.19**
16	275	92	464	576	0.5927	**0.1597**	62888.72	**70284.81**
17	308	94	496	566	0.6210	**0.1661**	61514.92	**73594.82**
18	273	77	470	551	0.5809	**0.1397**	66370.62	**72033.78**
19	285	85	434	579	0.6567	**0.1468**	63651.26	**70284.81**
20	269	113	512	516	0.5254	**0.2190**	69896.12	**72033.78**
avg.	264	97.7	466	553.8	0.5669	**0.1769**	63509.22	**71025.68**

proves the convergence properties of the algorithm. Especially when dealing with very large search spaces, the exploitation of local information around a solution is valuable in the evolutionary process since it drives the evolutionary process faster towards the most promising solutions. On the other hand, the other variation operators are also very important when using such local information since a strong local search process can stall into a small number of local optima which are only a few points compared to a large feasible, Pareto-optimal set. So a hybrid approach is preferable.

In addition to the results presented above, we also tested the performance of the HMOEA and the non-hybrid MOEA without a capital budget restriction for the respective portfolios. This means, we considered the unconstrained cases, too. We do not discuss them in detail here since the comparison of the HMOEA and the non-hybrid algorithm revealed similar results for all portfolio sizes: The average set coverage metric values of the hybrid approach were always better and even the average maximum spread values of the HMOEA were always equal or higher than these performance metric values of the non-hybrid approach. In all unconstrained cases, the HMOEA benefits strongly from its local search variation operator that enforces a higher quality of the discovered solutions and beyond that, leads to the discovery of the extreme solutions at the margins of the objective function spaces, which are not restricted in the unconstrained cases. Thus, the hybrid approach is also favourable in this problem setting.

5 Conclusion and Outlook

In this article we have formally defined a constrained multi-objective portfolio selection problem based on investment alternatives which incorporate credit risk. This problem consists of two conflicting objective functions, the aggregated net return from a portfolio and the aggregated downside risk, and an additional capital budget restriction. We have analysed the structure of the problem from a computational perspective and proved the NP-hardness of its associated decision problem.

For the approximation of a large set of feasible, global non-dominated solutions from the feasible, global Pareto-efficient set of solutions to our portfolio problem, we have proposed a hybrid multi-objective evolutionary algorithm framework that combines concepts from different multi-objective evolutionary algorithm schemes with a problem specific local search operator. The framework is not restricted to linear or convex objective functions and also flexible concerning the constraints. A proper genetic modelling of portfolio credit risk problems has been derived in general, and a fast greedy algorithm as a preprocessing stage to support evolutionary algorithms for portfolio credit risk problems has been developed. Further aspects of the algorithm have been considered, particularly with respect to computational complexity.

We have described the CreditRisk$^+$ portfolio credit risk model and derived a local search operator that exploits specific model features. This basis has been used for an implementation of our hybrid algorithm framework, and we have presented empirical results of a test using different portfolios. The results have indicated that our genetic modelling proposed for portfolio credit risk problems is successful since even a non-hybrid MOEA that used our preprocessing algorithm yielded good results for different problem instances. Moreover, the empirical results of different test portfolios showed that the quality of the MOEA could be improved significantly concerning the convergence speed towards the feasible, global Pareto-efficient set by applying the additional local search variation operator that has been developed in this article. Particularly for the medium and larger cases that we have considered, the hybridisation of the MOEA and the local search algorithm has yielded a better quality of the solutions found at a defined population step as well as a higher spread of the solutions in the objective function space both on average and in the majority of the single, independent algorithm runs. The additional computational cost of the local search variation operator are low compared to the advantages, and the user can decide about the amount of additional computational cost to be invested in favour of a higher convergence speed by setting a single parameter, the probability p_{local} for the application of the local search variation operator to each individual. To support this decision, we have carried out a sample analysis of the influence of this parameter on the performance of the algorithm for one of our test data sets.

Although our implementations of the non-hybrid MOEA and the HMOEA have been running on a single standard desktop PC, the algorithms have found approximations of many feasible, Pareto-optimal solutions in different problem instances within a few minutes. Remembering the fact that EAs are well suited for parallel implementation (see e. g. Schmeck et al. [27]) there are good perspectives for improving the speed of future implementations of our framework by using more than one CPU at least for some parts of the algorithms in each population step.

Further research from the viewpoint of risk modelling can e. g. extend the framework presented here by exploiting the latest developments in the CreditRisk[1] context published in Buergisser et al. [4] to include severity variations concerning the net exposures or use an alternative way of calculating the loss percentiles as proposed by Gordy [14]. Of course, the system can be extended to other credit risk exposure types, e. g. by embedding it into a mark-to-market model context. Due to the flexibility of our framework, many further constraints of practical interest can be easily integrated into our framework, e. g. the simultaneous use of different capital budgets or Credit-Value-at-Risk based limits per rating category and/or industry in the optimisation process. Even more sophisticated restrictions can be handled, e. g. a minimum overall quality of the parts of a portfolio to be sold in an Assed Backed Security transaction which is itself calculated using a non-linear pricing model.

Finally, the system presented in this paper can be integrated into a larger decision support system for risk-return optimisation in a financial institution that supports human portfolio risk-return managers and traders using software agent technology as proposed in Schlottmann & Seese [25].

The authors like to thank Tobias Dietrich, Michael Lesko, Thomas Stuempert and Stephan Vorgrimler for interesting discussions. Furthermore, support from GILLARDON AG financial software, Germany and from an anonymous German bank is hereby gratefully acknowledged.

6 Appendix

6.1 Proof of Lemma 2

The decision problem that corresponds to Problem 1 has the following form:

Problem 3 *Are there numbers $x_i \in \{0, e_i\}$, $x := (x_i)_{i=1,\ldots,m}$ for given non-negative rational numbers $e_i, r_i, p_i, w_i, K, R, Z$ and a rational function h such that the following three inequalities are satisfied:*

$$ret(x, p, r) \geq R \tag{28}$$

$$risk(x, p; h) \leq Z \tag{29}$$

$$cap(x, w) \leq K \tag{30}$$

Now consider an instance of the following decision problem which is known to be NP-complete:

Problem 4 *[0/1 KNAPSACK]. Given are a finite set U, a rational size $s(u) > 0$, a rational value $v(u) > 0$ for each element $u \in U$ and positive rational numbers V, W. Is there an assignment of an integer value $c(u) \in \{0, 1\}$ to each $u \in U$ such that the following two conditions are satisfied:*

$$\sum_{u \in U} c(u) v(u) \geq V \tag{31}$$

$$\sum_{u \in U} c(u) s(u) \leq W \tag{32}$$

We can construct an equivalent instance of Problem 3 for a given [0/1 KNAPSACK] problem instance by using a polynomial time calculable 1-1 function $f : U \to N$ that assigns a subsequent natural number to each element $u \in U$ starting from $f(u) := 1$ for the first element in U and by setting $m := |U|, R := V, K := W$ and

$$\forall i \in \{1, ..., m\} : e_i := 1, r_i := v(f^{-1}(i)), w_i := s(f^{-1}(i)).$$

Furthermore, we can set $\forall i \in \{1, \ldots, m\} : p_i \equiv 0$ for the given instance of [0/1 KNAPSACK] in our instance of Problem 3 so that the inequality (29) is not binding for any given positive rational number Z since $risk(x, p; h) \equiv 0$ according to Definition 2.5 in this case. Thus, we have to consider only inequalities (28) and (30) in our construction of the equivalent Problem 3 instance.

For $p_i \equiv 0$ the inequality (28) simplifies to

$$ret(x, p, r) = \sum_{i=1}^{m} x_i (r_i - p_i) \stackrel{p_i \equiv 0}{=} \sum_{i=1}^{m} x_i r_i \geq R. \tag{33}$$

So a solution to our constructed instance of Problem 3 has to satisfy the following conditions:

$$\sum_{i=1}^{m} x_i r_i \geq R \tag{34}$$

$$\sum_{i=1}^{m} x_i w_i \leq K \tag{35}$$

By construction of the variables e_i, r_i, w_i, R and K a solution $x = (x_i)_{i=1,\ldots,m}$ is a solution to this instance of Problem 3 if and only if

$$\forall i \in \{1,\ldots,m\} : c(f^{-1}(i)) := x_i$$

is a solution to the given instance of [0/1 KNAPSACK].

So we have found a polynomial time reduction from [0/1 KNAPSACK] to our Problem 3. Since [0/1 KNAPSACK] is known to be NP-complete, our Problem 3 is NP-hard. □

6.2 Proof of Formula (24)

Given is a portfolio specified by the vectors x, p, σ, Θ, r.
The function f is defined as follows:

$$f(x, p, \sigma, \Theta, r) := \frac{\sum_{i=1}^{m} x_i (r_i - p_i)}{q_{pf}^{0.99}(x, p, \sigma, \Theta) - \sum_{i=1}^{m} x_i p_i} \tag{36}$$

If we calculate a constant multiplier for the given portfolio data

$$\xi_{pf} := \frac{q_{pf}^{0.99}(x, p, \sigma, \Theta) - \mu_{pf}(x, p)}{\sigma_{pf}(x, p, \sigma, \Theta)} \tag{37}$$

which can be abbreviated by

$$\xi_{pf} := \frac{q_{pf}^{0.99} - \mu_{pf}}{\sigma_{pf}} \tag{38}$$

in analogy to CreditSuisse Financial Products [6], p. 63, the 99th percentile function can be reformulated by

$$q_{pf}^{0.99} = \mu_{pf} + \xi_{pf} \sigma_{pf}. \tag{39}$$

By substituting the 99th percentile function in formula (36) according to (39) we obtain:

$$\frac{\sum_{i=1}^{m} x_i (r_i - p_i)}{\mu_{pf} + \xi_{pf} \sigma_{pf} - \sum_{i=1}^{m} x_i p_i} \tag{40}$$

Taking into account that

$$\mu_{pf} := \sum_{i=1}^{m} x_i p_i \tag{41}$$

formula (40) can be simplified to

$$\frac{\sum_{i=1}^{m} x_i (r_i - p_i)}{\xi_{pf}\sigma_{pf}} \tag{42}$$

The partial derivative of f is calculated by deriving (42) using quotient rule:

$$d_j := \frac{\partial}{\partial x_j} f(x, p, \sigma, \Theta, r)$$

$$= \frac{(r_j - p_j)(\xi_{pf}\sigma_{pf}) - \left(\sum_{i=1}^{m} x_i(r_i - p_i)\right)\left(\frac{\partial}{\partial x_j}(\xi_{pf}\sigma_{pf})\right)}{(\xi_{pf}\sigma_{pf})^2} \tag{43}$$

For $x_j \neq 0$ formula (43) is equivalent to

$$\frac{x_j(r_j - p_j)(\xi_{pf}\sigma_{pf}) - x_j\left(\sum_{i=1}^{m} x_i(r_i - p_i)\right)\left(\frac{\partial}{\partial x_j}(\xi_{pf}\sigma_{pf})\right)}{x_j(\xi_{pf}\sigma_{pf})^2}$$

$$= \frac{x_j(r_j - p_j)(\xi_{pf}\sigma_{pf}) - \left(\sum_{i=1}^{m} x_i(r_i - p_i)\right)\left(x_j\frac{\partial}{\partial x_j}(\xi_{pf}\sigma_{pf})\right)}{x_j(\xi_{pf}\sigma_{pf})^2}$$

$$= \frac{x_j(r_j - p_j)(\xi_{pf}\sigma_{pf}) - \left(\sum_{i=1}^{m} x_i(r_i - p_i)\right)\left(\xi_{pf}x_j\frac{\partial}{\partial x_j}\sigma_{pf}\right)}{x_j(\xi_{pf}\sigma_{pf})^2} \tag{44}$$

Finally, remembering that

$$RC_j^{\sigma} := x_j \frac{\partial \sigma_{pf}}{\partial x_j}$$

the substitution of the partial derivative yields

$$d_j = \frac{x_j(r_j - p_j)(\xi_{pf}\sigma_{pf}) - \left(\sum_{i=1}^{m} x_i(r_i - p_i)\right)(\xi_{pf}RC_j^{\sigma})}{x_j(\xi_{pf}\sigma_{pf})^2}. \tag{45}$$

□

6.3 Specification of Portfolio m20n2

The following table shows the data of portfolio m20n2.

Note that the variables p_i, r_i and w_i are calculated on a basis of e_i. The variation coefficient for the second sector was set to $\omega_2 := 0.75$ in analogy to real-world variation coefficients of default rates.

Table 6. Specification of portfolio m20n2

i	e_i	Θ_{i1}	Θ_{i2}	p_i	r_i	w_i
1	12700	89%	11%	2.0%	4.72%	12.60%
2	15000	73%	27%	2.0%	3.33%	10.00%
3	3500	71%	29%	4.0%	2.86%	8.57%
4	19800	54%	46%	3.0%	5.05%	12.63%
5	30100	29%	71%	2.0%	8.31%	12.96%
6	30600	75%	25%	6.0%	7.52%	12.09%
7	43000	37%	63%	3.0%	4.19%	9.30%
8	22800	68%	32%	6.0%	7.02%	14.04%
9	23500	53%	47%	5.0%	5.11%	8.51%
10	9200	39%	61%	4.0%	14.13%	15.22%
11	40800	32%	68%	4.0%	6.13%	9.07%
12	26200	58%	42%	7.0%	4.20%	10.69%
13	42100	24%	76%	4.0%	5.46%	8.79%
14	27200	39%	61%	5.0%	7.72%	11.40%
15	1900	44%	56%	6.0%	5.26%	10.53%
16	34700	27%	73%	5.0%	4.03%	8.65%
17	40900	22%	78%	5.0%	8.80%	9.29%
18	28000	14%	86%	5.0%	6.43%	8.93%
19	32200	8%	92%	5.0%	2.80%	8.70%
20	4800	7%	93%	5.0%	4.17%	8.33%

References

1. Frederick Andersson and Stanislaus Uryasev. Credit risk optimization with conditional value-at-risk criterion. 1999. University of Gainesville, working paper.
2. Philippe Artzner, Freddy Delbaen, Jean-Marc Eber, and David Heath. Coherent measures of risk. *Mathematical Finance*, 9(3):203–228, 1999.
3. Basel Committee for Banking Supervision. The new basel capital accord. 2001. Bank for International Settlements, Basel, www.bis.org/publ/bcbsca03.pdf.
4. Peter Buergisser, Alexander Kurth, and Armin Wagner. Incorporating severity variations into credit risk. *Journal of Risk*, 3(4):5–31, 2001.
5. T. Chang, N. Meade, J. Beasley, and Y. Sharaiha. Heuristics for cardinality constrained portfolio optimisation. *Computers & Operations Research*, 27:1271–1302, 2000.
6. CreditSuisse Financial Products. CreditRisk+ (TM) - a credit risk management framework. 1997. www.csfp.co.uk/creditrisk/assets/creditrisk.pdf.
7. Kalyanmoy Deb. *Multi-objective optimisation using evolutionary algorithms*. John Wiley & Sons, Chichester, 2001.
8. Matthias Ehrgott. *Multicriteria optimisation*. Springer, Heidelberg, 2000.
9. Edwin Elton and Martin Gruber. *Modern portfolio theory and investment analysis*. John Wiley & Sons, Chichester, 1995.

10. David Fogel and Zbriginiew Michalewicz. *How to solve it: Modern heuristics.* Springer, Heidelberg, 2000.
11. Ruediger Frey and Alexander McNeil. Modelling dependent defaults. 2001. ETH Zuerich, working paper, www.math.ethz.ch/~frey/credit-paper.pdf.
12. Manfred Gilli and Evis Kellezi. Portfolio selection, tail risk and heuristic optimisation. 2000. University of Geneva, working paper.
13. David Goldberg. *Genetic algorithms for search, optimisation and machine learning.* Addison-Wesley, Reading, 1989.
14. Michael Gordy. Calculation of higher moments in CreditRisk+ with applications. 2001. Board of Governors of the Federal Reserve System, Washington, working paper.
15. Greg Gupton, Christopher Finger, and Mickey Bhatia. CreditMetrics (TM) technical document. 1997. JPMorgan Inc., New York.
16. John Holland. *Adaptation in natural and artificial systems.* Michigan University Press, Ann Arbor, 1975.
17. Stephen Kealhofer. Portfolio management of default risk. 1998. KMV Corporation, San Francisco.
18. Frank Lehrbass. Rethinking risk-adjusted returns. *Credit Risk Special Report, Risk*, (4):35–40, 1999.
19. Harry Markowitz. Portfolio selection. *Journal of Finance*, 7:77ff, 1952.
20. Harry Panjer. Recursive evaluation of a family of compound distributions. *ASTIN Bulletin*, 12:22–26, 1981.
21. Harry Panjer and Gordon Willmot. *Insurance risk models.* Society of Actuaries, Schaumburg, 1992.
22. Georg Pflug. Some remarks on the value-at-risk and the conditional value-at-risk. In S. Uryasev, editor, *Probabilistic constrained optimization*, pages 272–281. Kluwer, Dordrecht, 2000.
23. Guenter Rudolph and Alexandru Agapie. Convergence properties of some multi-objective evolutionary algorithms. In A. Zalzala, editor, *Proceedings of the 2000 Congress on Evolutionary Computation*, pages 1010–1016. IEEE Press, Piscataway, 2000.
24. Jonathan Schaffer. *Some experiments in machine learning using vector evaluated genetic algorithms.* PhD thesis, Vanderbilt University, Nashville, 1984.
25. Frank Schlottmann and Detlef Seese. An agent-based architecture for the management and trading of portfolio credit risk. 2001. University of Karlsruhe, working paper (in German).
26. Frank Schlottmann and Detlef Seese. A hybrid genetic-quantitative method for risk-return optimisation of credit portfolios. In Carl Chiarella and Eckhard Platen, editors, *Quantitative Methods in Finance 2001 Conference abstracts*, page 55. University of Technology, Sydney, 2001. Full paper: www.business.uts.edu.au/finance/resources/qmf2001/ Schlottmann_F.pdf.
27. Hartmut Schmeck, Juergen Branke, and Udo Kohlmorgen. Parallel implementations of evolutionary algorithms. In A. Zomaya, F. Ercal, and S. Olariu, editors, *Solutions to parallel and distributed computing problems*, pages 47–68. John Wiley & Sons, Chichester, 2001.
28. Detlef Seese and Frank Schlottmann. The building blocks of complexity: a unified criterion and selected problems in economics and finance. 2002. Presented at the Sydney Financial Mathematics Workshop 2002, www.qgroup.org.au/SFMW.

29. Detlef Seese and Frank Schlottmann. Large grids and local information flow as a reason for high complexity. In Gerry Frizelle and Huw Richards, editors, *Tackling industrial complexity: the ideas that make a difference*, pages 193–207. University of Cambridge, Cambridge, UK, 2002.
30. Detlef Seese and Frank Schlottmann. Structural reasons for high complexity: A survey on results and problems. pages 1–160, 2002. University of Karlsruhe, working paper.
31. William Sharpe. Capital asset prices. a theory of market equilibrium under conditions of risk. *Journal of Finance*, 19:425–442, 1964.
32. Thomas Wilson. Portfolio credit risk (I). *Risk*, (9):111–119, 1997.
33. Thomas Wilson. Portfolio credit risk (II). *Risk*, (10):56–61, 1997.
34. Eckart Zitzler. *Evolutionary algorithms for multiobjective evolutionary optimization: methods and applications.* PhD thesis, Eidgenoessisch-Technische Hochschule, Zuerich, 1999.

Credit Risk Modelling and Estimation via Elliptical Copulae

Rafael Schmidt

Abteilung Zahlentheorie und Wahrscheinlichkeitstheorie, Universität Ulm, Germany

Summary. Dependence modelling plays a crucial role within internal credit risk models. The theory of copulae, which describes the dependence structure between a multi-dimensional distribution function and the corresponding marginal distributions, provides useful tools for dependence modelling. The difficulty in employing copulae for internal credit risk models arises from the appropriate choice of a copula function. From the practical point of view the dependence modelling of extremal credit default events turns out to be a desired copula property. This property can be modelled by the so-called tail dependence concept, which describes the amount of dependence in the upper-right-quadrant tail or lower-left-quadrant tail of a bivariate distribution. We will give a characterization of tail dependence via a tail dependence coefficient for the class of elliptical copulae. This copula class inherits the multivariate normal, t, logistic, and symmetric general hyperbolic copula. Further we embed the concepts of tail dependence and elliptical copulae into the framework of extreme value theory. Finally we provide a parametric and non-parametric estimator for the tail dependence coefficient.

Introduction

The New Basel Capital Accord [16] will be the next stepping-stone for regulatory treatment of credit risk. Within the Internal Ratings-Based approach (IRB) [15] the dependence structure, specifically the default correlation, of credit risky exposure represents a primary input for regulatory capital requirements. The IRB approach utilizes a single factor model for credit risk modelling. In particular, the asset portfolio is modelled by a multivariate normal distribution. However, multivariate normal distributions encounter two major insufficiencies leading to short comings in multivariate asset return modelling: On the one hand, their normal distributed margins are not flexible enough, and on the other hand, the copula related to the normal distribution (normal copula) does not possess the tail dependence property. This property enables the modelling of dependencies of extremal credit default events. To be more precise: The tail dependence concept describes the amount of dependence in the upper-right-quadrant tail or lower-left-quadrant tail of a bivariate distribution. In this paper we propose substituting the normal copula by an elliptical copula which possesses the tail dependence property. Further we show that this property remains valid after a change of margins of the corresponding asset return random vector, i.e. the tail dependence concept is a copula property.

This paper is organized as follows: After an outline of the dependence structure modelling within The New Basel Capital Accord we present the theory of copulae as a general framework for modelling dependence. In Section 3 we introduce the tail dependence concept as a copula property and give a characterization of tail dependence for elliptical copulae in Section 4. Section 5 embeds the concepts of tail dependence and elliptical copulae into the framework of extreme value theory. In the last section we provide a parametric and non-parametric estimator for the tail dependence coefficient.

1 Credit Risk Modelling Within the New Basel Capital Accord

The 1988 Basel Capital Accord is a current benchmark for many national regulatory laws related to "economic capital" on commercial bank lending businesses. This Accord requires banks to keep an 8% capital charge of the loan face value for any commercial loan in order to cushion losses from an eventual credit default. An increasing problem arises from the overall 8% capital charge which does not include the financial strength of the borrower and the value of the collateral. This led to an off-balance-sheet movement of low-risk credits and a retainment of high-risk credits. To overcome the insufficiency of credit-risk differentiation and other inadequacies, the Basel Committee launched the New Basel Capital Accord - Basel II which is expected to become national law in 2006. Basel II is divided into the Standard approach and the Internal Ratings-Based approach (IRB). Despite the Standard approach, which basically reflects the 1988 Basel Accord, the IRB approach calculates the "economic capital" by using a credit risk portfolio model. By credit-risk portfolio model we understand a function which maps a set of instrument-level and market-level parameters (cf. Gordy, [9], p. 1) to a distribution for portfolio credit losses over a specified horizon. In this context "economic capital" denotes the Value at Risk (VaR) of the portfolio loss distribution. Following the IRB-approach banks are now required to derive probabilities of default per loan (or exposure) via estimation or mapping. Together with other parameters the required regulatory capital increases with increasing probability of default and decreasing credit quality, respectively. Furthermore banks have an incentive to raise the "economic capital" in order to improve their own credit rating.

The IRB approach utilizes a single-factor model as credit-risk model which describes credit defaults by a two-state Merton model. This model can be compared to a simplified framework of CreditMetrics©. In particular, borrower i is linked to a random variable X_i which represents the normalized return of its assets, i.e.

$$X_i = \omega_i Z + \sqrt{1 - \omega_i^2}\varepsilon_i, \qquad i = 1, \ldots, n, \tag{1}$$

where Z is a single common systematic risk factor related to all n borrowers and ε_i, $i = 1, \ldots, n$, denotes the borrower-specific risk. The random variables Z and ε_i, $i = 1, \ldots, n$, are assumed to be standard normally distributed and mutually independent. The parameters ω_i, $i = 1, \ldots, n$, are called factor loadings and regulate the sensitivity towards the systematic risk factor Z.

The simplicity of the above single factor model has a significant advantage: It provides portfolio-invariant capital charges, i.e. the charges depend only on the loan's own properties and not on the corresponding portfolio properties. According to Gordy and Heitfeld [8], p. 5, this is essential for an IRB capital regime.

Observe that the IRB credit risk portfolio model (1) makes use of a multi-dimensional normal distribution, and thus the dependence structure of the portfolio's asset returns is that of a multi-dimensional normal distribution. Many empirical investigations ([10], [19], [6] and others) reject the normal distribution because of its inability to model dependence of extremal events. For instance, in the bivariate normal setting the probability that one component is large given that the other component is large tends to zero. In other words, the probability that the VaR in one component is exceeded given that the other component exceeds the VaR tends to zero. The concept of tail dependence, which we define soon, describes this kind of dependence for extremal events.

In the following we substitute the dependence structure of the above multivariate normal distribution by the dependence structure of elliptically contoured distributions, which contains the multi-dimensional normal distribution as a special case. This class, to be defined later, inherits most of the properties established in the IRB credit risk portfolio model. As additional advantage we can characterize dependence structures of elliptically contoured distributions which model dependencies of extremal default events.

Before continuing we clarify the general dependence concept of multivariate random vectors in the next section. Therefore the theory of copulae is needed.

2 The Theory of Copulae

The theory of copulae investigates the dependence structure of multi-dimensional random vectors. On the one hand, copulae are functions that join or "couple" multivariate distribution functions to their corresponding marginal distribution functions. On the other hand, a copula function itself is a multivariate distribution function with uniform margins on the interval $[0, 1]$.

Copulae are of interest in credit-risk management for two reasons: First, as a way of studying the dependence structure of an asset portfolio irrespective of its marginal asset-return distributions; and second, as a starting point for constructing multi-dimensional distributions for asset portfolios, with a view to simulation. First we define the copula function in a common way (Joe [12], p. 12).

Definition 2.1 *Let $C : [0,1]^n \to [0,1]$ be an n-dimensional distribution function on $[0,1]^n$. Then C is called a copula if it has uniformly distributed margins on the interval $[0,1]$.*

The following theorem gives the foundation for a copula to inherit the dependence structure of a multi-dimensional distribution.

Theorem 2.2 (Sklar's theorem) *Let F be an n-dimensional distribution function with margins $F_1, \ldots, F_n$. Then there exists a copula C, such that for all $x \in \mathbb{R}^n$*

$$F(x_1, \ldots, x_n) = C(F_1(x_1), \ldots, F_n(x_n)). \qquad (2)$$

If $F_1, \ldots, F_n$ are all continuous, then C is unique; otherwise C is uniquely determined on $\text{Ran} F_1 \times \cdots \times \text{Ran} F_n$. Conversely, if C is a copula and $F_1, \ldots, F_n$ are distribution functions, then the function F defined by (2) is an n-dimensional distribution function with margins $F_1, \ldots, F_n$.

We refer the reader to Sklar [22] or Nelsen [14] for the *proof.*

An immediate Corollary shows how one can obtain the copula of a multi-dimensional distribution function.

Corollary 2.3 *Let F be an n-dimensional continuous distribution function with margins $F_1, \ldots, F_n$. Then the corresponding copula C has representation*

$$C(u_1, \ldots, u_n) = F(F_1^{-1}(u_1), \ldots, F_n^{-1}(u_n)), \quad 0 \le u_1, \ldots, u_n \le 1,$$

where $F_1^{-1}, \ldots, F_n^{-1}$ denote the generalized inverse distribution functions of $F_1, \ldots, F_n$, i.e. for all $u_i \in (0,1) : F_i^{-1}(u_i) := \inf\{x \in \mathbb{R} \mid F_i(x) \ge u_i\}$, $i = 1, \ldots, n$.

According to Schweizer and Wolff [21]: "... the copula is invariant while the margins may be changed at will, it follows that it is precisely the copula which captures those properties of the joint distribution which are invariant under a.s. strictly increasing transformations" and thus the copula function represents the dependence structure of a multivariate random vector. We add some more copula properties needed later.

Remarks.

1. A copula is increasing in each component. In particular the partial derivatives $\partial C(u)/\partial u_i$, $i = 1 \ldots n$, exist almost everywhere.

2. Consequently, the conditional distributions of the form

$$C(u_1, \ldots, u_{j-1}, u_{j+1}, \ldots, u_n \mid u_j), \quad j = 1, \ldots, n, \qquad (3)$$

exist.
3. A copula C is uniformly continuous on $[0,1]^n$.

For more details regarding the theory of copulae we refer the reader to the monographs of Nelsen [14] and Joe [12].

3 Tail Dependence: A Copula Property

Now we introduce the concepts of tail dependence and regularly varying (multivariate) functions. We will embed the tail dependence concept within the copula framework. Recall that multivariate distributions possessing the tail dependence property are of special practical interest within credit portfolio modelling, since they are able to incorporate dependencies of extremal credit default events. According to Hauksson et al. [10], Resnick [19], and Embrechts et al. [6] tail dependence plays an important role in extreme value theory, finance, and insurance models. Tail dependence models for multivariate distributions are mostly related to their bivariate marginal distributions. They reflect the limiting proportion of exceedence of one margin over a certain threshold given that the other margin has already exceeded that threshold. The following approach represents one of many possible definitions of tail dependence.

Definition 3.1 (Tail dependence, Joe [12], p. 33) *Let $X = (X_1, X_2)'$ be a 2-dimensional random vector. We say that X is tail dependent if*

$$\lambda := \lim_{v \to 1^-} \mathbb{P}(X_1 > F_1^{-1}(v) \mid X_2 > F_2^{-1}(v)) > 0; \qquad (4)$$

where the limit exists and F_1^{-1}, F_2^{-1} denote the generalized inverse distribution functions of X_1, X_2. Consequently, we say $X = (X_1, X_2)'$ is tail independent if λ equals 0. Further, we call λ the (upper) tail dependence coefficient.

Remark. Similarly, one may define the lower tail dependence coefficient by

$$\omega := \lim_{v \to 0^+} \mathbb{P}(X_1 \leq F_1^{-1}(v) \mid X_2 \leq F_2^{-1}(v)).$$

The following Proposition shows that tail dependence is a copula property. Thus many copula features translate to the tail dependence coefficient, for example the invariance under strictly increasing transformations of the margins.

Proposition 3.2 *Let X be a continuous bivariate random vector, then*

$$\lambda = \lim_{u \to 1^-} \frac{1 - 2u + C(u,u)}{1 - u}, \qquad (5)$$

where C denotes the copula of X. Analogous $\omega = \lim_{u \to 0^+} \frac{C(u,u)}{u}$ holds for the lower tail dependence coefficient.

Proof. Let F_1 and F_2 be the marginal distribution functions of X. Then

$$\lambda = \lim_{u \to 1^-} \mathbb{P}(X_1 > F_1^{-1}(u) \mid X_2 > F_2^{-1}(u))$$

$$= \lim_{u \to 1^-} \frac{\mathbb{P}(X_1 > F_1^{-1}(u), X_2 > F_2^{-1}(u))}{\mathbb{P}(X_2 > F_2^{-1}(u))}$$

$$= \lim_{u \to 1^-} \frac{1 - F_2(F_2^{-1}(u)) - F_1(F_1^{-1}(u)) + C(F_1(F_1^{-1}(u)), F_2(F_2^{-1}(u)))}{1 - F_2(F_2^{-1}(u))}$$

$$= \lim_{u \to 1^-} \frac{1 - 2u + C(u,u)}{1 - u}.$$

□

Although we provided a simple characterization for upper and lower tail dependence by the last proposition, it will be still difficult and tedious to verify certain tail dependencies if the copula is not a closed-form expression, as in the case for most well-known elliptically contoured distributions. Therefore, the following Theorem gives another approach calculating tail dependence. We restrict ourselves to the upper tail.

Proposition 3.3 *Let X be a bivariate random vector with differentiable copula C. Then the (upper) tail dependence coefficient λ can be expressed using conditional probabilities if the following limit exists:*

$$\lambda = \lim_{v \to 1^-} \Big(\mathbb{P}(U_1 > v | U_2 = v) + \mathbb{P}(U_2 > v | U_1 = v) \Big), \qquad (6)$$

where (U_1, U_2) are distributed according to the copula C of X.

Proof. Let C denote the copula of X which is assumed to be differentiable on the interval $(0,1)^2$. Therefore we may write $\mathbb{P}(U_1 \leq v | U_2 = u) = \partial C(u,v)/\partial u$ and $\mathbb{P}(U_1 > v | U_2 = u) = 1 - \partial C(u,v)/\partial u$, respectively. The rule of L'Hospital implies that

$$\lambda = \lim_{u \to 1^-} \frac{1 - 2u + C(u,u)}{1 - u} = \lim_{u \to 1^-} \left(-\left(-2 + \frac{dC(u,u)}{du} \right) \right)$$

$$= \lim_{u \to 1^-} \left(2 - \frac{\partial C(x,u)}{\partial x}\bigg|_{x=u} - \frac{\partial C(u,y)}{\partial y}\bigg|_{y=u} \right)$$

$$= \lim_{u \to 1^-} \Big(\mathbb{P}(U_1 > u | U_2 = u) + \mathbb{P}(U_2 > u | U_1 = u) \Big).$$

□

Some of the following results for copulae of elliptically contoured distributions are characterized by regularly varying or O-regularly varying functions and multivariate regularly varying random vectors, which are defined as follows.

Definition 3.4 *(Regular and O-regular variation of real valued functions)*

1. A measurable function $f : \mathbb{R}_+ \to \mathbb{R}_+$ is called *regularly varying (at ∞)* with index $\alpha \in \mathbb{R}$ if for any $t > 0$

$$\lim_{x \to \infty} \frac{f(tx)}{f(x)} = t^\alpha.$$

2. A measurable function $f : \mathbb{R}_+ \to \mathbb{R}_+$ is called *O-regularly varying (at ∞)* if for any $t \geq 1$

$$0 < \liminf_{x \to \infty} \frac{f(tx)}{f(x)} \leq \limsup_{x \to \infty} \frac{f(tx)}{f(x)} < \infty.$$

Thus, regularly varying functions behave asymptotically like power functions.

Definition 3.5 (Multivariate regular variation of random vectors)
An n-dimensional random vector $X = (X_1, \ldots, X_n)^T$ and its distribution are said to be *regularly varying with limit measure ν* if there exists a function $b(t) \nearrow \infty$ as $t \to \infty$ and a non-negative Radon measure $\nu \neq 0$ such that

$$t \, \mathbb{P}\left(\left(\frac{X_1}{b(t)}, \ldots, \frac{X_n}{b(t)} \right) \in \cdot \right) \xrightarrow{v} \nu(\cdot) \tag{7}$$

on the space $E = [-\infty, \infty]^n \setminus \{0\}$.

Notice that convergence $\xrightarrow{v}$ stands for vague convergence of measures, in the sense of Resnick [18], p. 140. It can be shown that (7) requires the existence of a constant $\alpha \geq 0$ such that for relatively compact sets $B \subset E$ (i.e. the closure $\overline{B}$ is compact in E)

$$\nu(tB) = t^{-\alpha} \nu(B), \quad t > 0. \tag{8}$$

Thus we say X is regularly varying with limit measure ν and index $\alpha \geq 0$, if (7) holds. Moreover, the function $b(\cdot)$ is necessarily regular varying with index $1/\alpha$. In the following we always assume $\alpha > 0$.

Later, when we consider elliptically contoured distributions, it turns out that polar coordinate transformations are a convenient way to deal with multivariate regular variation. Denote by $\mathbb{S}^{n-1} := \{x \in \mathbb{R}^n : ||x|| = 1\}$ the $(n-1)$-dimensional unit sphere for some arbitrary norm $||\cdot||$ in $\mathbb{R}^n$. Then the polar coordinate transformation $T : \mathbb{R}^n \setminus \{0\} \to (0, \infty) \times \mathbb{S}^{n-1}$ is defined by

$$T(x) = \left(||x||, \frac{x}{||x||} \right) =: (r, a).$$

Observe, the point x can be seen as being distance r from the origin 0 away with direction $a \in \mathbb{S}^{n-1}$. It is well-known that T is a bijection with inverse transform $T^{-1} : (0, \infty) \times \mathbb{S}^{n-1} \to \mathbb{R}^n \setminus \{0\}$ given by $T^{-1}(r, a) = ra$. For notational convenience we denote the euclidian-norm by $\|\cdot\|_2$ and the related unit sphere by $\mathbb{S}_2^{n-1}$. The next proposition, stated essentially in Resnick [19], Proposition 2, characterizes multivariate regularly varying random vectors under polar-coordinate transformation.

Proposition 3.6 *The multivariate regular variation condition (7) is equivalent to the existence of a random vector Θ with values in the unit sphere $\mathbb{S}^{n-1}$ such that for all $x > 0$*

$$t\, \mathbb{P}\left(\left(\frac{\|X\|}{b(t)}, \frac{X}{\|X\|}\right) \in \cdot\right) \xrightarrow{v} c\nu_\alpha\, \mathbb{P}(\Theta \in \cdot), \quad \text{as}\quad t \to \infty, \tag{9}$$

where $c > 0$, ν_α is a measure on Borel subsets of $(0, \infty]$ with $\nu_\alpha((x, \infty]) = x^{-\alpha}$, $x > 0, \alpha > 0$, and $\|\cdot\|$ denotes an arbitrary norm in $\mathbb{R}^n$. We call $S(\cdot) := \mathbb{P}(\Theta \in \cdot)$ the spectral measure of X.

Remark. According to Stărică [23], p. 519, multivariate regular variation condition (7) is also equivalent to

$$\frac{\mathbb{P}(\|X\| > tx, X/\|X\| \in \cdot)}{\mathbb{P}(\|X\| > t)} \xrightarrow{v} x^{-\alpha}\, \mathbb{P}(\Theta \in \cdot), \quad \text{as}\quad t \to \infty, \tag{10}$$

where $\|\cdot\|$ denotes an arbitrary norm in $\mathbb{R}^n$ and $S(\cdot) := \mathbb{P}(\Theta \in \cdot)$ is the spectral measure of X. Observe that regular variation of random variables is equivalent to regular variation of its distribution's tail function.

Notice that the latter proposition also implies that the multivariate regular variation property (9) does not depend on the choice of the norm. For more details regarding regular variation, O-regular variation, and multivariate regular variation we refer the reader to Bingham, Goldie, and Teugels [1], pp. 16, pp. 61, and pp. 193 and Resnick [18], pp. 12, pp. 250.

4 Tail Dependence of Elliptical Copulae

Elliptically contoured distributions (in short: elliptical distributions) play a significant role in risk management due to many properties which fit very well in the Value at Risk and Markowitz framework. The best known elliptical distributions are the multivariate normal distribution, the multivariate t-distribution, the multivariate logistic distribution, and multivariate symmetric general hyperbolic distribution.

Definition 4.1 (Elliptical distribution) *Let X be an n-dimensional random vector and $\Sigma \in \mathbb{R}^{n \times n}$ be a symmetric positive semi-definite matrix.*

If $X - \mu$, for some $\mu \in \mathbb{R}^n$, has a characteristic function of the form $\phi_{X-\mu}(t) = \Phi(t^T \Sigma t)$, then X is said to be elliptically distributed with parameters μ, Σ, and Φ. Let $E_n(\mu, \Sigma, \Phi)$ denote the class of elliptically contoured distributions with the latter parameters. We call Φ the characteristic generator.

Definition 4.2 (Elliptical copulae) *We say C is an elliptical copula, if it is the copula of an elliptically contoured distribution.*

Remark. The density function, if it exists, of an elliptically contoured distribution has the following form:

$$f(x) = |\Sigma|^{-1/2} g((x-\mu)^T \Sigma^{-1} (x-\mu)), \quad x \in \mathbb{R}^n, \tag{11}$$

for some function $g : \mathbb{R} \to \mathbb{R}_+$, which we call the *density generator*. Observe that the name "elliptically contoured" distribution is related to the elliptical contours of the latter density.

Examples. In the following we give some examples of density generators for n-dimensional elliptical distributions. Here C_n denotes a normalizing constant depending only on the dimension n.

1. Normal distribution: $g(u) = C_n \exp(-u/2)$.
2. t-distribution: $g(u) = C_n (1 + \frac{t}{m})^{-(n+m)/2}, m \in \mathbb{N}$.
3. logistic distribution: $g(u) = C_n \exp(-u)/(1 + \exp(-u))^2$.
4. Symmetric generalized hyperbolic distribution:
   $g(u) = C_n K_{\lambda - \frac{n}{2}}(\sqrt{\psi(\chi + u)})/(\sqrt{\chi + u})^{\frac{n}{2} - \lambda}$, $u > 0$, where $\psi, \chi > 0$, $\lambda \in \mathbb{R}$, and K_ν denotes the modified Bessel function of the third kind (or Macdonald function).

A characteristic property of elliptical distributions is that all margins are elliptically distributed with the same characteristic generator or density generator, respectively. However, in most risk models one encounters the problem that the margins of the multivariate asset-return random vector are empirically not of the same distribution type. As a solution we propose joining appropriate marginal distributions with an elliptical copula, because of its well-known and statistically tractable properties. One important issue is of course the estimation of the copula or the copula parameters, respectively. According to Theorem 2.15 in Fang, Kotz, and Ng [7], elliptical copulae C corresponding to elliptically distributed random vectors $X \in E_n(\mu, \Sigma, \Phi)$ with positive-definite matrix Σ, are uniquely determined up to a positive constant by the matrix $R_{ij} = \Sigma_{ij}/\sqrt{\Sigma_{ii}\Sigma_{jj}}$, $1 \leq i,j \leq n$ and the characteristic generator Φ or the density generator g, respectively. Uniqueness is obtained by setting $|\Sigma| = 1$ without loss of generality. Observe that R corresponds to the linear correlation matrix, if it exists. Embrechts et al. [6] propose the following robust estimator for R via Kendall's Tau τ. This estimator is based on the relationship $\tau(X_i, X_j) = \frac{2}{\pi} \arcsin(R_{ij})$, $1 \leq i,j \leq n$, for X_i and X_j having continuous distributions:

$$\hat{R} = \sin(\pi\hat{\tau}/2) \quad \text{with} \quad \hat{\tau} = \frac{c-d}{\binom{n}{2}}, \tag{12}$$

where c and d denote the number of concordant and discordant tuples of a bivariate random sample. The characteristic generator Φ or the density generator g can be estimated via non-parametric estimators as they are discussed in Bingham and Kiesel [2].

An immediate question arises: Which elliptical copula should one choose? The previous discussion showed that tail dependence represents a desired copula property in the context of credit-risk management. It is well-known that the Gaussian copula with correlation coefficient $\rho < 1$ does not inherit tail dependence (see Schmidt [20] for more details). Therefore we seek a characterization of elliptical copulae possessing the latter property. The following stochastic representation turns out to be very useful.

Let X be an n-dimensional elliptically distributed random vector, i.e. $X \in E_n(\mu, \Sigma, \Phi)$, with parameters μ and symmetric positive semi-definite matrix Σ, $\text{rank}(\Sigma) = m$, $m \leq n$. Then

$$X \stackrel{d}{=} \mu + R_m A' U^{(m)}, \tag{13}$$

where $A'A = \Sigma$ and the univariate random variable $R_m \geq 0$ is independent of the m-dimensional random vector $U^{(m)}$. The random vector $U^{(m)}$ is uniformly distributed on the unit sphere $\mathbb{S}_2^{n-1}$ in $\mathbb{R}^m$. In detail, R_m represents a radial part and $U^{(m)}$ represents an angle of the corresponding elliptical random vector X. We call R_m the *generating variate* of X. The above representation is also applicable for fast simulation of multi-dimensional elliptical distributions and copulae. Especially in risk-management practice, where large exposure portfolios imply high-dimensional distributions, one is interested in fast simulation technics.

Although tail dependence is a copula feature we will state, for the purpose of generality, the next characterization of tail dependence for elliptical distributions. For this we need the following condition, which is easy to check in the context of density generators.

Condition 4.3 *Let $h : \mathbb{R}^+ \to \mathbb{R}^+$ be a measurable function eventually decreasing such that for some $\varepsilon > 0$*

$$\limsup_{x \to \infty} \frac{h(tx)}{h(x)} \leq 1 - \varepsilon \quad \text{uniformly } \forall\, t > 1.$$

Theorem 4.4 *Let $X \in E_n(\mu, \Sigma, \Phi)$, $n \geq 2$, with positive-definite matrix Σ. If X possesses a density generator g then*
α) all bivariate margins of X possess the tail dependence property if g is regularly varying, and
β) if X possesses a tail dependent bivariate margin and g satisfies Condition 4.3, then g must be O-regularly varying.

The *Proof* and examples are given in Schmidt [20].

Remark. Although we cannot show the equivalence of tail dependence and regularly varying density generator, most well-known elliptical distributions and elliptical copulae are given either by a regularly varying or a not 0-regularly varying density generator. That justifies a restriction to the class of elliptical copulae with regularly varying density generator if one wants to incorporate tail dependence.

Additionally we can state a closed form expression for the tail dependence coefficient of an elliptically contoured random vector $(X_1, X_2)' \in E_2(\mu, \Sigma, \Phi)$ with positive-definite matrix Σ, and the corresponding elliptical copula, having a regular varying density generator g with index $-\alpha/2 - 1 < 0$:

$$\lambda = \lambda(\alpha, \rho) = \frac{\int_0^{h(\rho)} \frac{u^\alpha}{\sqrt{1-u^2}} du}{\int_0^1 \frac{u^\alpha}{\sqrt{1-u^2}} du}, \quad (14)$$

with $\rho := \frac{\sigma_{12}}{\sqrt{\sigma_{11}\sigma_{22}}}$ and $h(\rho) := \left(1 + \frac{(1-\rho)^2}{1-\rho^2}\right)^{-1/2}$ (see also Figure 1). This formula has been developed in the proof of Theorem 5.2 in Schmidt [20], p. 20. Note, that ρ corresponds to the correlation coefficient when this exists (see Fang, Kotz, and Ng [7], p. 44, for the covariance formula of elliptically contoured distributions). We remark that the (upper) tail dependence coefficient λ coincides with the lower tail dependence coefficient and depends only on the "correlation" coefficient ρ and the regular variation index α.

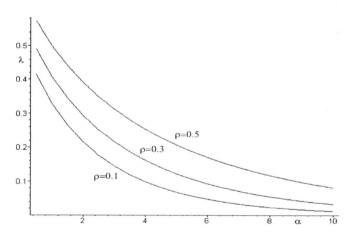

Fig. 1. Tail dependence coefficient λ versus regular variation index α for $\rho = 0.5, 0.3, 0.1$

For completeness we state the following generalization of Theorem 4.4. The proof can also be found in Schmidt [20].

Theorem 4.5 *Let $X \in E_n(\mu, \Sigma, \Phi)$, $n \geq 2$, with positive-definite matrix Σ and stochastic representation $X \stackrel{d}{=} \mu + R_n A' U^{(n)}$. Denote by H_n the distribution function of R_n.*
α) If X has a tail dependent bivariate margin, then the tail function $1 - H_n$ of R_n must be O-regularly varying.
β) If X has a tail dependent bivariate margin, then the tail function $1 - F_i$ must be O-regularly varying, where F_i denote the distribution functions of the univariate margins of X_i, $i = 1, \ldots, n$.
γ) Suppose the distribution function H_n of R_n has a regularly varying tail. Then all bivariate margins are tail dependent.

5 Tail Dependence: A Tool of Multivariate Extreme Value Theory

In this section we embed the concepts of tail dependence and elliptical copulae, introduced in Section 3 and 4, into the framework of multivariate extreme value theory. Extreme value theory is the natural choice for inferences on extremal events of random vectors or the tail behavior of probability distributions. Usually one approximates the tail of a probability distribution by an appropriate extreme value distribution. In the one-dimensional setting the class of extreme value distributions has a solely parametric representation, so it suffices to apply parametric estimation methods. By contrast, multi-dimensional extreme value distributions are characterized by a parametric and a non-parametric component. This leads to more complicated estimation methods. First we provide the necessary background for our purpose. Let $X, X^{(1)}, X^{(2)}, \ldots, X^{(m)}$, $m \in \mathbb{N}$ be independent multivariate random vectors with common continuous distribution function F. We say X or its distribution is in the domain of attraction of a multivariate extreme value distribution G if there exists a sequence of normalizing constants $(a_{mi})_{m=1}^{\infty}$, $(b_{mi})_{m=1}^{\infty}$ with $a_{mi} > 0$ and $b_{mi} \in \mathbb{R}$, $i = 1, \ldots, n$ such that

$$\mathbb{P}\left(\frac{\max_{1 \leq i \leq m} X_1^{(m)} - b_{m1}}{a_{m1}} \leq x_1, \ldots, \frac{\max_{1 \leq i \leq m} X_n^{(m)} - b_{mn}}{a_{mn}} \leq x_n\right) \quad (15)$$

converges to the limit distribution function G with non-degenerate margins as $m \to \infty$. In particular, the latter is equivalent to

$$\lim_{m \to \infty} F^m(a_{m1} x_1 + b_{m1}, \ldots, a_{mn} x_n + b_{mn}) = G(x_1, \ldots, x_n).$$

Before turning to elliptical copulae we prove the following theorem for elliptical distributions.

Credit Risk Modelling and Estimation via Elliptical Copulae 279

Theorem 5.1 *Let $X \in E_n(\mu, \Sigma, \Phi)$ with stochastic representation $X \stackrel{d}{=} \mu + R_n A' U^{(n)}$ and positive-definite matrix Σ. If the generating variate R_n possesses a regularly varying tail function, then X lies in the domain of attraction of an extreme value distribution.*

Proof. Suppose $X \in E_n(\mu, \Sigma, \Phi)$ with stochastic representation $X \stackrel{d}{=} \mu + R_n A' U^{(n)}$ and positive-definite matrix Σ. We start by showing that a regularly varying R_n requires X to be in the class of multivariate regularly varying random vectors, introduced in Definition 3.5. Consider first the case $\mu = 0$ and $\Sigma = I$, i.e. $X \stackrel{d}{=} R_n U^{(n)}$. We need the following characterization of vague convergence stated in Resnick [18], Proposition 3.12, p. 142: A sequence of Radon measures ν_m on some space $\mathbb{E}$ converges vaguely to a Radon measure ν on $\mathbb{E}$ if and only if $\lim_{m \to \infty} \nu_m(B) = \nu(B)$ for all relatively compact Borel sets $B \in \mathbb{E}$ (i.e. the closure $\overline{B}$ is compact in $\mathbb{E}$) with $\nu(\partial B) = 0$. Note that the Borel sets $(x, \infty] \times C$, $x > 0$, of $(0, \infty] \times \mathbb{S}_2^{n-1}$ represent a generating Π-system of the class of relatively compact sets of $(0, \infty] \times \mathbb{S}_2^{n-1}$. Thus it suffices to consider for $x > 0$ and $0 < b(t) \nearrow \infty$

$$t \mathbb{P}\left(T\left(\frac{X}{b(t)}\right) \in (x, \infty] \times C\right) = t \mathbb{P}\left(\left(\frac{\|X\|_2}{b(t)}, \frac{X}{\|X\|_2}\right) \in (x, \infty] \times C\right)$$
$$= t \mathbb{P}\left(\frac{R_n}{b(t)} > x\right) \mathbb{P}(U^{(n)} \in C) \to x^{-\alpha} \mathbb{P}(U^{(n)} \in C) \quad \text{as } t \to \infty,$$

since R_n and $U^{(n)}$ are stochastically independent and R_n is regularly varying. Further the latter convergence is locally uniformly due to locally uniform convergence of regularly varying functions (see Resnick [18], Proposition 0.5, p. 17) and the absolute continuity of $U^{(n)}$. Applying Proposition 3.6 yields the existence of $b(t) \nearrow \infty$ with

$$t \mathbb{P}\left(\frac{X}{b(t)} \in \cdot\right) \xrightarrow{v} \nu(\cdot) \tag{16}$$

for a Radon measure ν on $E = [-\infty, \infty]^n \setminus \{0\}$ and locally uniform convergence transfers, because T and its inverse are continuous functions on E.

Let now $\mu \in \mathbb{R}^n$ and $\Sigma \in \mathbb{R}^{n \times n}$, positive-definite, be arbitrary. Set $A \in \mathbb{R}^{n \times n}$ such that $\Sigma = A' A$ and A is a regular matrix. Define the Radon measure $\rho(\cdot) := \nu(A' \cdot)$ on E. Then

$$t \mathbb{P}\left(\frac{X}{b(t)} \in \cdot\right) = t \mathbb{P}\left(\frac{A' R_n U^{(n)}}{b(t)} + \frac{\mu}{b(t)} \in \cdot\right) \xrightarrow{v} \rho(\cdot) \tag{17}$$

holds because of the locally uniform convergence property. Further the sets $A'B := \{A'x \mid x \in B \text{ relatively compact in } E\}$ are relatively compact on E and $\nu(\partial(A'B)) = 0$ if $\rho(\partial B) = \nu(A' \partial B) = 0$, since $\partial(A'B) \subset A'(\partial B)$ holds for regular matrixes A' $((A')^{-1} x$ is a continuous function on E). Again Proposition 3.6 yields

$$t\,\mathbb{P}\left(\frac{||X||}{b(t)} > x, X/||X|| \in \cdot\right) \xrightarrow{v} cx^{-\alpha}\,\mathbb{P}(\Theta \in \cdot),\ c > 0, \tag{18}$$

for some spectral measure $S(\cdot) = \mathbb{P}(\Theta \in \cdot)$ on the unit sphere $\mathbb{S}_2^{n-1}$. We refer the reader to Hult and Lindskog [11] for explicit calculations of the spectral measure with respect to the euclidian and the max-norm. Finally (18) and Corollary 5.18 in Resnick [18], p. 281, require X to be in the domain of attraction of an extreme value distribution. □

Typically, elliptically contoured distributions are given by their density function or their density generator, respectively. Thus, the next Corollary turns out to be helpful.

Corollary 5.2 *Let $X \in E_n(\mu, \Sigma, \Phi)$ be an elliptically contoured distribution with regularly varying density generator g introduced in (11). Then X is in the domain of attraction of an extreme value distribution.*

Proof. According to Proposition 3.7 in Schmidt [20], p. 10, a regularly varying density generator implies a regularly varying density function of the generating variate R_n. In particular the latter proposition yields the existence of a density function of R_n. By Karamata's Theorem (see Bingham, Goldie, and Teugels [1], p. 26) regular variation is transferred to the tail function of R_n. The corollary follows by Theorem 5.1. □

The following calculation clarifies the relationship between the spectral measure arising from multivariate regular variation of random vectors (see (9)) and extreme value distributions. According to Corollary 5.18 in Resnick [18], p. 281, every multivariate regularly varying random vector with associated spectral measure $S(\cdot)$ is in the domain of attraction of an extreme value distribution G with representation

$$G(x) = \exp\left(-\nu([-\infty, x]^c)\right), \quad x \in \mathbb{R}_+^n, \tag{19}$$

where $\nu(\{x \in \mathbb{R}^n \setminus \{0\} : ||x|| > t, x/||x|| \in \cdot\}) = t^{-\alpha} S(\cdot)$ and $[-\infty, x]^c := \{y \in E \mid y_i > x \text{ for some } i = 1, \ldots, n\}$. In the literature ν is referred to as the *exponent measure* and ν_T is the corresponding measure under polar coordinate transformation $T(x) = (||x||, x/||x||)$. In particular ν_T represents a product measure and $S(\cdot) = \nu(\{x \in \mathbb{R}^n \setminus \{0\} : ||x|| > 1, x/||x|| \in \cdot\})$. Recall that T is a bijection on E and $T^{-1}(r, a) = ra$ holds. Set $\overline{\mathbb{S}}^{n-1} := \mathbb{S}^{n-1} \setminus (-\infty, 0]^n$. Then for $x \in \mathbb{R}_+^n$

$$\begin{aligned}\nu([-\infty, x]^c) &= \nu_T(T([-\infty, x]^c)) \\ &= \nu_T(\{(r, a) \in \mathbb{R}_+ \times \mathbb{S}^{n-1} \mid ra_i > x_i \text{ for some } i = 1, \ldots, n\}) \\ &= \nu_T((r, a) \in \mathbb{R}_+ \times \mathbb{S}^{n-1} \mid a \in \overline{\mathbb{S}}^{n-1}, r > \min\{\frac{x_i}{a_i}, i \in I_a\} =: g(a))\end{aligned}$$

$$= \int_{\mathbb{S}^{n-1}} \int_{g(a)}^{\infty} \frac{1}{\alpha+1} \frac{1}{r^{\alpha+1}} \, dr \, S(da) = \int_{\mathbb{S}^{n-1}} \frac{1}{g(a)^{\alpha}} S(da)$$

$$= \int_{\mathbb{S}^{n-1}} \frac{1}{[\min\{\frac{x_i}{a_i}, i \in I_a\}]^{\alpha}} S(da)$$

$$= \int_{\mathbb{S}^{n-1}} [\max\{\frac{a_i}{x_i}, i \in I_a\}]^{\alpha} S(da),$$

where $I_a = \{j \in \{1, \ldots, n\} \mid a_j > 0\}$. We summarize the above results in the following proposition.

Proposition 5.3 *Let X be a multivariate regularly varying random vector according to Definition 3.5. Then X is in the domain of attraction of a multi-dimensional extreme value distribution*

$$G(x) = \exp\Big(-\int_{\mathbb{S}^{n-1}} [\max\{\frac{a_i}{x_i}, i \in I_a\}]^{\alpha} S(da)\Big), \quad x \in \mathbb{R}_+^n, \quad (20)$$

with spectral measure $S(\cdot)$ living on the unit sphere $\mathbb{S}^{n-1}$.

In general, multi-dimensional extreme value distributions are characterized by an extreme value index and a finite measure, which is commonly referred to as the spectral or angular measure. According to the latter proposition, a multivariate regularly varying random vector is in the domain of attraction of an extreme value distribution with spectral measure coinciding with that of Definition 3.5.

For the family of elliptically contoured distributions the spectral measure is given in closed form. Especially for an elliptical random vector $X \in E_n(0, I, \Phi)$ with stochastic representation $X \stackrel{d}{=} R_n U^{(n)}$ and R_n having a regularly varying tail function with index α we obtain

$$G(x) = \exp\Big(-\int_{\mathbb{S}_2^{n-1}} [\max\{\frac{a_i}{x_i}, i \in I_a\}]^{\alpha} \, da\Big), \quad x \in \mathbb{R}_+^n, \quad (21)$$

with $\mathbb{S}_2^{n-1}$ denoting the unit sphere with respect to the euclidian norm $\|\cdot\|_2$. Recall that the spectral measure $S(\cdot)$ is uniformly distributed on the unit sphere $\mathbb{S}_2^{n-1}$. Moreover, in the bivariate setup straightforward calculation yields

$$G(x_1, x_2) = \exp\Big(-\frac{1}{2\pi}\Big(\frac{\sqrt{\pi}}{2}\frac{\Gamma((1+\alpha)/2)}{\Gamma(1+\alpha/2)}\Big)\Big(\frac{1}{x_1^{\alpha}} + \frac{1}{x_2^{\alpha}}\Big) \quad (22)$$

$$+ \frac{1}{x_1^{\alpha}} \int_0^{\tan^{-1}(x_2/x_1)} \cos^{\alpha}\theta \, d\theta + \frac{1}{x_2^{\alpha}} \int_{\tan^{-1}(x_2/x_1)}^{\pi/2} \sin^{\alpha}\theta \, d\theta\Big)\Big).$$

Having established the connection among elliptically contoured distributions and multivariate extreme value theory, we now turn towards the relationship between the tail dependence coefficient, elliptical copulae, and bivariate extreme value theory. In the following we will always assume that the bivariate random vector X is in the domain of attraction of an extreme value distribution. Recall, for a bivariate random vector X with distribution function F in the domain of attraction of an extreme value distribution G there must exist constants $a_{mi} > 0$ and $b_{mi} \in \mathbb{R}$, $i = 1, 2$, such that

$$\lim_{m \to \infty} F^m(a_{m1}x_1 + b_{m1}, a_{m2}x_n + b_{m2}) = G(x_1, x_2).$$

Transformation of the margins of G to so-called standard Frechét margins yields

$$\lim_{m \to \infty} F_*^m(mx_1, mx_2) = G^*(x_1, x_2), \qquad (23)$$

where F_* and G^* are the standardized distributions of F and G, respectively, i.e. $G_{*i}(x_i) = \exp(-1/x_i)$, $x_i > 0$, $i = 1, 2$, and

$$F_*(x_1, x_2) = F\left(\left(\frac{1}{1-F_1}\right)^{-1}(x_1), \left(\frac{1}{1-F_2}\right)^{-1}(x_2)\right).$$

This standardization does not introduce difficulties as shown in Resnick [18], Proposition 5.10, p. 265. Moreover the following continuous version of (23) can be shown:

$$\lim_{t \to \infty} F_*^t(tx_1, tx_2) = G^*(x_1, x_2), \qquad (24)$$

or equivalently

$$\lim_{t \to \infty} t(1 - F_*(tx_1, tx_2)) = -\log(G^*(x_1, x_2)). \qquad (25)$$

Summarizing the above facts, we obtain

$$\lim_{t \to \infty} t\left(1 - F\left(\left(\frac{1}{1-F_1}\right)^{-1}(tx_1), \left(\frac{1}{1-F_2}\right)^{-1}(tx_2)\right)\right)$$
$$= -\log G\left(\left(\frac{1}{-\log G_1}\right)^{-1}(x_1), \left(\frac{1}{-\log G_2}\right)^{-1}(x_2)\right)$$
$$= -\log G^*(x_1, x_2). \qquad (26)$$

Thus the tail dependence coefficient, if it exists, can be expressed as

$$\lambda = \lim_{v \to 1^-} \mathbb{P}(X_1 > F_1^{-1}(v) \mid X_2 > F_2^{-1}(v))$$
$$= \lim_{v \to 1^-} \mathbb{P}(X_1 > F_1^{-1}(v), X_2 > F_2^{-1}(v))/(1-v)$$
$$= \lim_{t \to \infty} t\, \mathbb{P}\left(X_1 > F_1^{-1}\left(1 - \frac{1}{t}\right), X_2 > F_2^{-1}\left(1 - \frac{1}{t}\right)\right).$$

Further, easy calculation shows that

$$-\log G^*(x_1, x_2) = \lim_{t \to \infty} t\left(1 - \mathbb{P}\left(X_1 \le \left(\frac{1}{1-F_1}\right)^{-1}(tx_1), X_2 \le \left(\frac{1}{1-F_2}\right)^{-1}(tx_2)\right)\right)$$
$$= \frac{1}{x_1} + \frac{1}{x_2} - \lim_{t \to \infty} \mathbb{P}\left(X_1 > F_1^{-1}\left(1 - \frac{1}{tx_1}\right), X_2 > F_2^{-1}\left(1 - \frac{1}{tx_2}\right)\right)$$

and hence
$$\lambda = 2 + \log G\left(\left(\frac{1}{-\log G_1}\right)^{-1}(1), \left(\frac{1}{-\log G_2}\right)^{-1}(1)\right). \qquad (27)$$

The latter equation shows how one could model the tail dependence coefficient by choosing an appropriate bivariate extreme value distribution.

Using the above results, for an elliptical random vector $X \in E_n(0, I, \Phi)$ with stochastic representation $X \stackrel{d}{=} R_n U^{(n)}$ and R_n having a regularly varying tail function with index α, we derive

$$\lambda = \frac{\int_{\pi/4}^{\pi/2} \cos^\alpha \theta \, d\theta}{\int_0^{\pi/2} \cos^\alpha \theta \, d\theta}. \qquad (28)$$

Observe that for $X \in E_n(0, I, \Phi)$ this formula coincides with formula (14) after a standard substitution.

Within the framework of copulae we can rewrite (5) and (27) to obtain

$$\lambda = 2 - \lim_{t \to \infty} t\left(1 - C\left(1 - \frac{1}{t}, 1 - \frac{1}{t}\right)\right) \qquad (29)$$
$$= 2 + \log\left(C_G\left(\frac{1}{e}, \frac{1}{e}\right)\right),$$

where C and C_G denote the copula of F and G, respectively. Using the notation of co-copulae (see Nelsen [14], p. 29) (29) implies

$$\lambda = 2 - \lim_{t \to \infty} t\, C_{co}\left(\frac{1}{t}, \frac{1}{t}\right) = 2 + \log\left(C_G\left(\frac{1}{e}, \frac{1}{e}\right)\right). \qquad (30)$$

The above results lead to the observation that a bivariate random vector inherits the tail dependence property if the standardized distribution F_* or the related copula function C (equals the copula of F) lies in the domain of attraction of an extreme value distribution which does not have independent margins. Consequently it is not necessary that the bivariate distribution function F itself is in the domain of attraction of some extreme value distribution. This is an important property for asset portfolio modelling.

Remark. In particular every bivariate regularly varying random vector with a spectral measure not concentrated on $(c, 0)^T$ and $(0, c)^T$ for some $c > 0$ possesses the tail dependence property according to Corollary 5.25 in Resnick [18], p. 292. This is in line with Theorem 4.4 and Theorem 4.5, since the spectral measure of a non-degenerated elliptical distribution is not concentrated on single points.

Based on the previous results we finish this section with an important theorem about elliptical copulae.

Theorem 5.4 *Let C be an elliptical copula corresponding to an elliptical random vector $X \stackrel{d}{=} \mu + R_n A' U^{(n)} \in E_n(\mu, \Sigma, \Phi)$, Σ positive-definite, with regularly varying generating variate R_n or regularly varying density generator. Then C is in the domain of attraction of some extreme value distribution and all bivariate margins of C possess the tail dependence property.*

Proof. Suppose $X \stackrel{d}{=} \mu + R_n A' U^{(n)} \in E_n(\mu, \Sigma, \Phi)$, Σ positive-definite, with regularly varying generating variate R_n or regularly varying density generator. Then Theorem 5.1 and Corollary 5.2 require X to be in the domain of attraction of some extreme value distribution. According to Proposition 5.10 in Resnick [18], p. 265, all margins X_i, $i = 1, \ldots, n$ and the standardized distribution F_*, i.e. $F_*(x_1, \ldots, x_n) = F((1/(1-F_1))^{-1}(x_1), \ldots, (1/(1-F_n))^{-1}(x_n))$ are in the domain of attraction of an extreme value distributions and $\lim_{m \to \infty} F_*^m(mx_1, \ldots, mx_n) = G(x_1, \ldots, x_n)$. Since $F_*(x_1, \ldots, x_n) = C(1 - 1/x_1, \ldots, 1 - 1/x_n) = C_*(x_1, \ldots, x_n)$, $x_1, \ldots, x_n \geq 1$ and uniform distributions on $[0,1]$ are in the domain of attraction of some extreme value distribution, we conclude again with Proposition 5.10 in Resnick [18] that C is in the domain of attraction of an extreme value distribution. Applying Proposition 3.1 in Schmidt [20], p.7, every bivariate margin of C is an elliptical copula with regularly varying generating variate or regularly varying density generator. Thus tail dependence for all bivariate margins of C follows by the results stated above this theorem. □

6 Estimating the Tail Dependence Coefficient for Elliptical Copulae

Suppose $X, X^{(1)}, \ldots, X^{(m)}$ are iid bivariate random vectors with distribution function F and elliptical copula C. Further we assume continuous marginal distribution functions F_i, $i = 1, 2$. There are several parametric and nonparametric estimation methodologies for the tail dependence coefficient of an elliptical copula available. We distinguish between two kinds of bivariate random vectors possessing an elliptical copula as dependence structure: Those which are elliptically distributed and those which are not elliptically distributed, i.e. the margins might follow different distributions. Statistics testing for tail dependence and tail independence are given in Ledford and Tawn [13].

At the presence of tail dependence Theorem 4.4 and Theorem 4.5 justify to consider only elliptical copulae with regularly varying generating variate or regularly varying density generator.

i) First, we consider the case of an elliptically contoured bivariate random vector X. In particular we consider an elliptical random vector $X \stackrel{d}{=} \mu + R_2 A' U^{(2)}$ with regularly varying generating variate R_2, i.e. for $x > 0$

$$\lim_{t \to \infty} \frac{\mathbb{P}(\|X\|_2 > tx)}{\mathbb{P}(\|X\|_2 > t)} = \lim_{t \to \infty} \frac{\mathbb{P}(R_2 > tx)}{\mathbb{P}(R_2 > t)} = x^{-\alpha}, \ \alpha > 0.$$

Then formula (14) shows that the tail dependence coefficient λ depends only on the tail index α and the "correlation" coefficient ρ, precisely $\lambda = \lambda(\alpha, \rho)$. A robust estimator $\hat{\rho}$ for ρ based on its relationship to Kendall's Tau was given in (12). Regarding the tail index α there are several well-known estimators obtainable from extreme value literature. Among these, the Hill estimator represents a natural one for the tail index α:

$$\hat{\alpha}_m = \left(\frac{1}{k} \sum_{j=1}^{k} \log \|X_{(j,m)}\|_2 - \log \|X_{(k,m)}\|_2\right)^{-1}, \tag{31}$$

where $\|X_{(j,m)}\|_2$ denotes the j-th order statistics of $\|X^{(1)}\|_2, \ldots, \|X^{(m)}\|_2$ and $k = k(m) \to \infty$ is chosen in an appropriate way; for a discussion on the right choice we refer the reader to Embrechts et al. [5], pp. 341.

ii) Now we consider the case that X is not elliptically distributed but is still in the domain of attraction of some extreme value distribution. Then we can estimate the tail dependence coefficient using the homogeneity property (8) and the spectral measure representation (20) arising from the limiting extreme value distribution. Einmahl et al. [3] and Einmahl et al. [4] propose a non-parametric and a semi-parametric estimator for the spectral measure of an extreme value distribution.

iii) Finally, if we have to reject that X follows an elliptical distribution and X is in the domain of attraction of an extreme value distribution, we propose the following estimator for λ which is based on the copula representation (29). Let C_m be the empirical copula defined by

$$C_m(u_1, u_2) = F_m(F_{1m}^{-1}(u_1), F_{2m}^{-1}(u_2)), \tag{32}$$

with F_m, F_{im} denoting the empirical distribution functions corresponding to F, F_i, $i = 1, 2$. Let $R_{m1}^{(j)}$ and $R_{m2}^{(j)}$ be the rank of $X_1^{(j)}$ and $X_2^{(j)}$, $j = 1, \ldots, m$, respectively. Then

$$\hat{\lambda}_m = 2 - \frac{m}{k}\left(1 - C_m\left(1 - \frac{k}{m}, 1 - \frac{k}{m}\right)\right) \tag{33}$$

$$= 2 - \frac{m}{k} + \frac{1}{k} \sum_{j=1}^{m} 1_{\{R_{m1}^{(j)} \leq m-k, R_{m2}^{(j)} \leq m-k\}}$$

with $k = k(m) \to \infty$ and $k/m \to 0$ as $m \to \infty$. The optimal choice of k is related to the usual variance-bias problem, which we address in a forthcoming work. The next theorem states the strong consistency property of $\hat{\lambda}_m$.

Theorem 6.1 *Let X be a bivariate random vector with elliptical copula C having a regularly varying density generator or regularly varying generating variate. Let $\hat{\lambda}_m$ be the tail dependence coefficient estimator given in (33). If $k = k(m) \to \infty$, $k/m \to 0$, $k/\log(\log m) \to \infty$ as $m \to \infty$ then*

$$\hat{\lambda}_m \to \lambda \quad \text{almost surely} \quad \text{as} \quad m \to \infty.$$

Proof. Let X possess an elliptical copula C with regularly varying density generator or regularly varying generating variate. According to Stute [24], p. 371, the distribution of C_m in (32) does not dependent on the marginal distributions F_1 and F_2 such that w.l.o.g we may assume that F_i, $i = 1, 2$, are uniform distributions on the unit interval and we are in the copula framework. Theorem 5.4 yields that C is in the domain of attraction of an extreme value distribution. The strong consistence is now a special case of Theorem 1.1 in Qi [17] because of the uniform convergence of $C^m(1 - 1/(mx_1), \ldots, 1 - 1/(mx_n))$ to its corresponding extreme value distribution. □

Asymptotic normality will be addressed in a forthcoming work.

The figures below graphically summarize the tail dependence properties of four financial data-sets. We provide the scatter plots of daily negative log-returns of the financial securities and compare them to the corresponding tail dependence coefficient estimate (33) for various k. Both plots give an intuition for the presence of tail dependence and the order of magnitude of the tail dependence coefficient. For modelling reasons we assume that that the daily log-returns are iid observations. All plots related to the estimation of the tail dependence coefficient show the typical variance-bias problem for various k. In particular, a small k comes along with a large variance of the estimate, whereas an increasing k results in a strong bias. In the presence of tail dependence, k is chosen such that the tail dependence coefficient estimate $\hat{\lambda}$ lies on a plateau between the decreasing variance and the increasing bias. Thus in Figure 2 one takes k between 80 and 110 to obtain the estimate $\hat{\lambda} = 0.28$.

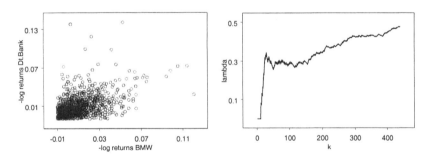

Fig. 2. Scatter plot of BMW versus Dt. Bank daily stock log-returns (2325 data points) and the corresponding tail dependence coefficient estimate $\hat{\lambda}$ for various k

Credit Risk Modelling and Estimation via Elliptical Copulae 287

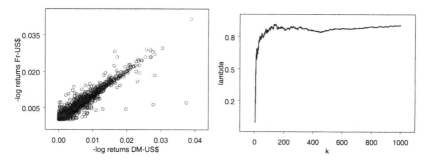

Fig. 3. Scatter plot of DM-US\$ versus FF-US\$ daily exchange rate log-returns (5000 data points) and the corresponding tail dependence coefficient estimate $\hat{\lambda}$ for various k

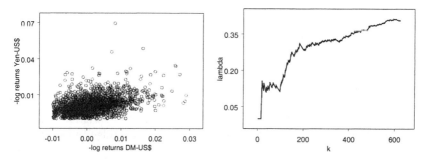

Fig. 4. Scatter plot of DM-US\$ versus Yen-US\$ daily exchange rate log-returns (3126 data points) and the corresponding tail dependence coefficient estimate $\hat{\lambda}$ for various k

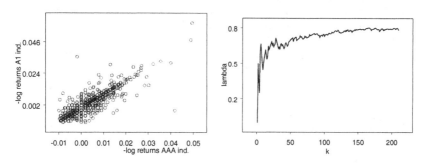

Fig. 5. Scatter plot of AAA.Industrial versus A1.Industrial daily log-returns (1043 data points) and the corresponding tail dependence coefficient estimate $\hat{\lambda}$ for various k

7 Conclusions

Summarizing the results, we have found that elliptical copulae provide appealing dependence structures for asset portfolio modelling within internal credit-risk management. We characterized those elliptical copulae which incorporate dependencies of extremal credit default events by the so-called tail dependence property. Further we showed that most elliptical copulae having the tail dependence property are in the domain of attraction of an extreme value distribution. Thus powerful tools of extreme value theory can be applied. Moreover, the application of elliptical copulae is recommended due to the existence of good estimation and simulation techniques.

Acknowledgment

The author wants to thank Prof. Dr. U. Stadtmüller and Prof. Dr. R. Kiesel for helpful comments and discussions.

References

1. N. H. Bingham, C. M. Goldie, and J. L. Teugels, *Regular variation*, Encyclopedia of Mathematics and Its Applications, vol. 27, Cambridge University Press, Cambridge, 1987.
2. N. H. Bingham and R. Kiesel, *Semi-parametric modelling in finance: theoretical foundation*, Quantitative Finance 2 (2002), 241–250.
3. J.H.J. Einmahl, L. de Haan, and V.I. Piterbarg, *Multivariate extremes estimation*, Ann. Statistics **19** (2001), 1401–1423.
4. J.H.J. Einmahl, L. de Haan, and A. K. Sinha, *Estimating the spectral measure of an extreme value distribution*, Stochastic Processes and their Applications **70** (1997), 143–171.
5. P. Embrechts, C. Klueppelberg, and T. Mikosch, *Modelling extremal events*, Applications of Mathematics, Stochastic Modelling and Applied Probability 33, Springer Verlag Berlin, 1997.
6. P. Embrechts, F. Lindskog, and A. McNeil, *Modelling dependence with copulas and applications to risk management*, preprint (2001).
7. K.T. Fang, S. Kotz, and K.W. Ng, *Symmetric multivariate and related distributions*, Monographs on Statistics and Applied Probability 36, Chapman and Hall, London, 1990.
8. M. Gordy and E. Heitfield, *Estimating factor loadings when rating performance data are scarce*, Technical Report, Board of Governors of the Federal Reserve System (2000).
9. M. B. Gordy, *A risk-factor model foundation for ratings-based bank capital rules*, Draft, Board of Governors of the Federal Reserve System (2001).
10. H.A. Hauksson, M. Dacorogna, T. Domenig, U. Mueller, and G. Samorodnitsky, *Multivariate extremes, aggregation and risk estimation*, Quantitative Finance **1** (2001), 79–95.

11. H. Hult and F. Lindskog, *Multivariate extremes, aggregation and dependence in elliptical distributions*, preprint (2001).
12. H. Joe, *Multivariate models and dependence concepts*, Monographs on Statistics and Applied Probabilty 73, Chapman and Hall, London, 1997.
13. A.W. Ledford and J.A. Tawn, *Statistics for near independence in multivariate extreme values*, Biometrika **83** (1996), 169–187.
14. R.B. Nelsen, *An introduction to copulas*, Lecture Notes in Statistics 139, Springer Verlag, 1999.
15. Basel Committee on Banking Supervision, *The internal ratings-based approach*, Bank of International Settlement (2001).
16. *The New Basel Capital Accord*, Bank of International Settlement (2001).
17. Y. Qi, *Almost sure convergence of the stable tail empirical dependence function in multivariate extreme statistics*, Acta Math. Appl. Sinica (English ser.) (1997), no. 13, 167–175.
18. S. Resnick, *Extreme values, regular variation, and point processes*, Springer New York, 1987.
19. *Hidden regular variation, second order regular variation and asymptotic dependence*, preprint (2002).
20. R. Schmidt, *Tail dependence for elliptically contoured distributions*, Math. Methods of Operations Research **55** (2002), no. 2, 301–327.
21. B. Schweizer and E.F. Wolff, *On nonparametric measures of dependence for random variables*, Ann. Statist. **9** (1981), 870–885.
22. A. Sklar, *Random variables, distribution functions, and copulas - a personal look backward and forward*, Distributions with Fixed Marginals and Related Topics, ed. by L. Rueschendorff, B. Schweizer, and M.D. Taylor (1996), 1–14.
23. C. Stărică, *Multivariate extremes for models with constant conditional correlation*, Journal of Empirical Finance **6** (1999), 515–553.
24. W. Stute, *The oscillation behavior of empirical processes: The multivariate case*, The Ann. of Probability **12** (1984), no. 2, 361–379.

Credit Risk Models in Practice - a Review

Stefan Trück and Jochen Peppel

Institut für Statistik und Mathematische Wirtschaftstheorie, Universität Karlsruhe, Germany

1 Introduction

Due to the changes in the new Basel Capital Accord for Credit Risk evaluation banks will have to concentrate more thoroughly on internal or external models to assess the risk adequately. Our paper reviews the current most popular models proposed by the industry not leaving out the theoretical background behind.

We start with the so-called structural approach to Credit Risk that was originally introduced by Robert Merton in 1974. In this approach default is considered to happen when the value of a firm's assets falls below the liabilities of the company. This approach was slightly modified and is applied by following described 'distance-to-default' methodology of the KMV company - one of the most popular models used in practice.

The second class of models are Credit Value-at-Risk methodologies. The probably most famous model of this category is applied by JP Morgan with CreditMetrics. Here the rating of a company is the decisive variable entering the model and historical transition matrices are used to determine default probabilities and the Value-at-Risk of bond portfolios within a given time horizon.

We also describe the third major approach in practice as it is proposed by Credit Suisse Financial Products (CSFP) with CreditRisk+. Here, actuarial methods from insurance mathematics are applied to model frequencies and severities of defaults which are assumed to follow exogenous Poisson processes.

We give a brief description of McKinsey's CreditPortfolioView which in addition to a transition matrix uses macroeconomic variables to capture the effect of the business cycle on credit risk.

Finally, we consider the one-factor model by Frye (2000) and Gordy (2000) which is the starting point for the approach suggested in the new Basel Capital Accord to determine Credit VaR.

2 The Merton Approach

The structural approach to credit risk was introduced by Robert Merton in 1974[1]. He applied the Black-Scholes option pricing formula to Contingent Claims Analysis as it was suggested by Black and Scholes themselves in their seminal paper from 1973.

The basic idea of the structural approach is that default occurs when the value of a company's assets falls below the value of its outstanding debt. So the probability of default is determined by the dynamics of the assets. If the assets exceed the firm's debt at maturity, the shareholders will pay back the debt in full. In structural models a stochastic process driving the dynamics of the assets is assumed. The firm is in financial distress when the stochastic process hits a certain lower boundary at (respectively before) maturity of the obligations. In this case – due to limited liability of the shareholders – the lenders are repaid only with the residual value of the assets.

The similarity to option price theory is quite obvious: the position of the shareholders can be described as having a call option on the firm's assets with a strike price equal to the face value of the outstanding debt.

It is clear that in this model all kinds of outstanding debt have to be valued simultaneously. Therefore, for companies with complex capital structures the valuation method for corporate debt can be quite difficult or impractical. Therefore, Merton simplified the structure of the model. He assumed the same conditions of an arbitrage free market as Black and Scholes (1973) did. Furthermore, four somehow critical assumptions were imposed as well:

ASSUMPTION 1: *The riskless interest rate is constant over time. The value of a sure dollar at time t payable at $T > t$ is therefore*

$$D(t,T) = e^{-r(T-t)}, \qquad (1)$$

where $r(t) = r$ *is the instantaneous riskless interest rate.*
ASSUMPTION 2: *The firm is in default when the value of its assets falls below the value of its debt.*
ASSUMPTION 3: *The default can exclusively occur at maturity of the bond. Low asset values during the lifetime of the bond do not necessarily force the firm into bankruptcy.*
ASSUMPTION 4: *The payouts in case of bankruptcy follow strict absolute priority*[2].

Imposing these assumptions we can now derive a closed form expression for valuing corporate debt.

Let $V(t)$ be the value of the firm's assets at time t whose dynamics are given by

---
[1] See Merton (1974).
[2] 'Strict absolute priority' means that in case of bankruptcy obligations are paid back in succession of their seniority.

$$dV(t) = \mu V(t)dt + \sigma V(t)dZ(t), \tag{2}$$

where μ and σ are constants and $Z(t)$ is a standard Wiener process[3].

B denotes the nominal value of a (zero-)bond[4]. At maturity time T the payoff to the shareholder $S(T)$ is either zero, if the firm defaults, or, otherwise, the difference between the assets and the debt:

$$S(T) = \max\{V(T) - B, 0\}. \tag{3}$$

On the other side the bondholder receives either the full nominal value, or the residual value of the assets in case of bankruptcy. His payoff at T is then

$$\begin{aligned} B(T) &= \min\{B, V(T)\} \\ &= B - \max\{B - V(T), 0\}. \end{aligned} \tag{4}$$

Since $\max\{B - V(T), 0\}$ is the payoff from a put option on the assets with strike price B, the value of the risky bond at time t ($t < T$) is equal to the difference between a discounted riskless bond and the price of the put option at t:

$$B(t, T) = Be^{-r(T-t)} - p(V(t), t). \tag{5}$$

The price of the put option $p(V(t), t)$ satisfies the put call parity

$$p(V(t), t) = c(V(t), t) - V(t) + Be^{-r(T-t)}, \tag{6}$$

where $c(V(t), t)$ is the price of the corresponding call.

The value $c(V(t), t)$ can then be calculated using the Black-Scholes formula:

$$c(V(t), t) = V(t)\Phi(d_1) - Be^{-r(T-t)}\Phi(d_2) \tag{7}$$

with

$$d_1 = \frac{\ln \frac{V(t)}{B} + (r + \frac{1}{2}\sigma^2)(T-t)}{\sigma\sqrt{T-t}},$$

$$d_2 = \frac{\ln \frac{V(t)}{B} + (r - \frac{1}{2}\sigma^2)(T-t)}{\sigma\sqrt{T-t}}$$

---

[3] A 'Wiener Process' (also called 'Brownian Motion') is a stochastic process $\{X(t), 0 \leq t < \infty\}$ with homogenous, independent and normally distributed increments that satisfies $\Pr(X_0 = 0) = 1$.

[4] In the original structural model there is no other class of outstanding debt, see Merton (1974).

Thus, we derived a closed form expression for the time t value of a risky zero bond. It clearly depends on the stochastic process driving the dynamics of the assets and the assumed term structure of interest rates.

$$B(t,T) = V(t)(1 - \Phi(d_1)) + Be^{-r(T-t)}\Phi(d_2) \tag{8}$$

There is no doubt that Merton's structural approach is convincing in its basic concept as it is based on modern theory of corporate finance. The approach can be applied to any public company and being based on stock market data rather than on accounting data, it can be seen as "forward looking".

However, its success in empirical investigations is rather doubtful[5]. In 1998 Jarrow and van Deventer (Kamakura Cooperation) set up a competition between various hedging strategies to see how the Merton model held up in practice[6]. The result was rather disillusioning as on a First Interstate Bancorp data set between 1987 and 1993 the hedge performance of Merton's model could not compete with a one-for-one hedge using US Treasuries. 20% of the time even no hedge at all would have been preferable to the structural model.

The shortcomings of the Merton model have two major sources. First of all, based on the APT, the model predicts credit spreads which are consistent with expected losses in a risk neutral world. Since investors tend to be risk-averse the prices of risky bonds are consequently overestimated.

It hardly makes sense to calculate unexpected losses with the Merton model. Different payoffs at maturity are only due to a varying recovery rate in case of default. The recovery itself depends exclusively on the value of the remaining assets. This combination of assumptions cannot satisfy real world conditions.

The second source of the model's shortcomings are its basic assumptions (assumption 1 - 4). Prior default experience suggests that a firm defaults long before its assets fall below the value of the debt (assumption 2). This is one reason why the analytically calculated credit spreads are much smaller than actual spreads from observed market prices.

Black and Cox (1976) improved this limitation by simplifying the bankruptcy condition of the model. They introduced an exogenously given lower boundary at which default occurs. Modelling default as the first passage time to this boundary also allows to relax assumption 3. The firm can now default at any time during the lifetime of a bond.

Besides more realistic credit spreads the simplified bankruptcy condition has another advantage. The debt structure of the firm does not matter anymore. It is treated as one homogenous class of debt. This makes the model tractable in the case of more complex capital structures. In Merton's approach all sorts of outstanding debt would have to be valued simultaneously which is an impractical task.

---

[5] The predictive power of Merton's model is tested e.g. in Jones, Mason and Rosenfeld (1983).

[6] See also Locke (1998).

2.1 Conclusion

One major weakness in empirical implementations of the model is that in general the structural approach predicts default probabilities and credit spreads too low. With the Black-Scholes-Model continuous time diffusion process for asset values, the probability for a default, respectively for the firm's asset value falling below its debt boundary declines substantially as the time horizon gets smaller. The implication would be that credit spreads should be zero close to maturity. However, observable short-term credit spreads are nonzero, what cannot only be due to liquidity or transaction cost effects. It is further observable that structural models generally underestimate the probability of default not only for short horizons. One reason for this underestimation is the underlying assumption of normally distributed asset values around some mean level.

In recent years some of the shortcomings of Merton's model have been overcome by model extensions. In the model of Longstaff and Schwartz (1995) the most obvious shortcoming of Merton's approach is eliminated. Longstaff-Schwartz introduce a structural model with stochastic riskless interest rates.

Another approach by Zhou (1997) attempts to address the underestimation of default probabilities by allowing for jumps in the asset value of the firm.

Finally Rachev et al (2000)[7] investigate the assumption of Gaussian asset returns for volatile companies and find that asset returns can rather be modelled by an alpha-stable distribution. Then a Levy motion instead of the standard Brownian motion is used for modelling the process driving the value of the assets.

The most famous model extension that is used in practice is probably the KMV model. In the KMV approach[8] the problem of underestimation default probabilities is overcome by introducing the so-called 'empirical EDF'. The empirical EDF e.g. for a one-year horizon is not defined by the calculated probability based on the model but using the fraction of firms that defaulted within a year among all firms having asset values with a certain distance $k\sigma$ from default boundary according to the structural model. Since this model is widely used in practice, we will have a closer look at the KMV model in the next section.

3 The KMV Approach

KMV extends the original Merton approach in several ways to a framework which is know as the Vasicek-Kealhofer model. Based on this model, KMV derives the actual individual probability of default for each obligor, which in KMV terminology is then called expected default frequency or EDF. Since

---
[7] See e.g. in Rachev, Schwartz and Khindanova (2000) or Trück and Rachev (2002)
[8] See e.g. Crosbie (1998) or the description in the next section.

EDFs are derived using observable variables like the firm's capital structure, the current market value of equities, etc., they can be calculated at any point of time. This property makes it possible to continuously observe changes in EDF values.

EDFs are probabilities of default for a given time horizon and most commonly quoted in basis points, thus representing a cardinal ranking of firms. By slicing the interval of possible EDF values into bands, a mapping of EDF values to default oriented rating systems is possible. Taking the structural approach setup proposed by Merton as a starting point, the dynamics of the issuer's asset value is considered to be a key driver for default. If the asset value falls below a certain level at the credit horizon, the issuer goes into default. This level is called the default point (DPT).

3.1 Deriving EDFs

Using the Vasicek Kealhofer model as a basis, there are essentially three steps to derive the EDF for a publicly traded company:

1. Estimation of the market value and the volatility of the firm's assets.
2. Calculation of the distance-to-default, which is an index measure of default risk.
3. Translation of the distance-to-default into the actual probability of default, which is also called EDF.

Estimation of the market value and the volatility of the firm's asset

The market value of the firm's asset is usually not directly observable. For this reason, the extended Merton approach is used to establish a link between the firm's equity market value and asset market value. Similar to the previous section the market value of the firm's assets, V_A, is assumed to follow the stochastic process

$$dV_A = \mu V_A dt + \sigma_A V_A dz, \tag{9}$$

where μ and σ_A are the drift rate and volatility of the assets, and dz is a standard Wiener process.

Assuming that the firm's liabilities consist of a single class of debt due at time T with book value X and a single class of equity with market value V_E, it is possible to consider the equity as a call option and to price it using the Black/Scholes option price formula:

$$V_E = V_A N(d_1) - e^{-rT} X N(d_2) \tag{10}$$

$$d_1 = \frac{\ln\left(\frac{V_A}{X}\right) + \left(r + \frac{\sigma_A^2}{2}\right)T}{\sigma_A\sqrt{T}}$$

$$d_2 = d_1 - \sigma_A\sqrt{T},$$

where r is the risk-free interest rate. A formula for the volatility of equity value can be derived by

$$\Delta V_E = \frac{\partial V_E}{\partial V_A}\Delta V_A$$

$$\frac{\Delta V_E}{V_E} = \underbrace{\frac{\partial V_E}{\partial V_A}}_{N(d_1)}\frac{V_A}{V_E}\frac{\Delta V_A}{V_A}$$

$$\sigma_E = N(d_1)\frac{V_A}{V_E}\sigma_A. \tag{11}$$

Thus the two functions

$$V_E = f(V_A, \sigma_A, X, r, T) \quad \text{and}$$
$$\sigma_E = g(V_A, \sigma_A, X, r, T) \tag{12}$$

can be solved simultaneously for V_A and σ_A. But there are two problems connected with this proceeding:

1. The equity volatility σ_E is not directly observable and relatively unstable.
2. Equation (12) holds only instantaneously.

For this reason, KMV uses a more complex iterative procedure[9] to solve for the asset volatility σ_A. An initial guess of the volatility is used to determine the asset value and to de-lever the equity return. The volatility of the resulting asset returns is then used as the input for the next iteration step, and so on, to produce a series of asset returns. Iteration is continued until convergence. Additionally, the volatility derived in this procedure is combined with country, industry and size averages in order to produce a more adequate estimate.

Calculation of the distance-to-default

In the KMV model, default occurs, if the firm's asset value is below some threshold, the so-called default point (DPT), at the time horizon T under consideration[10]. From empirical observations, KMV has found that this point

---

[9] Unfortunately, the available information is not very precise at this point. See for instance Crosbie(1998).
[10] Note that this way of modelling default allows the asset value process to drop below this threshold before T without any influence on the outcome. The decisive point of time under consideration is T.

lies somewhere between the value of total liabilities and the value of short term debt. Thus, they define the default point as the par value of current liabilities, including short term debt (STD) to be serviced over the time horizon plus half the long term debt (LTD)

$$\text{DPT} = \text{STD} + \frac{1}{2}\text{LTD}.$$

From the default point and the asset value process, an index measure is calculated that captures the distance between the expected asset value at T and the default point in units of asset volatilities and is, therefore, called distance-to-default:

$$\text{DD} = \frac{E(V_A) - \text{DPT}}{\sigma_a} \qquad (13)$$

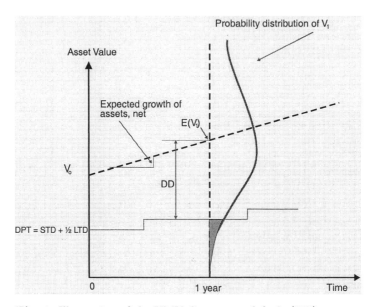

Fig. 1. Illustration of the KMV distance-to-default (DD) concept

Figure 1 illustrates[11] the distance-to-default by assuming a time horizon of one year. The asset value starts in $t = 0$ at V_0 and develops according to some process to V_1 at $t = 1$, which translates into the probability distribution of V_1. The difference between the expected value of V_1 in $t = 1$, $E(V_1)$, and the DPT, expressed in units of the asset return standard deviation at $t = 1$, is the distance to default.[12]

---

[11] The figure has been taken from Crouhy, Galai and Mark (2000), page 90.
[12] Note that it is not specified in the figure that the DD is measured in units of the asset return standard deviation at the credit horizon.

This first set up of DD is not yet directly applicable to our Black/Scholes world, because there, the set of consistent variables would be

- $E_T(\ln(V_A))$ since we consider the process of logarithmic asset values
- $\ln(DPT_T)$ to be consistent with the logarithmic modelling, and
- $\sigma_A\sqrt{T}$ as σ_A is the volatility for the unit time interval of the stochastic asset value process under consideration, we match time horizons by multiplying with $\sqrt{T}$

Thus

$$DD = \frac{E_T(\ln(V_A)) - \ln(DPT_T)}{\sigma_A\sqrt{T}} \qquad (14)$$

and since

$$E_T(\ln(V_A)) = \ln(V_{A,t=0}) + \left(\mu - \frac{\sigma_A^2}{2}\right)T$$

equation (14) yields

$$DD = \frac{\ln(V_{A,t=0}) + \left(\mu - \frac{\sigma_A^2}{2}\right)T - \ln(DPT_T)}{\sigma_A\sqrt{T}}$$

$$= \frac{\ln\left(\frac{V_{A,t=0}}{DPT_T}\right) + \left(\mu - \frac{\sigma_A^2}{2}\right)T}{\sigma_A\sqrt{T}}. \qquad (15)$$

Translating DD to EDF

The probability of default p_d is the probability that the asset value at time T is lower than the default point when starting at time $t = 0$ with a total asset value of $V_A^0 = V_A$

$$p_d = P\left(V_A^T \leq DPT_T \mid V_A^0 = V_A\right)$$
$$= P\left(\ln V_A^T \leq \ln DPT_T \mid V_A^0 = V_A\right).$$

Assuming the process for the change of the firm's asset values described by (9) the logarithm of the asset value at T is

$$\ln V_A^T = \ln V_A + \left(\mu - \frac{\sigma_A^2}{2}\right)T + \sigma_A\sqrt{T}\,\epsilon, \qquad (16)$$

where $\epsilon \sim N(0,1)$ is the random component of the firm's return. Thus, the default probability can be transformed into

$$p_d = P\left[\ln V_A + \left(\mu - \frac{\sigma_A^2}{2}\right)T + \sigma_A\sqrt{T}\epsilon \leq DPT_T\right]$$

$$= P\left[\epsilon \leq -\frac{\ln\left(\frac{V_A}{DPT_T}\right) + \left(\mu - \frac{\sigma_A^2}{2}\right)T}{\sigma_A\sqrt{T}}\right]$$

$$= \Phi\left[-\frac{\ln\left(\frac{V_A}{DPT_T}\right) + \left(\mu - \frac{\sigma_A^2}{2}\right)T}{\sigma_A\sqrt{T}}\right]. \quad (17)$$

Note that the term in brackets is just $-DD$ and thus $p_d = \Phi(-DD)$. Having calculated the distance to default, this would allow a direct solution for the default probability or EDF, respectively. But there are some inaccuracies involved in this proceeding.

Firstly, the normal distribution is a rather poor choice when defining the probability of default, because the default point is, in reality, also a random variable. Surely firms will adjust their liabilities as they head towards default and use undrawn commitments. Furthermore, the assumption of normality of the asset return distribution may capture the normal development of a firm quite well, but may not be able to capture extreme events correctly. Then some simplifying assumptions were made about the capital structure of the firm in order to set up a simple model of liability and asset dependence.

Being well aware of the problems, KMV does not use the distance-to-default measure as a direct means for default probability calculation. Instead, based on historical observations of firms which were grouped according to their DD values, KMV has elaborated a mapping between distance-to-default and EDF values. A DD of four would yield, taking the normal distribution assumption, a default rate of essentially zero, whereas KMV's EDF mapping yields an EDF of approximately 100 bp.

The idea behind EDF is that based on historical information on a large sample of firms, one can estimate the proportion of firms of a given ranking, say e.g. $DD = 4$, which actually defaulted after 1 year. This proportion then is the EDF for firms with a given ranking.

$$EDF = \frac{\text{Firms actually defaulted with certain DD (e.g. DD=4)}}{\text{Total population of firms with certain DD (e.g. DD=4)}} \quad (18)$$

3.2 Conclusion

Obviously with its more empirically based view on credit risk the KMV approach is able to overcome some of the disadvantages of the classical structural approach models. Mapping distance-to-default to an empirical EDF leads to more realistic default probabilities and credit spreads as they can be observed in the market. KMV tries to combine the strengths of the basic concept of the structural approaches being based on modern theory of corporate finance with an empirical mapping to obtain results based on "real-world" figures.

Also the idea of calculating an explicit probability of default instead of just using the rating may give additional information about a company's risk. In many models the rating is the decisive variable as it comes to determining default frequencies and all loans within a certain rating grade are assigned the same default probability. In reality there are substantial differences of default rates. So comparing its model to other approaches for KMV the question is: 'Do we yield better results by increasing the overall number of categories using EDFs as an additional information?'. Miller (1998) tested whether using EDFs in addition to general rating categories assigned by S&P or Moody's could improve the default prediction power of a model. He found that the KMV ratings seem to have a significant predictive value especially for time horizons shorter than two years. As one would expect, the ability to predict default decreases with longer time horizons, however the predictive power of his model was still better than just using a company's rating class.

Despite this advantages the KMV approach is still a structural model. Thus, several of the problems of this model class remain like the difficulties of estimating correctly the value and volatility of a company's assets or also the calculation of its default point. Also taking the value of a firm as the decisive variable for estimating default probabilities may be critical, especially using Brownian Motion as the stochastic process driving the value of the assets may be unrealistic. In many cases default happened as a sudden and unforeseen event and no convergence of the asset value to a default point could be observed before. These problems lead to another kind of approach to credit risk, the so-called reduced form models as they are implemented in CreditRisk+ or CreditMetrics. In the next two sections we will have a look at these models.

4 The CreditMetrics Model

4.1 Building the Model

The *CreditMetrics* approach departs from the assumption that the market value of a bond or loan is a function of the variables as they were used in firm value models: time to maturity τ, face value F, the risk free rate r.

As a risk management tool the model must be applicable to all kinds of financial instruments with inherent credit risk, and the valuation procedure must be consistent with actual market prices[13]. Therefore *CreditMetrics* uses empirically derived bond prices for the valuation. CreditMetrics is the theoretical framework for J.P.Morgan's risk management software *CreditManager*.

---

[13] For more detailed information see Gupton, Finger and Bhatia (1997): Credit Metrics - Technical Document.

It is further assumed that all variables, except the current rating state of the issuer π_t, behave deterministically over time. Thus the value of the bond or loan at the risk horizon T is essentially dependent on the rating state of the issuer at this point of time, π_T.

- In the case of default, the recovery rate is simulated by a beta distribution whose mean and standard deviation are calibrated to fit the parameters of the historically observed recovery rate that corresponds to the seniority of the item.[14]
- If the issuer is not in a state of default at the risk horizon, the value of the bond or loan is determined by discounting the outstanding cash flows using credit spreads over r that correspond to the rating state π_T of the issuer in T.

The distribution of bond or loan values in T is thus given by the probabilities $P(\pi_T)$ of the different rating states in T, together with the corresponding values of the bond $V_T(\,\cdot\,,\pi_T)$.

To calculate the price of the bond at a future time t (risk horizon), we then divide the space of time from present to maturity date ($[0,T]$) into two phases:

- The first phase ($[0,t]$) is the risk horizon. During this time the bond can go to any rating category according to an estimated transition matrix.
- Then, for the second time interval the bond is discounted ($(t,T]$) with an appropriate discount function (calculated implicitly from observed market prices).

We will illustrate the *CreditMetrics* model considering a four-year corporate bond issued by an industrial company. The face value of the bond is 1,000 Euro and coupons of 6% are paid yearly. The seniority class of the bond is "senior unsecured" and the company initial S&P's rating for long-term debt is BBB.

Table 1. Cash flows from a BBB-bond for each year $t \in [0, 4]$.

t	0	1	2	3	4
Coupon C_t (Euro)	0	60	60	60	60
Principal B (Euro)	0	0	0	0	1000

---

[14] For a table of historical observed recovery rates and their standard deviation see for example Carty and Lieberman (1996), table 5 on page 16.

4.2 Rating Transitions

In the first stage of the model we determine the distribution of ratings the exposure can take on at the end of a given risk horizon t. This is done with the help of a transition matrix P. An exemplary transition matrix is given in Table 2.

Table 2. Exemplary one-year transition matrix (%) (source: Standard&Poor's)

Initial Rating	Rating at year-end							
	AAA	AA	A	BBB	BB	BB	CCC	Default
AAA	87.74	10.93	0.45	0.63	0.12	0.09	0.02	0.02
AA	0.84	88.23	7.47	2.16	1.11	0.12	0.05	0.02
A	0.27	1.59	89.05	7.40	1.48	0.12	0.06	0.03
BBB	1.84	1.89	5.00	84.21	6.51	0.32	0.16	0.07
BB	0.08	2.91	3.29	5.53	74.68	8.05	4.14	1.32
B	0.21	0.36	9.25	8.29	2.31	63.87	10.13	5.58
CCC	0.06	0.25	1.85	2.06	12.34	14.87	39.97	18.60
Default	0	0	0	0	0	0	0	100

This table provides the probabilities for a rating change within one year. Suppose that the initial rating of the exposure at time 0 is $i \in \{1, 2, \ldots K\}$. This initial setting can be represented by the vector

$$p_{i\cdot}(0) = \delta_i.$$

To obtain the distribution of possible ratings at t we have to multiply the initial rating vector with a t-step transition matrix. If the risk horizon is longer than one year then it is suggested to compute the required vector of transition probabilities $p_{i\cdot}(t)$ either with a multiple of a one-year transition matrix

$$p_{i\cdot}(t) = \delta_i \cdot P(1)^t,$$

or, if available, with a directly estimated t-year transition matrix

$$p_{i\cdot}(t) = \delta_i \cdot P(t).$$

Thus, we obtain all possible future ratings at time t and the corresponding transition probabilities.

Rating at t	1	2	$\ldots$	$K-1$	K
Migration probability	$p_{i1}(t)$	$p_{i2}(t)$	$\ldots$	$p_{i(K-1)}(t)$	$p_{iK}(t)$

Thus, assuming a constant transition matrix in our example to obtain the distribution of all possible ratings that our bond can take on at the end of the two year risk horizon we have to compute

$$p^{(2)}_{BBB \cdot} = \delta_{BBB} \cdot P^2$$

$$= (0,0,0,1,0,0,0,0) \cdot$$

$$\left(\frac{1}{100}\right) \begin{pmatrix} 87.74 & 10.93 & 0.45 & 0.63 & 0.12 & 0.09 & 0.02 & 0.02 \\ 0.84 & 88.23 & 7.47 & 2.16 & 1.11 & 0.12 & 0.05 & 0.02 \\ 0.27 & 1.59 & 89.05 & 7.40 & 1.48 & 0.12 & 0.06 & 0.03 \\ 1.84 & 1.89 & 5.00 & 84.21 & 6.51 & 0.32 & 0.16 & 0.07 \\ 0.08 & 2.91 & 3.29 & 5.53 & 74.68 & 8.05 & 4.14 & 1.32 \\ 0.21 & 0.36 & 9.25 & 8.29 & 2.31 & 63.87 & 10.13 & 5.58 \\ 0.06 & 0.25 & 1.85 & 2.06 & 12.34 & 14.87 & 39.97 & 28.60 \\ 0 & 0 & 0 & 0 & 0 & 0 & 0 & 100 \end{pmatrix}^2 \quad (19)$$

$$= \tfrac{1}{100}(3.20, 3.73, 9.06, 71.72, 10.47, 1.29, 0.50, 0.03)$$

For example, the probability that our BBB-rated bond will have climbed up one category to A in two years is 9.06%.

4.3 Forward Prices

Now we will derive for each rating state a risk-adjusted forward price. The case of default and non-default states are considered separately.

The remaining cash flows from t to T in non-default categories are discounted with state specific forward rates. The forward zero curve for each rating category can be found by calibrating forward rates to observed credit spreads of different maturities.

Table 3. Forward zero curve (US industrial and treasury bonds)

(%)			$t = 2$			
	1	2	3	5	7	10
Treasury	5.60	5.84	5.99	6.12	6.32	6.43
AAA	5.80	6.08	6.22	6.37	6.49	6.65
AA	5.84	6.12	6.26	6.42	6.56	6.71
A	5.91	6.20	6.32	6.49	6.64	6.79
BBB	6.08	6.36	6.50	6.64	6.84	7.05
BB	6.53	6.96	7.20	7.53	7.75	8.01
B	7.14	7.53	7.83	8.23	8.60	8.93
CCC	8.19	8.90	10.00	11.12	12.19	13.30

In the case of non-default agreed payments before t will be fully received and can be added – including the earned interest until $t-$ to the risk-adjusted time t value.

$$B_j(t,T) = \sum_{k=1}^{t} C_k(1+f^*(k,t))^{t-k} + \sum_{k=t+1}^{T} \frac{C_k}{(1+f_j(t,k))^{k-t}} + \frac{B}{(1+f_j(t,T))^{T-t}} \quad (20)$$

with

C_k the nominal coupon in year k
B the nominal principal
f^* the riskless forward rate
f_j the forward rate for j-rated bonds.

In case that the bond defaults before t a recovery payment is assigned:

$$B_K(t,T) = R \cdot (\sum_{k=1}^{T} C_k + B) \quad (21)$$

where R is the expected fraction of the bond's nominal cash flows that is paid back. R is estimated as the average return in prior default experience and depends on the seniority class of the bond. Unfortunately, recovery payments are highly uncertain. Therefore, later calculations of expected and unexpected losses also have to take the variance of recovery rates into consideration. We assume the recovery rate to be $R = 0.5113$ which is the average historical recovery rate as it is provided by Moody's for Senior Unsecured bonds.

$$B_{AAA}(2,4) = 60 \cdot 1.04 + 60 + \frac{60}{1,0580} + \frac{60+1000}{1.0608^2} = 1121.08€,$$
$$B_{AA}(2,4) = 60 \cdot 1.04 + 60 + \frac{60}{1,0584} + \frac{60+1000}{1,0612^2} = 1120.35€,$$
$$\ldots = \ldots$$
$$B_K(2,4) = 0.5113(60+60+60+60+1000) = 634.01€.$$

Compared to these values the value of a riskless bond with the same cash flows is

$$B^*(2,4) = 60 \cdot 1.04 + 60 + \frac{60}{1,0560} + \frac{60+1000}{1.0584^2} = 1125.47€.$$

With these bond prices and the mass distribution we will now use the so-called *distribution of values* to calculate value-at-risk numbers for the bond.

4.4 The Distribution of Values

We regard the bond prices B_j as a random variable. Since the mass distribution of this random variable is given by the vector $p_i.$, we obtain the Distribution of Values (DoV) for a given initial rating and risk horizon. Credit risk measures like the Expected or Unexpected Loss can be derived from the DoV. It maps the predicted rating specific bond price at the risk horizon to the probability of the transition to this rating.

Every DoV of a risky bond can be described by some characteristic properties:

- $\Pr(B_j < 0) = 0$. The worst case that can happen is that a bond defaults and the recovery rate equals zero.
- $\Pr(B_j \geq B^*) = 0$. A risky bond is always worth less than a riskless bond.
- There is a large probability of earning a small profit (riskless interest rate + risk premia), because default probabilities are quite low.
- There is a small probability of losing a large amount of investment in the case of default.

For our example we get the following distribution of values - they are illustrated in Figure 2.

Table 4. Distribution of values for our exemplary bond

Rating Category	Migration probability	Risk-adjusted bond value
AAA	3.20%	1121.08 €
AA	3.73%	1120.35 €
A	9.06%	1118.90 €
BBB	71.71%	1115.98 €
BB	10.47%	1105.26 €
B	1.29%	1095.14 €
CCC	0.50%	1071.68 €
D	0.03%	634.01 €

One reason why Credit Risk cannot be handled with established VAR methods (developed for market risk measurement) is the shape of the DoV. Most models in finance assume normal distributed returns when they are affected by market risk (e.g. changes in stock prices, interest rates, etc.) However, as you can see in Figure 2 a symmetric distribution is unsuitable for returns subject to credit risk. The distributions are highly skewed to the right with a "fat tail" on the loss side.

Still, the distinction between market risk and credit risk is not always clear. For example, the value of a credit exposure can change due to FX

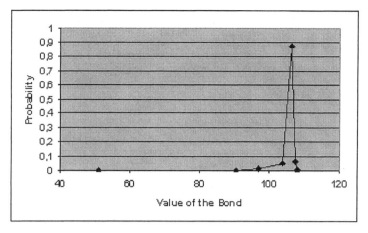

Fig. 2. 'Distribution' of values for our exemplary bond

moves without a change in the credit quality of the obligor. Or the value of a credit exposure can change due to a change of credit quality of the obligor which itself is caused by sources of market risk. Therefore, sometimes credit risk is regarded as a function of market risk[15].

Expected Loss

Keeping the new Basel capital accord in mind a very important question as it comes to measuring Credit Risk is the average return from a bond, respectively the average loss that a bank may suffer from the default of a bond. This average loss is the minimum amount of capital that should be set aside for compensation.

The Expected Return (ER) of an i-rated bond to the risk horizon t is defined as the expectation value of its DoV:

$$ER_i(t) = \sum_{j=1}^{K} p_{ij}(t) B_j(t, T) \qquad (22)$$

In risk management tools it is more common to express future values in terms of losses.

The Expected Loss (EL) is defined as the difference between the Expected Return and the value of a riskless bond with the same cash flows (or simply the expected value of the credit spreads):

---

[15] See for example in Das (1997).

$$EL_i(t) = B^*(t,T) - ER_i(t) \qquad (23)$$
$$= \sum_{j=1}^{K} p_{ij}(t)(B^*(t,T) - B_j(t,T))$$

Thus, for our example we get:

$$ER_{BBB}(2) = \frac{1}{100} \begin{pmatrix} 3.20 \\ 3.73 \\ 9.06 \\ 71.72 \\ 10.47 \\ 1.29 \\ 0.50 \\ 0.03 \end{pmatrix}^T \begin{pmatrix} 1121.08 \\ 1120.35 \\ 1118.90 \\ 1115.98 \\ 1105.26 \\ 1095.14 \\ 1071.68 \\ 634.01 \end{pmatrix} € = 1114.81€$$

Then the EL is the difference between the ER and the price of the riskless bond

$$EL_{BBB}(2) = 1125.47€ - 1114.81€ = 10.66€,$$

Unexpected Loss

The Unexpected Loss (UL) is an indicator of the amount of money that can be lost in a worst case scenario. There are two possibilities to measure the UL: the standard deviation and percentile levels of the DoV.

The UL of an i-rated exposure can be defined as the standard deviation of its DoV.

$$UL_i(t) = \sqrt{\sum_{j=1}^{K} p_{ij}(t) B_j(t,T)^2 - ER_i(t)^2}. \qquad (24)$$

As we mentioned before recovery rates are always highly uncertain. Therefore it is recommended that the volatility of the recovery rate enters into the calculation of ULs[16]:

$$UL_i(t) = \sqrt{\sum_{j=1}^{K} p_{ij}(t)(B_j(t,T)^2 + \sigma_j^2) - ER_i(t)^2} \qquad (25)$$

with

---

[16] For a derivation of this formula refer to Appendix D of the *CreditMetrics* Technical Document (1997).

$$\sigma_j^2 = 0 \qquad j \neq K.$$

As a second measure for the unexpected loss of an i-rated exposure to a given confidence level α we can define the α-percentile level of the DoV

$$UL_i^* = Q_\alpha(DoV). \tag{26}$$

By these UL definitions the uncertainty of future bond values comes only from potential rating transitions and the volatility of recovery rates. One might yield more realistic results by taking the volatility of predicted bond values in each rating category into account.

We get:

$$UL_{BBB}(2) = (\tfrac{1}{100} \begin{pmatrix} 3.20 \\ 3.73 \\ 9.06 \\ 71.72 \\ 10.47 \\ 1.29 \\ 0.50 \\ 0.03 \end{pmatrix}^T \cdot \begin{pmatrix} 1121.08^2 \\ 1120.35^2 \\ 1118.90^2 \\ 1115.98^2 \\ 1105.26^2 \\ 1095.14^2 \\ 1071.68^2 \\ 634.01^2 \end{pmatrix} - 1114.81^2)^{1/2} \text{\euro}$$

$$= \sqrt{98.34}\,\text{\euro}$$
$$= 9.92\,\text{\euro}.$$

4.5 Conclusion

Obviously *CreditMetrics* offers a quite different approach to measuring credit risk than the firm value models presented in the sections before. Firstly, *CreditMetrics* departs from the assumption that the market value of a bond or loan is a function of the variables firm value or volatility, time to maturity, face value and the risk free rate. Also the idea of using empirically derived bond prices for the valuation is completely different to the methodology of the structural approaches. One should be aware that these differences are mainly due to the fact that *CreditMetrics* is a rather empirical Value-at-Risk approach to measuring credit risk that should be consistent with actual market prices and is rather interested in the question "how much will I lose" in worst-case scenarios. Here historical transition matrices and forward prices are more important than the value of the firm.

However, we should not forget to mention that as it comes to deriving joint transition matrices for two or more individual companies the company's asset value is considered as the key driver of rating changes[17]. For measuring asset return correlations the issuers' equity returns from publicly available quotations are used. Therefore, *CreditMetrics* cannot be considered as a reduced

---
[17] See Credit Metrics - technical Document (1997)

form model only but is rather a hybrid model that uses both structural and reduced form approaches to measure VaR figures for bond or loan portfolios.

5 An Actuarial Model: CreditRisk$^+$

Credit Suisse First Boston's CreditRisk$^+$ model applies the actuarial way of proceeding commonly used for insurance matters to credit risk modelling. Thus, default as the elementary event that drives credit risk is modelled directly by assuming a Bernoullian default game for every firm. Unlike in other frameworks discussed earlier on, no assumptions are made about the causes of default, and downgrade risk is not captured.

For a better understanding of the CR$^+$ approach, a first simple way of modelling will be considered in section 5.1 that reveals the basic ideas and analyzes problems arising. These deficiencies are then tackled in the next section by introducing the extended CR$^+$ model.

5.1 The First Modelling Approach

The derivation of the default loss distribution in the CR$^+$ model comprises the following steps:

- Modelling the frequencies of default for the portfolio.
- Modelling the severities in the case of default.
- Linking these distributions together to obtain the default loss distribution.

Frequencies of default in the portfolio

For every firm, a single stationary default probability for the risk horizon under consideration, p_i, is assigned. Furthermore, default events are assumed to be independent of each other and the number of defaults is assumed to be independent over periods.

Thus, departing from the probability generating function (pgf) for a single obligor i

$$F_i(z) = (1 - p_i)\, z^0 + p_i z^1 = 1 + p_i(z - 1),$$

the pgf for the portfolio yields

$$F(z) = \prod_i (1 + p_i(z - 1)). \qquad (27)$$

Using the approximation

$$1 + p_i(z - 1) \simeq e^{p_i(z-1)} \qquad (28)$$

and setting $\mu = \sum_i p_i$, equation (27) becomes

$$F(z) = e^{\sum_i p_i(z-1)} = e^{\mu(z-1)} = e^{-\mu}e^{\mu z} = \sum_{n=0}^{\infty} \frac{e^{-\mu}\mu^n}{n!} z^n. \qquad (29)$$

Hence the probability for n defaults in the portfolio is

$$P(n \text{ defaults}) = \frac{e^{-\mu}\mu^n}{n!}, \qquad (30)$$

which is the well-known Poisson distribution with parameter μ, mean μ and standard deviation $\sqrt{\mu}$. Note that the approximation in (28) is, from a statistical point of view, equivalent to approximating the sum of independent Bernoulli draws by the Poisson distribution.

Modelling severities

In the case of default, a loss arises that is equal to the (gross) exposure minus the recovery rate. The CR$^+$ model assumes that the recovery rates are exogenously given and for that reason requires adjusted (net) exposures as an input. Thus, talking about exposures in this section always relates to the remaining part of exposure once netted with the recoveries.

For computational efficiency, equal exposures are grouped into bands. This is done by assuming some base unit of exposure L and expressing all exposures in rounded multiples of L.

Let

- ν_j be the common exposure in band j in units of L;
- ε_j be the expected loss in band j in units of L;
- μ_j be the expected number of defaults in exposure band j; and
- n_j be the actual number of defaults in exposure band j.

Since default events are assumed to be independent, one can treat every band j as an independent portfolio. The Loss in one of those subportfolios expressed in units of L is then $n_j \nu_j$.

Assume for example N=100 loans in a band with loss exposure of L=20.000 Euro. Let further p=0.03 be the default rate assumed to follow a Poisson distribution

Then we get for μ_j, the expected number of defaults in band j

$$\mu = N \cdot p = 100 \cdot 0.03 = 3 \qquad (31)$$

and for the probability of k defaults in band j

$$P(k \text{ defaults}) = \frac{e^{-3} 3^n}{n!} \qquad (32)$$

Figure 3 shows the obtained loss distribution in band j.

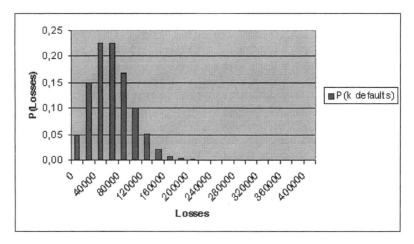

Fig. 3. Loss distribution in band j with N=100, L=20000 Euro and $\mu_j = 3$.

Obtaining the default loss distribution

Using the results of the last two sections, one can easily derive the distribution of default losses through the following steps:

- Identify the pgf $G_j(z)$ for the default loss distribution of band j.
- Combine the pgfs into a single pgf.
- Find a computationally efficient way of calculating the actual distribution of credit losses on the basis of the combined pgf.

The pgf for a loss of $n_j \nu_j$ units of L in band j is

$$G_j(z) = \sum_{n_j=0}^{\infty} P(\text{loss of } n_j \nu_j) \, z^{n_j \nu_j}$$

$$= \sum_{n_j=0}^{\infty} P(\text{number of defaults is } n_j) \, z^{n_j \nu_j}$$

and using equation (30)

$$G_j(z) = \sum_{n_j=0}^{\infty} \frac{e^{-\mu_j} \mu_j^{n_j}}{n_j!} \, z^{n_j \nu_j} \tag{33}$$

$$= e^{-\mu_j + \mu_j z^{\nu_j}} \tag{34}$$

gives the pgf for the default loss distribution of band j. Since independence is assumed between the bands, the combined pgf is just the product over the exposure band pgfs.

$$G(z) = \prod_{j=1}^{m} G_j(z)$$

$$= \prod_{j=1}^{m} e^{-\mu_j + \mu_j z^{\nu_j}}$$

$$= e^{-\sum_{j=1}^{m} \mu_j + \sum_{j=1}^{m} \mu_j z^{\nu_j}}. \tag{35}$$

The final task is now to find a computationally efficient way of calculating the actual distribution of credit losses on the basis of this combined pgf. Since equation (35) cannot be transformed into the standard pgf form that offers a closed solution for $P(n)$, the following useful property of a pgf is used.

For an arbitrary pgf $F(z) = \sum_{n=0}^{\infty} P(n) z^n$ holds:

$$\frac{\partial^k F(z)}{\partial z^k} = \sum_{n=k}^{\infty} P(n) \frac{n!}{(n-k)!} z^{n-k}$$

$$\left. \frac{\partial^k F(z)}{\partial z^k} \right|_{z=0} = P(k) k!$$

$$\Rightarrow P(n) = \frac{1}{n!} \left. \frac{\partial^k F(z)}{\partial z^k} \right|_{z=0} \tag{36}$$

Using this property, CSFB[18] derives a recurrence relationship for the probability of an overall portfolio loss of n units of L:

$$P(n) = \sum_{j:\, \nu_j \leq n} \frac{\varepsilon_j}{n} P(n - \nu_j)$$

$$P(0) = e^{-\sum_{j=1}^{m} \frac{\varepsilon_j}{\nu_j}} \tag{37}$$

These probabilities can be expressed in closed form and obviously only depend on the parameters ε_j and ν_j.

5.2 Shortcomings of the First Modelling Approach

In the last section the distribution of the number of defaults in a portfolio, departing from some initial assumptions, was inferred to be Poisson with parameter μ. This would include the hypothesis that – in the long term – the average observed volatility should converge towards $\sqrt{\mu}$. However, empirical evidence[19] shows that the observed standard deviation of the number of defaults is significantly larger than $\sqrt{\mu}$.

To overcome this problem, CR$^+$ assumes the default rate itself to be stochastic. While the expected number of default events remains the same,

---
[18] See CSFB (1997), section A4.1.
[19] See for example Carty and Lieberman (1996).

the distribution becomes significantly skewed and fat-tailed when introducing default rate volatility. This can be seen in Figure 4, which was taken from the CR^+ technical document[20].

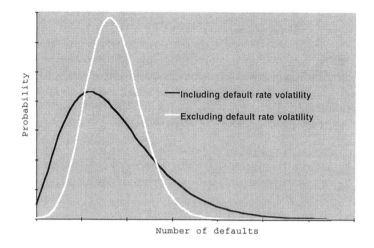

Fig. 4. Distribution of the number of defaults excluding and including default rate volatility.

Another questionable core assumption is the independence of default events. Though there might be no direct dependence between individual events, there is a clear link between the co-movements of default probabilities or default rates respectively. In an economic recession for example, one can observe a significantly higher number of defaults than on average. Since the state of the economy is surely not the only explanatory variable for the co-movement of default probabilities, CR^+ introduces a multi-factor model that influences the variable default rates of the obligors, and thus induces interdependence amongst them.

5.3 Extensions in the CR^+ Model

As mentioned above the basic model has some limitations. Some of them are overcome by the extensions in the CR^+ model. For example stochastic default rates are incorporated, a factor model is introduced that explains the dependence between the variation of the stochastic default rates of the different obligors by attributing their default rate processes to common factors.

For the matter of simplification, we make the assumption that every obligor is driven by exactly one factor. Later on, a generalization of the model will be introduced that allows for arbitrary factor affiliation for the different obligors.

---

[20] The graph was taken from CSFB(1997), page 18.

Allocating obligors to one of several factors

Define a set of n independent background factors that allow the decomposition of the set of obligors into disjoint subsets, each of which is affiliated to exactly one background factor. In CR$^+$ terminology, the subsets are also called *sectors*. Each background factor influences the expected default rate and the standard deviation of default rate of its sector.

Let therefore X_k be a random variable with mean μ_k and standard deviation σ_k that represents the average default rate of the kth factor. Let every obligor A in sector k have an actual default probabiliy of X_A with mean p_A and standard deviation σ_A and denote the set of obligors that belong to sector k as $\mathbb{K}$.

At this point, the CR$^+$ model assumes that X_A is modelled proportional to x_k by

$$X_A = p_A \frac{X_k}{\mu_k} \qquad (38)$$

and sets

$$\mu_k = \sum_{A \in \mathbb{K}} p_A.$$

For any proposed relationship between X_k and X_A, we will demand

$$E(\sum_{A \in \mathbb{K}} X_A) = \mu_k$$

and

$$VAR(\sum_{A \in \mathbb{K}} X_A) = \sigma_k^2,$$

as X_k is supposed to represent the average default rate for sector k.

The actual sector parameters are thus obtained by estimating the parameters p_A and σ_A for every obligor which determines the sector parameters as follows:

$$\mu_k = \sum_{A \in \mathbb{K}} p_A$$

and

$$\sigma_k = \frac{\sigma_k}{\mu_k} \left(\sum_{A \in \mathbb{K}} p_A \right) = \sum_{A \in \mathbb{K}} \frac{p_A}{\mu_k} \sigma_k = \sum_{A \in \mathbb{K}} \sigma_A$$

which shows that the sector standard deviation is just the sum over the standard deviations of all obligors which belong to the sector.

The pgf for the number of defaults

Having modelled the background factor dependency, the next step is to develop the pgf for the number of defaults in the kth factor. The way of proceeding is analogous to the case of a fixed default rate which was discussed in section 5.1. Recall that, assuming a fixed default rate μ, the pgf as stated in equation (29) turned out to be

$$F(z) = e^{\mu(z-1)}.$$

Since the random variable X_k represents the average default rate for sector k, it follows that conditional on the value of X_k, the pgf remains the same.

$$F_k(z \mid X_k = x) = e^{x(z-1)}$$

To obtain the unconditional pgf for sector k we integrate over the density function of X_k which is supposed to be $f_k(x)$

$$\begin{aligned} F_k(z) &= \sum_{n=0}^{\infty} P(\text{n defaults in sector k}) z^n \\ &= \sum_{n=0}^{\infty} z^n \int_{x=0}^{\infty} P(\text{n defaults in sector k} \mid x) f_k(x) dx \\ &= \int_{x=0}^{\infty} F_k(z \mid X_k = x) f_k(x) dx = \int_{x=0}^{\infty} e^{x(z-1)} f_k(x) dx. \end{aligned} \quad (39)$$

To continue solving the integral, a key assumption about the density function $f_k(x)$ is needed. Here, the CR$^+$ model assumes a Gamma distribution $\Gamma(\alpha, \beta)$ with mean μ_k and standard deviation σ_k. Therefore the parameters of the Gamma distribution are

$$\alpha_k = \frac{\mu_k^2}{\sigma_k^2} \quad \text{and} \quad \beta_k = \frac{\sigma_k^2}{\mu_k}.$$

After resolving the integral,[21] setting an additional auxiliary variable

$$p_k := \frac{\beta_k}{1 + \beta_k}$$

and expanding the obtained pgf

$$F_k(z) = \left(\frac{1 - p_k}{1 - p_k z}\right)^{\alpha_k}$$

in its Taylor series, the following explicit formula for the probability of n defaults in sector k is obtained

---
[21] See CSFB(1997), section A8.3.

$$P(\text{n defaults in sector k}) = (1 - p_k)^{\alpha_k} \binom{n + \alpha_k - 1}{n} p_k^n, \qquad (40)$$

which is the probability density function of the Negative Binomial distribution.

To obtain the pfg for the number of defaults in the whole portfolio comprising n sectors, recall that the sectors are assumed to be independent. Thus, the pgf is just the product over the sector pfgs

$$F(z) = \prod_{k=1}^{n} F_k(z) = \prod_{k=1}^{n} \left(\frac{1 - p_k}{1 - p_k z} \right)^{\alpha_k}. \qquad (41)$$

Note that the corresponding default event distribution is, accordingly, the sum of the independent Negative Binomial sector distributions and therefore **not** Negative Binomial in general.

The pgf for the default loss distribution

Analogous to the previous section, results already obtained for the fixed default rate case in section 5.1 can be used here by utilizing conditional probabilities. Let $G_k(z)$ be the pgf for the default loss distribution of sector k.

$$G_k(z) = \sum_{n=0}^{\infty} P(\text{loss of nL in sector } k) z^n$$

$$= \sum_{n=0}^{\infty} z^n \int_{x=0}^{\infty} P(\text{loss of nL in sector } k \mid x_k) f_k(x_k) \, dx_k \qquad (42)$$

This pgf has a similar integral form like the pgf obtained in the previous section in equation (39) and can be solved similarly. The result is[22]

$$G_k(z) = \left(\frac{1 - p_k}{1 - \frac{p_k}{\mu_k} \sum_{A \in \mathbb{K}} p_A z^{\nu_A}} \right)^{\alpha_k}. \qquad (43)$$

Since the sectors are defined to be independent, the overall pgf for the default loss distribution is just the product over the sector pgfs.

$$G(z) = \prod_{k=1}^{n} G_k(z)$$

Again, a recurrence relation is derived that allows a numerical approximation of the underlying distribution[23].

---

[22] For detailed calculation see CSFB(1997), section A9.
[23] See CSFB(1997), section A10.1

Generalization of obligor allocation

So far, we have assumed that every obligor can be assigned to one of several mutually independent sectors. Every sector is again driven by one underlying background factor. In another extension (see CSFB 1997) this assumption can also be relaxed and the default rate of an obligor can now depend on a weighted subset of the set of background factors. The weights $\theta_{A,k}$ are combined to an obligor-specific weight vector θ_A that represents the extent to which the default probability of obligor A is affected by the background factors k. One obligor's weights have to sum up to the unit over all sectors.

The default loss distribution

After estimation of the means and standard deviations of the obligors, and of the weights $\theta_{A,k}$ representing the obligors' sensitivity to the background factors, one can use the above relationships to calculate the mean μ_k and standard deviation σ_k for all factors.

The equations developed in section 5.3 are still valid and can be applied directly using the modified parameters. Note that the incorporation of firm specific factors is directly possible in the factor set-up just discussed. Since firm specific factors are idiosyncratic and uncorrelated to other factors and other firms' idiosyncratic components, it is sufficient to assign them to one common factor with variance zero. This causes the covariance of this specific factor, with all other factors, to be zero.

5.4 Conclusion

CreditRisk+ is an approach derived from an actuarial science framework that is rather easy to implement. Only few inputs - the probability of default and the exposure for each instrument - are needed. Providing closed form solutions for the probability of portfolio loan losses and the implementation of marginal risk contributions to obligors makes the model also attractive from the computational point of view.

However, there are also some limitations of the approach. CreditRisk+ is a so-called default mode model that does not incorporate migration or market risk of an obligor. Therefore, in comparison to so-called *mark-to-market* models like *CreditMetrics* CreditRisk+ is a so-called *default mode* model. The exposure for each obligor is fixed and cannot depend on changes on the credit quality of an issuer. This rather unrealistic assumption is not overcome even in the most general form of the model.

6 Credit Portfolio View by McKinsey

6.1 The Macroeconomic Factor Model

Credit Portfolio View proposed by McKinsey & Company, and described in the paper of Wilson (1997) consider the assumption that default probabilities follow business cycles. One may assume that the macroeconomic effects may only have a small influence on whether an individual bond will default or not. However, the importance of macroeconomic variables especially in the aggregate is considerable. Therefore, instead of taking historical default probability averages, where data over many business cycles is being used, a macro-economic multi-factor model is set up. It is worth mentioning that one should not evaluate bond prices and risks without a sense of how aggregate economic performance affects the assumptions that underlie especially risky bond prices.

Deriving conditional default probabilities

The original setup comprises index values $Y_{j,t}$ capturing the state of the economy for each speculative grade, each industrial sector and each country:

$$Y_{j,t} = \beta_{j,0} + \beta_{j,1} X_{j,1,t} + \ldots + \beta_{j,m} X_{j,m,t} + \nu_{j,t}$$

where $X_{j,i,t}$ are the values of the macroeconomic variables influencing the jth index value. $\beta_{j,i}$ are the coefficients to be estimated, $\nu_{j,t}$ the error terms which are assumed to be independent of $X_{j,t}$ and $N(0, \sigma_j)$ distributed.

Further, each macro-economic variable is assumed to follow a univariate, auto-regressive model of order 2:

$$X_{j,i,t} = \gamma_{j,i,0} + \gamma_{j,i,1} X_{j,i,t-1} + \gamma_{j,i,2} X_{j,i,t-2} + e_{j,i,t}$$

where $\gamma_{j,i,x}$ are coefficients and $e_{j,i,t} \sim N(0, \sigma_{e_{j,i,t}})$ is the error term.

Using the index values as a basis, default probabilities are modelled as logit functions

$$p_{j,t} = \frac{1}{1 + e^{-Y_{j,t}}}.$$

Possible macro-economic factors are for example the unemployment rate, the growth in GDP, foreign exchange rates, the level of long-term interest rate, public disbursement and the aggregate savings rate. Macro-economic time series can then be used to calibrate the multi-factor model and to forecast conditional default probabilities.

Conditional rating migration matrices

Investigating historical defaults, downgrades and upgrades shows that there is an obvious link between these numbers and the macroeconomy. Both default and credit downgrade are more likely to occur in recessions. Therefore, estimations of rating transition matrices can only reflect historical averages of transition probabilities. Instead, transition matrices conditional on the current state of the economy can be used for more accurate calculations. Therefore, the estimators in historical rating transition matrices are used as a basis to calculate actual transition matrices conditional on the state of the economy to date.

From the natural feeling that default probabilities should be higher during times of recession and downward migration should become more likely than in an upturn the following ratios are defined:

$$\frac{p_{j,t}}{p_{j,avg}} > 1 \text{ in economic recession and}$$

$$\frac{p_{j,t}}{p_{j,avg}} < 1 \text{ in economic expansion.}$$

$p_{j,t}$ in this equation denotes the simulated default probability for a speculative obligor in rating class j and $p_{j,avg}$ denotes the unconditional average probability of default from the agency's rating transition matrix.

CPV then generates a conditional rating transition matrix by means of these ratios. If the ratio is greater than one more probability mass is shifted into the downgraded and defaulted states and vice versa. This situation is supported by the following empirical correlations:

Let M_{avg} denote the average historical transition matrix. At time t, the matrix

$$M_t = M(p_{j,t}/p_{j,avg})$$

represents the desired transition matrix conditional on the current state of the economy. M_t is obtained by shifting probability mass in matrix M_{avg} from the lower left corner in the direction of the upper right corner in times of higher default rate than average. Then, downgrades and defaults become more probable. In times when the default rate is lower than on average, probability mass is shifted in the opposite direction, which results in a higher probability for upgrades and a lower probability of default. Conditional multi-year transition matrices can be generated by multiplying conditional one-year transition matrices.

$$M_T = \prod_{t=1,...T} M(p_{j,t}/p_{j,avg})$$

An illustrative example

We will illustrate this with an example taken from Saunders and Allen (2002). Therefore we consider a transition matrix with only four states. Let's further assume that the unconditional default probability for C rated debt is

$p_{CD,avg} = 0.15$ while e.g. the migration probability from state C to state B is $p_{CB,avg} = 0.04$ as it is denoted in Table 5.

Now suppose that based on current macroeconomic conditions the conditional value of the default probability for rating state C is $p_{CD,t} = 0.174$. In this case without adjusting the transition matrix we were likely to underestimate the VAR of a loan portfolio and especially the default probability of a C-rated loan.

Table 5. Unconditional transition matrix

	A	B	C	D
A	...	...	...	...
B	...	...	...	...
C	0.01	0.04	0.80	0.15

With $\Delta p_{CD} = p_{CD,t} - p_{CD,avg} = 0.174 - 0,15 = 0,024$ we get the so-called diffusion term or shift parameter. This parameter is then used to change the respective row in the unconditional transition matrix to obtain the conditional transition matrix.[24] Clearly the shift in transition probabilities must be diffused throughout the row in a way ensuring that the sum of all probabilities equals one. The procedure aims $\Delta p_{CC} = -0.0204$, $\Delta p_{CB} = -0.006$ and $\Delta p_{CA} = +0.0024$ and we get conditional transition probabilities as they are denoted in Table 6 .

Table 6. Conditional transition matrix

	A	B	C	D
A	...	...	...	...
B	...	...	...	...
C	0.0124	0.034	0.7796	0.174

To determine the complete conditional transition matrix this procedure is repeated for each row of the unconditional transition matrix.

Using the auto-regressive model of order 2 one can simulate the macro variables and therefore, also the index $Y_{j,t}$ over any time horizon $t = 1, ..., T$. Then we can also generate multi-period transition matrices for these time periods. Simulating many times the transition matrix one can finally generate the distribution of the cumulative conditional default probability for any rating. With the same methodology conditional cumulative migration probabilities over any time horizon can be produced.

---

[24] For a more detailed description of the procedure we refer to Saunders and Allen (2002)

6.2 Conclusion

The main idea of CreditPortfolioView is the empirical observation that default and transition probabilities can vary over time. The approach is an extension of the classical reduced form transition matrix model and designed to evaluate cumulative default or value-at-risk figures for a portfolio. CreditPortfolioView proposes a methodology which links macroeconomic factors to default and migration probabilities. According to empirical observations of high volatilities of default frequencies especially in lower rating categories this approach seems to be quite convincing. Even small changes in the transition matrix may have larger effects on cumulative default probabilities for a portfolio in the longer run.

However, the calibration of this model clearly depends on reliable default data for each country, and possibly - if conditional migration probabilities are calculated for sectors as well - for each industry sector within each country. Also the forecasts for the macro index $Y_{j,t}$ depend on the assumption that the macroeconomic factors can be modelled by an auto-regressive model, which is somehow questionable. What remains, however, is the idea of a model that links migration probabilities to the current state of the economy which seems to be favorable to just taking the average of historical transition probabilities.

7 The Basel II Model

In this last section we will describe the approach of the Basel Committee on Banking Supervision (2001) to determine risk weights for corporate defaults as it is used in the Internal Rating Based (IRB) approach of the new Basel capital accord. Due to the importance of the new accord for all financial institutions this model can also be considered as one of the most attended and discussed within the last years.

The risk weight function of the second consultative document is based on a rather simple credit risk model that was derived by Michael Gordy (2001). The risk parameters, entering the model are the probability of default (PD), the loss given default (LGD), the exposure at default (EAD), and the maturity (M) of a loan. They are supposed to be estimated by the financial institution for each exposure individually.

The model is based on the assumption that the assets of all debtors are influenced by only one common systematic factor. The systematic risk factor can be considered for example as the current state of the economy that next to firm-specific risks has an influence on the fact whether a firm defaults or not. Such models are in the literature generally called *one-factor models.*[25]

---

[25] See *A one-parameter representation of credit risk and transition matrices*, Belkin, B., Forest, L.R., Suchower, S., CreditMetrics Monitor, Third Quarter, 1998 as well as *The one-factor CreditMetrics model in the New Basel Capital Accord*, Finger, C.C., RiskMetrics Journal 2, 2001.

The committee emphasizes that this approach provides a reasonable approximation to two of the industry-standard credit risk models CreditMetrics and CreditRisk+ that were described in the previous sections.

7.1 Deriving the Model

The borrowers' asset-value Y_i is divided into two components: A systematic (X) and a idiosyncratic (U_j) factor whereby the idiosyncratic risk factor is in contrast to the systematic risk factor dependent on every single borrower j (with j = 1,...,K). Concerning these two variables X and U_j we assume

$X, U_1, ..., U_K \sim N(0,1)$ and further

$\forall$ j = 1,...,K: X and U_j to be stochastically independent as well as all U_i and U_j for $\forall$ $i, j - 1, ..., K$ with $i \neq j$.

Due to these assumptions, X and U_j (for j = 1,...,K) are standard normal distributed and in pairs stochastically independent but not the sum of X and U_j. Therefore, a weight-factor w (with $w \in [0;1)$) has to be introduced with w_j for the systematic risk factor and with $\sqrt{1 - w_j^2}$ for the idiosyncratic risk factor.

Finally we get for all borrowers $j = 1, ..., K$ the modelling relationship between the borrowers' asset-value Y_j, the systematic risk factor X and the idiosyncratic risk factors U_j as follows:

$$Y_j = \sqrt{w_j} \cdot X + \sqrt{1 - w_j^2} \cdot U_j \qquad (44)$$

In addition, if there are identically correlations ρ for all borrowers with $\rho = \rho(Y_j, Y_k) = w_j \cdot w_k > 0$, equation 44 can be simplified to the following:[26]

$$Y_j = \rho \cdot X + \sqrt{1 - \rho} \cdot U_j \qquad (45)$$

From this assumption follows, that all Y_i are standard normal distributed with the same multi-normal distribution and the same correlation ρ. Concerning the correlation ρ, the fifth and last assumption is met. It is assumed that the asset return correlation ρ is constant over time, independent of any risk factor and assumed to be 0.2. This asset return correlation assumption is underpinning the proposed corporate credit risk charges and is - according to the Basel committee - consistent with the overall banks' experience.

Given all these assumptions, a *default-point model* is introduced. In this model, a default for borrower j happens exactly when Y_j is equal or less than a certain default point d_j which – in turn – is defined as:

---

[26] See *Die Risikogewichte der IRB-Ansätze: Basel II und "schlanke" Alternativen*, Wehrsohn, U. et al, risknews, 11/2001.

$$PD_j = P(D_j = 1) = P(Y_j \leq d_j) \qquad (46)$$

where: d_j is the default point with: $d_j = \Phi^{-1}(PD_j)$
D_j is a bernoulli distributed default indicator with:

$$D_j = \begin{cases} 1 &, \text{ for default by obligor j} \\ 0 &, \text{ otherwise} \end{cases}$$

To go back to the modelling relationship, it is necessary to abstract from a single borrower to the whole homogeneous portfolio. Therefore, in the event of a default – i.e $Y_j \leq d_j$ – the conditional PD for a fixed realisation x of the same risk factor X is used:

$$PD_j(x) := P(D_j = 1 | X = x)$$

$$= P\left[\sqrt{\rho} \cdot x + \sqrt{1-\rho} \cdot U_j \leq \Phi^{-1}(PD_j)\right]$$

$$= P\left[U_j \leq \frac{\Phi^{-1}(PD_j) - \sqrt{\rho} \cdot x}{\sqrt{1-\rho}}\right] \qquad (47)$$

With the assumptions above, all U_j are standard normal distributed and the average asset correlation coefficient ρ is 0.20. Furthermore, the committee decided to calibrate the coefficients within this expression to an assumed value-at-risk loss coverage target of 99.5%. This corresponds with choosing the 1 - 99.5% = 0.5% quantile of $N(0,1)$ for the systematicc risk factor X. Thus, we get: $\Phi^{-1}(0.005) = -2.576$ and therefore, $PD_j(x)$ can be calculated as follows:

$$PD_j(z) := \Phi\left(\frac{\Phi^{-1}(PD_j) - \sqrt{0.20} \cdot (-2.576)}{\sqrt{0.80}}\right)$$

$$= \Phi\left(1.118 \cdot \Phi^{-1}(PD_j) + 1.288\right) \qquad (48)$$

To calculate the Benchmark Risk Weights there is also made an adjustment to reflect that the BRW_C are calibrated to a three-year average maturity:

$$b(PD) = \left(1 + 0.047 \cdot (1 - PD)/(PD^{0.44})\right) \qquad (49)$$

This construction is based on survey evidence and simulation results which were pooled judgementally to develop a smooth functional relationship between the values of PD and b(PD).

In addition to this relation a constant scaling factor 976.5 is introduced to calibrate the BRW_C to 100% for the average default probability of PD = 0.7% and LGD = 50%.

Finally, we get for the relationship between a corporate borrower's PD and the associated risk weight for a benchmark the follwoing formula. This formula is valid for an "average" borrower having a three-year maturity and LGD equal to 50%. The formula presented here is the BRW function that was usually suggested in the second consultive document of the Basel committee:

$$BRW_C(PD) = 976.5 \cdot \Phi\left(1.118 \cdot \Phi^{-1}(PD) + 1.288\right) \cdot \left(1 + 0.047 \cdot \frac{1 - PD}{PD^{0.44}}\right)$$
(50)

A graphical depiction of the IRB risk weights for a hypothetical corporate exposure (with LGD equal to 50% and without explicit maturity dimension) is presented in Figure 5. It is obvious that with the BRW as it was stated in the second consultive document especially for more risky companies the assigned risk weights are quite high. This lead to severe criticism of the suggestions while especially small and medium-sized companies (SMEs) were afraid of higher capital costs for banks that would lead to worse credit conditions for these companies. This lead to some revisions of the BRW function. Since some of these revisions are still under discussion we will only describe the most significant changes below.[27]

Changes in the capital accord and the new BRW function

As a consequence of the comment on the accord and especially on the risk weight function assigned in the IRB approach the Basel committee will suggest a refined version of the accord. The most significant changes are concerned with the following issues:

Due to the criticism considering the assignment of quite high capital charges to SMEs and in recognition of the unique characteristics of national markets, supervisors will have the option of exempting smaller domestic firms from the maturity framework.

Further, banks that manage small-business-related exposures in a manner similar to retail exposures will be permitted to apply the less capital requiring retail IRB treatment to such exposures, provided that the total exposure of a bank to an individual SME is less than Euro 1 Million.

Another concern was the potential gap between the capital required under the foundation and advanced IRB approaches. To modestly narrow this gap, the average maturity assumption in the foundation approach will be modified

---

[27] For a detailed description of the criticism and the possible changes in the Capital Accord see e.g. Benzin and Trück (2002).

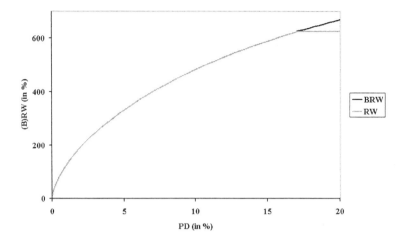

Fig. 5. (Benchmark) risk weights as a function of PD

from 3 years to 2.5 years, and the majority of the supervisory "loss-given-default" (LGD) values in the foundation IRB approach will now be reduced by five percentage points (e.g. for senior unsecured exposures from 50% to 45%).

Finally, the modified formula relating probability of default (PD) to capital requirements (BRW) differs from the formula proposed before in several ways.

One major change is that there is no explicit scaling factor in the formula anymore. Also the confidence level that was implicit in the formula has been increased from 0.995 to 0.999, to cover some of the elements previously dealt with by the scaling factor. The CP2 formula incorporated an implicit assumption that asset correlation is equal to 0.20 while the new formula assumes that asset correlation declines with PD according to the following formula:

$$\rho(PD) = 0.1 \cdot \frac{1 - e^{-50 \cdot PD}}{1 - e^{-50}} + 0.2 \cdot (1 - \frac{(1 - e^{-50 \cdot PD})}{1 - e^{-50}}) \qquad (51)$$

For the lowest PD value this function is equal to 0.20 and for the highest PD value it is equal to 0.10. The modified formula can be computed by first calculating the correlation value that corresponds to the appropriate PD value. This value enters then the main formula for the capital requirement. Capital requirements and risk-weighted assets are related in a straightforward manner. With the given confidence level and $\Phi^{-1}(0.999) = 3.090$, the resulting formula is the following:

$$BRW_C(PD) = \Phi\left(\frac{\Phi^{-1}(PD) + \sqrt{\rho} \cdot 3.090}{\sqrt{1-\rho}}\right)\left(1 + 0.047 \cdot \frac{1-PD}{PD^{0.44}}\right)$$
(52)

In Figure 6 the original CP2 BRW function is plotted against the revised function. Obviously, the combined impact of the changes is a risk-weight curve that is generally lower and flatter than that proposed in January. With this altered risk weight functions assigned BRWs are clearly lower than before and especially for more risky loans the required capital will decrease by up to 40%.

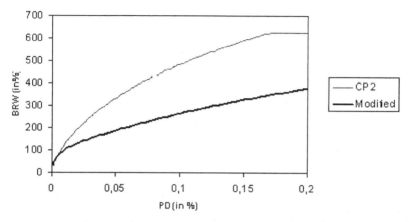

Fig. 6. Modified corporate risk weight curve compared to CP2

7.2 Conclusion

The approach of the Basel Committee on Banking Supervision (2001) as it is used in the Internal Rating Based (IRB) approach is clearly a Value-at-Risk approach. It is based on the assumption that the assets of all debtors are influenced by a systematic risk factor that next to firm-specific risks has an influence on the fact whether a firm defaults or not. The idea of the committee was to provide a quite simple and general model that provides a reasonable approximation to industry-standard credit risk models like e.g. the KMV approach, CreditMetrics and CreditRisk+. Gordy (2001) shows that the approximation to at least two of these models is quite good. The problem of assigning too high risk weights to more risky exposures could be solved by some minor changes and a re-calibration of the model. Therefore, despite the severe criticism one can consider the Basel II model as a favorable effort that provides a understandable value-at-risk methodology and also tries to reconcile the most important credit risk models in practice.

References

1. Aleksiejuk, A. and Holyst, J.A. (2001). A simple model of bank bankruptcies. Working Paper.
2. Basel Committee on Banking Supervision (2001) Potential Modifications to the Committee's Proposal.
3. Basel Committee on Banking Supervision (1999). Credit Risk Modelling: Current Practices and Applications.
4. Basel Committee on Banking Supervision (2001). Consultative Document – Overview of the New Basel Capital Accord.
5. Basel Committee on Banking Supervision (2001). Consultative Document – The Internal-Ratings-Based Approach to Credit Risk.
6. Basel Committee on Banking Supervision (2001). Consultative Document – The Standardised Approach to Credit Risk.
7. Basel Committee on Banking Supervision (2001). Consultative Document – The New Basel Capital Accord.
8. Bathia, M. and Finger, C.C. and Gupton, G.M. (1997). CreditMetrics-Technical Document. J.P. Morgan.
9. Belkin, B. and Forest, L.R. and Suchower, S. (1998). A one-parameter representation of credit risk and transition matrices. CreditMetrics Monitor.
10. Belkin, B. and Forest, L.R. and Suchower, S. (1998). The effect of systematic credit risk on loan portfolio value-at-risk and loan pricing. CreditMetrics Monitor.
11. Benzin, A., Trück, S. and Rachev, S. (2003). Approaches to Credit Risk in the New Basel Capital Accord, in: Bol et al: Credit Risk – Measurement, Evaluation and Management.
12. Black, F. and Cox, J.C. (1976). Valuing Corporate Securities: Some Effects of Bond Indenture Provisions. The Journal of Finance (31).
13. Bohn, J.R. (1998). A Survey of Contingent-Claims Approaches to Risky Debt Valuation. KMV. available from: http://www.kmv.com.
14. Das, S. (1997). *Risk Management and Financial Derivatives: A Guide to the Mathematics*, McGraw-Hill.
15. Carty, L.V. and Lieberman, D. (1996). Defaulted Bank Loan Recoveries. Moody's Investor Service, Global Credit Research.
16. Carty, L.V. and Lieberman, D. (1996). Corporate Bond Defaults and Default rates 1938–1995. Moody's Investor Service. Global Credit Research.
17. Credit Suisse Financial Products (1997). Credit Risk$^+$: A credit risk management framework. available from: http://www.csfb.com/creditrisk
18. Crosby, P.J. and Bohn, J.R. (2001). Modelling Default Risk. KMV Corporation. available from: http://www.kmv.com.
19. Crouhy M., Galai, D. and and Mark, R. (2000). A comparative analysis of current credit risk models. Journal of Banking and Finance (24).
20. Duffie, D. and Singleton, K. (1999). Simulating Correlated Defaults. available from: http://www-leland.stanford.edu/ duffie/
21. Entrop, O. and Völker, J. and Wilkens, M. (2001). Strukturen und Methoden von Basel II – Grundlegende Veränderung der Bankenaufsicht. Kreditwesen.
22. Finger, C.C. (2000). A comparison of stochastic default rate models. RiskMetrics Journal 1.
23. Finger, C.C. (1999). Conditionla approaches for CreditMetrics portfolio distributions. CreditMetrics Monitor.

24. Finger, C.C. (2001). The one-factor CreditMetrics model in the new Basel Capital Accord. RiskMetrics Journal 2.
25. Frye, John (2000). Depressing Recoveries. Working Paper.
26. Gordy, M. B. (2000). A comparative anatomy of credit risk models. Journal of Banking & Finance 24.
27. Gordy, M. B. (2001). A risk-factor model foundation for ratings-based bank capital rules. Working Paper.
28. Gupton, G.M., Finger, C.C. and Bhatia, M. (1997). Credit Metrics – Technical document. J.P. Morgan & Co. Incorporated. available from: http://www.riskmetrics.com/research.
29. Jarrow, R. and van Deventer, D. (1998) Practical Usage of Credit Risk Models in Loan Portfolio and Counterparty Exposure Management. Kamakura Cooperation.
30. Jones, E.P., Mason, S.P. and Rosenfeld, E. (1984). Contingent Claims Analysis of Corporate Capital Structures: an Empirical Investigation. The Journal of Finance (39).
31. Kealhofer, S. (2001). Portfolio Management of Default Risk. KMV Corporation. available from: http://www.kmv.com.
32. J.P. Morgan & Co.Incorporated (1997). Credit Metrics.
33. Link, T. and Rachev, S. and Trück, S. (2001). New Tendencies in Rating SMES with Respect to Basel II. Informatica (4/2001).
34. Locke, J. (1998). Credit Check. Risk Magazine (Sep 1998).
35. Longstaff, F.A. and Schwartz, E.S. (1995). A Simple Approach to Valuing Risky Fixed and Floating Rate Debt. The Journal of Finance (50).
36. Merton, R.C. (1974). On the Pricing of Corporate Debt: The Risk Structure of Interest Rates. Journal of Finance (29).
37. Miller, R. M. (1998). A Nonparametric Test For Credit Rating Refinements. Working Paper. KMV Corporation.
38. Rachev, S.T., Schwartz, E. and Khindanova, I. (2000). Stable Modeling of Credit Risk. Working Paper.
39. Saunders, A. and Allen, L. (2002). Credit Risk Measurement. John Wiley.
40. Sellers, M., Vasicek, O. and Levison, A. (2000). The KMV EDF Credit Measure and Probabilites of Default. KMV. available from: http://www.kmv.com.
41. Sundaram, R.K. (2001). The Merton/KMV Approach to Pricing Credit Risk. KMV Corporation.
42. Trück, S. (2001). Basel II and the Consequences of the IRB Approach for Capital Requirements. Proceedings of the METU International Conference in Economics, Ankara, Turkey.
43. Trück, S. and Rachev, S.T. (2002). A structural approach to Default Risk using the alpha-stable Distribution. Working Paper.
44. Wehrspohn, U. and Gersbach, H. (2001). Die Risikogewichte der IRB-Ansätze: Basel II und "schlanke" Alternativen. Risknews.
45. Wilde, T.(2001). IRB approach explained. Risk.
46. Wilde, T.(2001). Probing Granularity. Risk.
47. Wilson, T.C. (1997). Portfolio Credit Risk I. Risk (10-9).
48. Wilson, T.C. (1997). Portfolio Credit Risk I. Risk (10-10).
49. Wilson, T.C. (1997). Measuring and Managing Credit Portfolio Risk. McKinsey & Company.
50. Zhou, C. (1997). A Jump-Diffusion Approach to Modelling Credit Risk and Valuing Defautable Securities. Research Paper. Federal Reserve Board Washington DC.

List of Authors

Keyvan Amir-Atefi
Department of Economics, University of California, Santa Barbara, CA 93106-9210, USA, E-Mail: keyvan@econ.ucsb.edu

Arne Benzin
Lehrstuhl für Versicherungswissenschaft, Universität Karlsruhe (TH), Postfach 6980, 76128 Karlsruhe, Germany

Dr. Christian Bluhm
HypoVereinsbank, Group Credit Portfolio Management, Sederanger 5, 80538 München, Germany, E-Mail: christian.bluhm@hypovereinsbank.de

Dylan D'Souza
Department of Economics, University of California, Santa Barbara, CA 93106-9210, USA, E-Mail: dylan@econ.ucsb.edu

Prof. Dr. Wolfgang Härdle
Institut für Statistik und Ökonometrie, Humboldt-Universität zu Berlin, Spandauer Str. 1, 10178 Berlin, Germany, E-Mail: haerdle@wiwi.hu-berlin.de

Dr. Christoph Heidelbach
Credit Risk Management, DaimlerChrysler Bank AG, Nordbahnhofstraße 147, 70191 Stuttgart, Germany

Prof. Dr. Alexander Karmann
Dresden University of Technology, Chair for Economics, especially Monetary Economics, 01062 Dresden, Germany

Professor Dr. Rüdiger Kiesel
Department of Financial Mathematics, University of Ulm, Helmholtzstr. 18, 89069 Ulm, Germany, E-Mail: kiesel@mathematik.uni-ulm.de

Dr. Rainer Klingeler
ifb AG, Neumarkt-Galerie, Neumarkt 2, 50667 Köln, Germany, E-Mail: Rainer.Kingeler@ifbAG.com

Daniel Kluge
Kreditmanagement, Deutsche Genossenschafts-Hypothekenbank AG, Rosenstr. 2, 20095 Hamburg, Germany, E-Mail: Daniel.Kluge@dghyp.de

Dr. Werner Kürzinger
Credit Risk Management, DaimlerChrysler Bank AG, Nordbahnhofstraße 147, 70191 Stuttgart, Germany

Dr. Frank B. Lehrbass
Kreditmanagement, Deutsche Genossenschafts-Hypothekenbank AG, Rosenstr. 2, 20095 Hamburg, Germany, E-Mail: Frank.Lehrbass@dghyp.de

Filip Lindskog
RiskLab, Departement of Mathematics, ETH Zentrum, HG F 42.3, CH-8092 Zürich, Switzerland, E-Mail: lindskog@math.ethz.ch

Dominik Maltritz
Dresden University of Technology, Chair for Economics, especially Monetary Economics, 01062 Dresden, Germany, E-Mail: Dominik.Maltritz@mailbox.tu-dresden.de

Peter Martin
ifb AG, Neumarkt-Galerie, Neumarkt 2, 50667 Köln, Germany E-Mail: Peter.Martin @ifbAG.com

Prof. Dr. Alexander McNeil
Department of Mathematics, ETH Zentrum, HG G 32.3, CH-8092 Zürich, Switzerland, E-Mail: mcneil@math.ethz.ch

PD Dr. Marlene Müller
Institut für Statistik und Ökonometrie, Humboldt-Universität zu Berlin, Spandauer Str. 1, 10178 Berlin, Germany, E-Mail: marlene@wiwi.hu-berlin.de

PD Dr. Ludger Overbeck
CIB/Credit Risk Management, Risk Analytics and Instruments, Risk Research and Development, Deutsche Bank AG, Taunusanlage 12, 60325 Frankfurt, Germany, E-Mail: ludger.overbeck@db.com

Jochen Peppel
Institut für Statistik und Mathematische Wirtschaftstheorie, Universität Karlsruhe, Postfach 6980, 76128 Karlsruhe, Germany

Prof. Dr. William Perraudin
Department of Financial Economics, Birkbeck College, University of London, Malet Street, Bloomsbury, London, WC1E 7HX, United Kingdom

Svetlozar T. Rachev
Institut für Statistik und Mathematische Wirtschaftstheorie, Universität Karlsruhe, Postfach 6980, 76128 Karlsruhe, Germany and Department of Statistics and Applied Probability, University of California, Santa Barbara CA 93106-3110, USA, E-Mail: rachev@pstat.ucsb.edu

Borjana Racheva-Jotova
BRAVO Risk Management Group, 440-F Camino del Remedio, Santa Barbara, CA 93110, USA, E-Mail: Borjana.Racheva@bravo-group.com

Thomas Rempel-Oberem
ifb AG, Neumarkt-Galerie, Neumarkt 2, 50667 Köln, Germany, E-Mail: Thomas.Rempel-Oberem@ifbAG.com

Ingo Schäl
zeb/rolfes.schierenbeck.associates, Hammer Strasse 165, 48153 Münster, Germany, E-Mail: ISchael@zeb.de

Frank Schlottmann
Institut AIFB, Universität Karlsruhe (TH), Postfach 6980, 76128 Karlsruhe, Germany, E-Mail: schlottmann@aifb.uni-karlsruhe.de

Rafael Schmidt
Universität Ulm, Abteilung Zahlentheorie und Wahrscheinlichkeitstheorie, Helmholtzstr. 18, 89069 Ulm, Germany, E-Mail: Rafael.Schmidt@mathematik.uni-ulm.de

PD Dr. Uwe Schmock
RiskLab, Department of Mathematics, ETH Zentrum, HG F42, CH-8092 Zürich, Switzerland, E-Mail: schmock@math.ethz.ch,

Prof. Dr. Detlef Seese
Institut AIFB, Universität Karlsruhe (TH), Postfach 6980, 76128 Karlsruhe, Germany, E-Mail: seese@aifb.uni-karlsruhe.de

Stoyan Stoyanov
BRAVO Risk Management Group, 440-F Camino del Remedio, Santa Barbara, CA 93110, USA, E-Mail: Stoyan.Stoyanov@bravo-group.com

Dr. Alex P. Taylor
Judge Institute of Management, University of Cambridge, Trumpington Street, Cambridge CB2 1AG, United Kingdom

Stefan Trück
Institut für Statistik und Mathematische Wirtschaftstheorie, Universität Karlsruhe, Postfach 6980, 76128 Karlsruhe, Germany, E-Mail: stefan@lsoe.uni-karlsruhe.de

Springer Finance

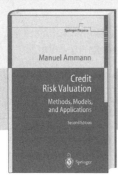

M. Ammann, University of St. Gallen, Switzerland

Credit Risk Valuation
Methods, Models, and Applications

Credit Risk Valuation offers an advanced introduction to the models of credit risk valuation. It concentrates on firm-value and reduced-form approaches and their applications in practice. Additionally, the book includes new models for valuing derivative securities with credit risk, focussing on options and forward contracts subject to counterparty default risk, but also treating options on credit-risky bonds and credit derivatives. The text provides detailed descriptions of the state-of-the-art martingale methods and advanced numerical implementations based on multi-variate trees used to price derivative credit risk. Numerical examples illustrate the effects of credit risk on the prices of financial derivatives.

2nd ed. 2001. Corr. 2nd printing 2002. X, 255, 17 figs., 23 tabs. Hardcover € **69.95**; sFr 116.50; £ 49 ISBN 3-540-67805-0

A. Ziegler, University of Lausanne, Switzerland

Incomplete Information and Heterogeneous Beliefs in Continuous-time Finance

This book considers the impact of incomplete information and heterogeneous beliefs on investor's optimal portfolio and consumption behavior and equilibrium asset prices. After a brief review of the existing incomplete information literature, the effect of incomplete information on investors' expected utility, risky asset prices, and interest rates is described. It is demonstrated that increasing the quality of investors' information need not increase their expected utility and the prices of risky assets. The impact of heterogeneous beliefs on investors' portfolio and consumption behavior and equilibrium asset prices is shown to be non-trivial.

XIII, 194 p. 51 illus. 2003. Hardcover € **64.95**; sFr 108; £ 45.50 ISBN 3-540-00344-4

Springer · Kundenservice
Haberstr. 7 · 69126 Heidelberg
Tel.: (0 62 21) 345 - 0 · Fax: (0 62 21) 345 - 4229
e-mail: orders@springer.de

Die €-Preise für Bücher sind gültig in Deutschland und enthalten 7% MwSt.
Preisänderungen und Irrtümer vorbehalten. d&p · BA 42333/2